De Gruyter Graduate

Sabbatini • Polymer Surface Characterization

Also of Interest

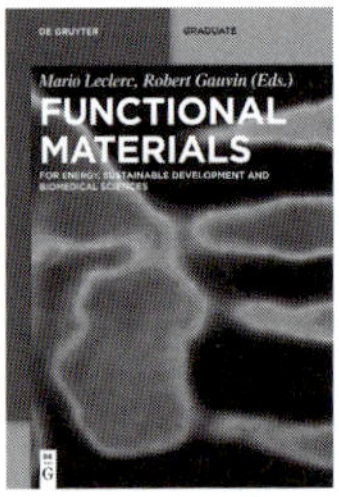

Functional Materials – For Energy, Sustainable Development, and Biomedical Sciences, 2014
Mario Leclerc, Robert Gauvin (Eds.)
ISBN 978-3-11-030781-8, e-ISBN 978-3-11-030782-5

Organic and Hybrid Solar Cells – An Introduction, 2015
Lukas Schmidt-Mende, Jonas Weickert
ISBN 978-3-11-028318-1, e-ISBN 978-3-11-028320-4

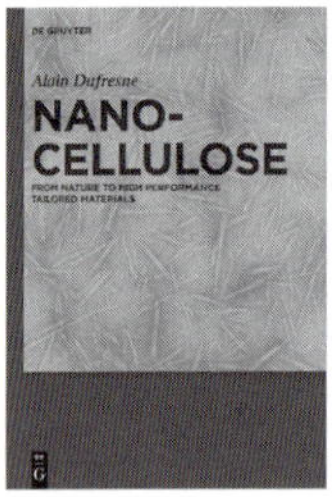

Nanocellulose – From Nature to High Performance Tailored Materials, 2012
Alain Dufresne
ISBN 978-3-11-025456-3, e-ISBN 978-3-11-025460-0,
Set-ISBN 978-3-11-219133-0

e-Polymers
Seema Agarwal, Andreas Greiner (Editor-in-Chief)
ISSN 1618-7229

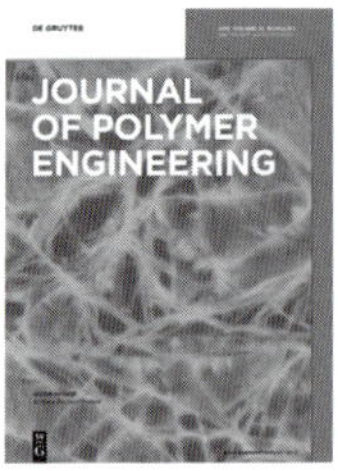

Journal of Polymer Engineering
Ica Manas-Zloczower (Editor-in-Chief)
ISSN 2191-0340

Polymer Surface Characterization

Edited by Luigia Sabbatini

DE GRUYTER

Editor
Prof. Dr. Luigia Sabbatini
University of Bari Aldo Moro
Chemistry Department
Via Orabona 4
70125 Bari
Italy
e-mail: luigia.sabbatini@uniba.it

ISBN 978-3-11-027508-7
e-ISBN 978-3-11-028811-7

Library of Congress Cataloging-in-Publication data
A CIP catalog record for this book has been applied for at the Library of Congress.

Bibliographic information published by the Deutsche Nationalbibliothek
The Deutsche Nationalbibliothek lists this publication in the Deutsche Nationalbibliografie; detailed bibliographic data are available in the Internet at http://www.dnb.de.

Typesetting: Compuscript Ltd.
Printing and Binding: Hubert & Co. GmbH & Co. KG, Göttingen
Cover image: “Green with Envy” by Waiken Wong. Permission granted by Materials Research Science and Engineering Center, University of Massachusetts, Amherst, MA, visual program supported by National Science Foundation.

∞ Printed on acid-free paper
Printed in Germany
www.degruyter.com

FSC www.fsc.org MIX Papier aus verantwortungsvollen Quellen FSC® C016439

Preface

Polymers permeate our everyday life more than any other materials. It is difficult to think of another material as versatile and irreplaceable as polymers in terms of applications both in consolidated and in emerging technological areas. Their use spans from the most trivial consumer goods to the most advanced microelectronics packaging, miniaturized pieces for artificial organs, high-performance coating in aerospace industry, etc.

Technological demand, moreover, requires materials not only with peculiar bulk properties but also with specific surface chemistry and structure, both generally different from the bulk one. The combination of these two requisites makes a polymeric material suitable and unique for the job it is called to do.

In this respect, surfaces in polymers, more than any other class of materials, can be quite easily handled and engineered to a high level of sophistication by fully exploiting their chemical reactivity. Furthermore, apart from the changes intentionally promoted on polymer surfaces to tailor new application-oriented properties, the importance of polymer dynamics occurring on surfaces whenever they are exposed to the environment (real world) or put in contact with another material (interface) is self-evident: dramatic changes of structural and thermodynamic properties in the confinement area can occur, and these dramatically reflect on properties such as bonding, friction, adsorption, wettability, inertness, etc. Moreover, aging modifies the surface chemistry of polymers (as well as of any other material), thus provoking a remarkable deterioration of their properties.

In all the above cases, knowledge of both the surface chemistry and the structure is mandatory to understand the behavior of polymeric materials, monitor the surface processing, and tailor and develop new surface modification strategies. A wealth of analytical techniques is available for the purpose; they are surface specific, and the analyst/researcher needs a basic knowledge of their principles and procedures in order to choose the most appropriate way to solve the specific problem. In this book, techniques have been selected that are well suited for characterization of surfaces/interfaces of thin polymer-based films but also of more general applicability in materials science. Basic principles, operative conditions, applications, performance, and limiting features are supplied, together with current advances in instrumental apparatus. Each chapter is devoted to one technique and is self-consistent; the end-of-chapter references would allow the reader a quick access to more detailed information.

After an introductory chapter, techniques that can interrogate the very shallow depth of a polymer surface, spanning from the top few angstroms in secondary ions mass spectrometry to 2–10 nm in X-ray photoelectron spectroscopy are discussed in Chapters 2 and 3, respectively. Their potential in the definition of surface chemical composition of polymers is highlighted and documented by selected application examples. Chapter 4 is dedicated to attenuated total reflection Fourier transform infrared spectroscopy, which provides information on polymer chemical structure with varying surface thicknesses, of the order of tenths of a micrometer. Chapter 5 deals with atomic force microscopy, a technique endowed with high-resolution capability to map a variety of material properties on the nanoscale. Evidence is supplied of how probing mechanical properties leads to the identification of important polymer surface processes such as phase segregation, continuity of phases, and dispersion, which play a major role in determining macroscopic properties. Characterization of surface morphology is also addressed in Chapter 6, which is focused on electron microscopy. Scanning electron microscopy is described in detail, and some general information on transmission electron microscopy is given. The point of strength of electron microscopy characterization, well highlighted in this chapter, is that the surface morphology of a polymer system is always closely connected with its chemical composition and processing conditions. Chapter 7 addresses the issue of wettability and its measurements; contact angle measurements and its relationships with surface energy and wetting behavior are clearly described as well as advanced technological processes to develop polymer surfaces with tunable hydrophobic/hydrophilic character. Chapter 8 ends the book with an insight on spectroscopic ellipsometry: after the description of the basics of the method, an investigation of optical, structural, and thermodynamic properties of polymers in the whole spectral range (UV-VIS-IR) is discussed. Some interesting industrial applications such as those in the quality control of roll-to-roll fabrication process or heterogeneity control of micropatterned films are also reported.

The editor gratefully acknowledges the invaluable contributions of the authors and, as it is behind every edited book, the kind invitation and precious support of the publisher, in particular, the enthusiasm of Karin Sora, Senior Editorial Director, Science and Technology, and the continuous, qualified, and friendly cooperation of Julia Lauterbach, Project Editor, Science, Technology & Medicine (STM), at De Gruyter.

Contents

Contributing authors

Eva Bittrich
Leibniz-Institut für Polymerforschung
Dresden e.V.
Hohe Strasse 6
01069 Dresden
Germany
e-mail: bittrich-eva@ipfdd.de

Stefania Cometa
Jaber Innovation s.r.l.
Via Calcutta 8
00144 Rome
Italy
e-mail: stefania.cometa@virgilio.it

Elvira De Giglio
Chemistry Department
University of Bari Aldo Moro
Via Orabona 4
70126 Bari
Italy
e-mail: elvira.degiglio@uniba.it

Rosa Di Mundo
Department of Mechanics Mathematics
and Management
Polytechnic University of Bari
V.le Japigia 182
70126 Bari
Italy
e-mail: rosa.dimundo@poliba.it

Nicoletta Ditaranto
Chemistry Department
University of Bari Aldo Moro
Via Orabona 4
70126 Bari
Italy
e-mail: nicoletta.ditaranto@uniba.it

Klaus-Jochen Eichhorn
Leibniz-Institut für Polymerforschung
Dresden e.V.
Hohe Strasse 6
01069 Dresden
Germany
e-mail: kjeich@ipfdd.de

Beat A. Keller
Swiss Federal Laboratories for Materials
Science and Technology (Empa)
CH-8600 Dübendorf
Switzerland
e-mail: beat.keller@empa.ch

František Lednický
Institute of Macromolecular Chemistry
Academy of Sciences of the Czech
Republic
Heyrovsky Sq. 2
162 06 Prague 6
Czech Republic
e-mail: ledn@imc.cas.cz

Filippo Mangolini
Department of Materials Science and
Engineering
University of Pennsylvania
Philadelphia, PA 19104
USA
e-mail: mfilippo@seas.upenn.edu

Fabio Palumbo
Chemistry Department
University of Bari Aldo Moro
Via Orabona 4
70126 Bari
Italy
e-mail: fabio.palumbo@cnr.it

Antonella Rossi
Dipartimento di Scienze Chimiche e Geologiche
Università degli Studi di Cagliari
INSTM Unit – Cittadella Universitaria di Monserrato
S.S. 554 Bivio per Sestu
I-09042 Monserrato, Cagliari
Italy
e-mail: rossi@unica.it

Luigia Sabbatini
Chemistry Department
University of Bari Aldo Moro
Via Orabona 4
70126 Bari
Italy
e-mail: luigia.sabbatini@uniba.it

Miroslav Šlouf
Institute of Macromolecular Chemistry
Academy of Sciences of the Czech Republic
Heyrovsky Sq. 2
162 06 Prague 6
Czech Republic
e-mail: slouf@imc.cas.cz

Taťana Vacková
Institute of Macromolecular Chemistry
Academy of Sciences of the Czech Republic
Heyrovsky Sq. 2
162 06 Prague 6
Czech Republic
e-mail: vackova@imc.cas.cz

Petr Wandrol
FEI
Podnikatelská 6
612 00 Brno
Czech Republic
e-mail: Petr.Wandrol@fei.com

Dalia Yablon
Surfacechar LLC
Sharon, MA 02067
USA
e-mail: dalia.yablon@surfacechar.com

Stefania Cometa and Luigia Sabbatini

1 Introductory remarks on polymers and polymer surfaces

1.1 Why polymers?

1.1.1 Generality

Polymers are everywhere in our everyday life. Both synthetic and natural polymers have unique properties (chemical, mechanical, electrical, thermal), which make them irreplaceable in the particular job they are called to do. The most impressive aspect in the world of polymers is the enormous variety of applications, which span from everyday consumer goods (house wares, toys, bottles, packaging, textiles, furniture), heavily used electrical and electronic items (computer, cable insulation, household appliances), construction (housing, pipes, adhesives, coatings, insulation), transportation (tires, appliance parts, hardware, carpeting), medical and hospital furniture and goods (gloves, syringes, catheters, bandage), to high-performance, sophisticated materials for aerospace, bulletproof vests, nonflammable fabric, artificial organs, degradable device for controlled release of drugs or chemicals, high-power-density batteries, high-strength cables for oceanic platforms, etc.

From a chemical point of view, polymers are high-molecular-weight compounds made up of simple repeating units called monomers. Molecular weights range from thousands to millions of atomic mass units, which makes writing a definitive molecular structure for these materials close to impossible. Polymer structures are therefore represented by enclosing the repeating unit (monomer) in brackets and placing "n" as subscript (Fig. 1.1a).

Polymers having more than one kind of repeating units are called copolymers; the units can be distributed either randomly in the chains or in an ordinate alternate sequence or in blocks. Moreover, the chains of the polymer structure may arrange in linear, branched, network, stars, ladder, or dendrimer fashion (Fig. 1.1b). Complex structures such as networks are formed when cross-linking occurs, i.e. the formation of covalent bonds between polymer chains; this process leads to a remarkable increase in molecular weight, and this in turn strongly influences chemical, physical, and mechanical properties of the material as well as its processability.

1.1.2 Synthesis

The first synthetic polymer that was produced on a commercial scale was Bakelite, a phenol-formaldehyde resin (by its inventor, the Belgian chemist Leo Baekeland [1]), at the beginning of 1900; however, it is the German chemist Hermann Staudinger who is

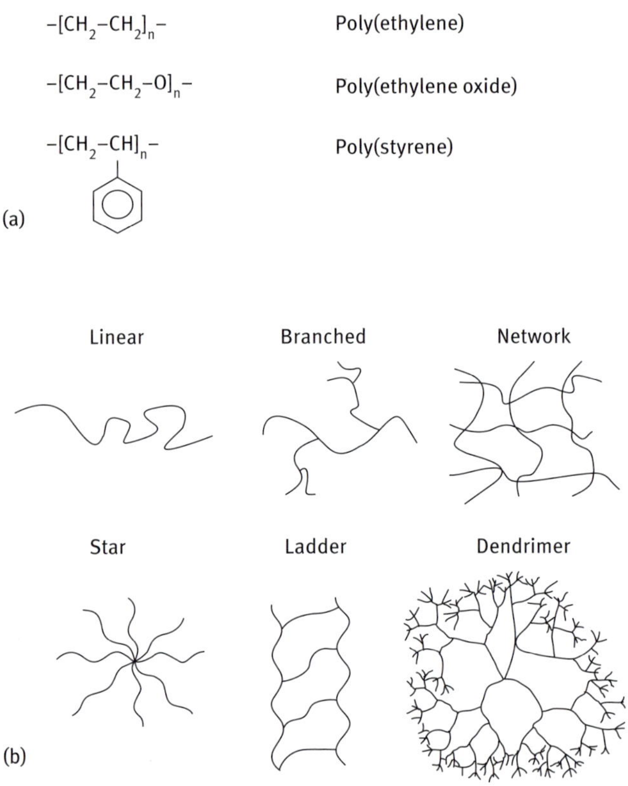

Fig. 1.1: (a) Conventional representation of polymers structures. (b) Arrangement of chains in the polymer structure.

rightly recognized as the father of the chemistry of macromolecules; for his studies on the relationship between structure of polymers and their properties, he received the Nobel Prize in Chemistry in 1953. During those years, polymer chemistry experienced the most significant advances and the beginning of a true revolution in the polymer industry, thanks to the work of Giulio Natta in Italy, who, for the first time, synthesized polymers having controlled stereochemistry [2] using the coordination catalysts developed by the German chemist Karl Ziegler [3]. The two scientists were awarded by the Nobel Prize in Chemistry in 1963. It was clearly demonstrated that polymers with similar chemical composition but exhibiting different molecular orientation have different morphology and their mechanical properties too may differ markedly. It is therefore vital, in view of commercial exploitation of polymer materials, to have full knowledge and control of the polymerization process. In brief, two reaction mechanisms are possible: step reaction (generally, a condensation reaction) and chain reaction (generally, an addition reaction). In the former, poly-functional monomers react randomly to give dimers, trimers, and oligomers, which can react with other monomers or among themselves: monomers are consumed rapidly, but polymer chains do not grow so much (i.e. the increase in molecular weight is slow). Polyamides (e.g. nylon 66) and polyesters (e.g. polyurethane) are produced by this reaction mechanism. In the chain reaction, however, an initiator (generally a free radical) starts the reaction by opening the double bond of a vinyl monomer and adding to it, with one electron remaining unpaired. The reaction propagates through the successive linking of a monomer to the end of the growing chain; finally, two free radicals react to end each other's growth activity (termination). Monomers are consumed more slowly with respect to step reactions, but there is a considerable quick increase of the chain molecular weight. Polytetrafluoroethylene (Teflon), polyethylene, and polystyrene (PS) are important examples of polymers obtained by chain reaction mechanism.

In polymer synthesis, it can be assumed that each chain reacts independently. Therefore, the bulk polymer is characterized by a wide distribution of molecular weights and chain lengths, and these factors primarily determine the properties of polymers. The degree of polymerization (DP) is the number of repeat units (included chain ends) in the chain and gives a measure of molecular weight; due to the presence of chains of varying lengths in any polymer, the average DP, $\overrightarrow{DP}$, is used more. Moreover, to characterize the distribution of polymer lengths in a sample, two parameters are defined: number average (M_n) and weight average (M_w) molecular weight. The number average is the sum of individual molecular weights divided by the number of polymers. The weight average is proportional to the square of the molecular weight. Therefore, the weight average is always larger than the number average (unless all molecules possess the same weight; then $M_n = M_w$).

1.1.3 Classification and nomenclature

Polymers are classified in many ways, the most common classifications being (1) based on source (natural, semisynthetic, or synthetic); (2) based on structure (linear, branched chain, cross-linked, or network); (3) based on molecular forces, which determine the mechanical properties (elastomers, fibers, thermoplastic polymers, or thermosetting polymers); (4) based on polymerization mode (addition and condensation polymers, which are currently referred to as chain growth polymers and step growth polymers, respectively).

The reported classifications, although useful from a practical point of view, are, in some cases, a rough simplification. For example, it must be emphasized that not all step reactions are condensation processes and not all the chain reactions are addition processes; this explains why IUPAC classified the polymerization processes [4] in a more comprehensive manner with respect to the above-simplified scheme.

At the same time, an IUPAC commission elaborated a series of rules for polymer nomenclature [5]. This is quite a complicated matter: some details and significant examples have been simply and clearly reviewed in many textbooks on polymers (see e.g. [6–9]), but this is outside the scope of this book. However, readers should be familiar with the most widely accepted terminology, and therefore, some general information is provided. Polymers (and copolymers) are grouped in families according to the functional group(s) present in the repeating units when their structure is easily identifiable: polyesters, polyethers, polyimides, polyamides, polyamideimides, etc. In the presence of structures that cannot be easily defined, (co)polymer nomenclature is based on the name of the starting monomer(s). This approach is referred to as source-based system and is more widely used than the structure-based IUPAC nomenclature. Moreover, when the starting monomers contain a carbon-carbon double bond, the resulting polymers are classified as vinyl polymers, in all the other cases, they are referred to as nonvinyl polymers: the nomenclature of latter may be quite complex, and the source-based approach once again is more generally used than IUPAC rules. Depending on the approach used, terminology may be quite different; to give an example, $(\text{-CH}_2\text{-CH}_2\text{-})_n$ is named polyethylene or poly(methylene) in the source-based and IUPAC approaches, respectively, and similarly, $(\text{-CH}_2\text{CHCOOH-})_n$ is named poly(acrylic acid) or poly(1-carboxylatoethylene), respectively. Given the complexity of polymer nomenclature, its abbreviations and/or trade terms have been and still are widely used both in the scientific literature and in the industrial world. Many of these terms are so popular and widely recognized that they are commonly used in the everyday language as well (PVC, PE, PET, HDPE, LDPE, PMMA, Teflon, etc.).

1.1.4 Morphology and properties

Chemical constitution and the way molecules self-assemble in a solid on the microscopic and macroscopic scales, together with molecular weight, are important factors in determining the ultimate mechanical and thermal properties of polymers, which in turn determine their applications and processability.

Regarding the assembly and orientation of molecules, polymers can be considered crystalline or amorphous materials. Crystalline materials have highly ordered structures with molecules arranged in repeating patterns. Amorphous materials, by contrast, show no long-range order, with molecules arranged randomly and the chains tangled and curved around one another. Polymers never achieve complete crystallinity, and they are more conveniently categorized as amorphous and semicrystalline, with the latter morphology being characterized by tangled and disordered regions surrounding crystalline zones (Fig. 1.2).

Semicrystalline polymers have true melting temperatures (Tm) at which the ordered regions break down and become disordered. In contrast, the amorphous regions soften over a relatively wide temperature range (always lower than Tm), known as the glass transition temperature (Tg). Fully amorphous polymers do not exhibit Tm, of course, but all polymers exhibit Tg. Above these temperatures, polymers are liquid-like.

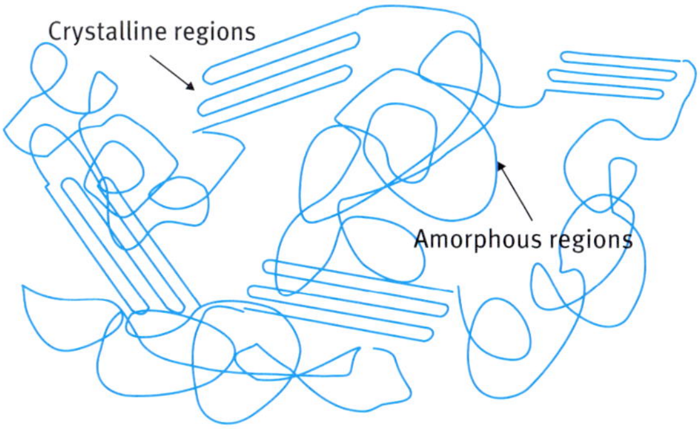

Fig. 1.2: Schematic representation of a semicrystalline polymer.

Polymers are usually in solid state at room temperature. Above a certain temperature, all polymers soften and, in the case of thermoplastics, are able to flow, having liquid-like order, i.e. they are disordered and in the melt state (for common polymers, this usually occurs in the range of 100°C–250°C).

When the temperature drops, their density increases. Depending on the polymer, there are two possibilities: (1) polymers with irregular molecular structure (atactic PS, atactic PP, PMMA) solidify, keeping their disordered microstructure and forming a stiff but brittle amorphous solid called polymer glass; (2) polymers that have a regular structure at the molecular scale (PE, PEO, isotactic PP, isotactic PS, PA, PTFE, PETP) partially crystallize.

In Fig. 1.3, which shows the specific volume *vs* temperature dependence, behavior 1 presents a change of slope in a continuous fashion at the specific temperature of the transition, and for this reason it is called glass transition temperature, Tg (left). Meanwhile, behavior 2 leads to a semicrystalline material in an abrupt manner at a characteristic temperature called crystallization temperature, Tc (right).

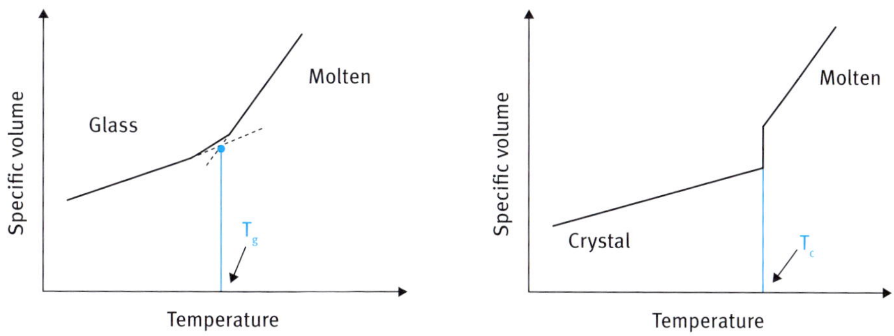

Fig. 1.3: Cooling down from the melt state of a polymer to glass or to crystal.

For polymers (unlike crystals of small molecules), the crystallization temperature Tc might differ from the melting temperature, Tm. Usually, Tc is lower than Tm by several degrees, due to the undercooling effect. Furthermore, both temperatures depend on the rate of cooling/heating. This is because long polymer chains, unlike small molecules, have decreased flexibility and are easily trapped in kinetically arrested states. For this reason, it is very difficult for them to attain their absolute thermodynamic equilibrium state of full crystallinity. High degrees of crystallinity can be attained in the case of very slow cooling rates. If they are cooled abruptly enough, they freeze in the fully disordered state, forming a polymer glass. Thermoset polymers and gels lack organization at all scales and are inherently amorphous.

It is worth noting that processes that use melting for polymer fabrication usually lead to a surface morphology different from that of the bulk of the material. The so-called skin-core effects, also observable in solution-prepared polymers as well as in natural polymers, are discussed in detail in Chapter 6.

Chemical structure and morphology are tightly related in polymers, and both relate to the main bulk properties: mechanical characteristics, thermal stability, chemical resistance, flame resistance, degradability, electrical conductivity, and nonlinear optical (NLO) properties.

A variety of polymer applications depend on their unique mechanical properties such as tensile, compressive, and flexural strength, elasticity, plasticity, and toughness, which are governed by intermolecular forces binding the polymer chains. Thermally stable is defined a polymer that does not decompose at temperatures lower than 400°C; thermal stability is related to bond energy, and this explains why, for example, high-performance materials for the aerospace industry that have to withstand extreme temperatures comprise aromatic compound-based polymers. Polymers with high halogen content exhibit both a good flame and chemical (solvent) resistance, but again, morphology plays a role. Durability is one of the "blessings" of polymers, but it is also a curse from the point of view of environmentalists; this justifies the many efforts spent in the last years by materials scientists in capturing the fundamentals of polymer degradation mechanisms, in order to design and synthesize materials with tuned degradability and capable of controlled release. An interesting class of polymers, characterized by an extended conjugated system and a high degree of crystallinity and which can be conveniently doped, exhibit electrical conductivity comparable to that of metals; this property, together with the much lower density and the higher processability with respect to metals, make such conducting polymers extremely attractive for many applications including low-weight batteries, light-emitting diodes, thin film transistors, etc. Polymers containing chromophores able to promote a change in the refractive index when interacting with a light-induced electric field exhibit the so-called NLO properties. These organic materials comprise one or more aromatic systems in conjugated positions, leading to charge transfer systems; functionalization of both ends of the π-conjugated system with appropriate electron donor and acceptor groups can increase the asymmetric electronic distribution in either or both the ground and excited states, thus leading to an increased optical nonlinearity. Organic NLO materials have potentially high nonlinearities and rapid response in electro-optic effect compared with inorganic NLO materials, foreseeing a great impact on information technology and industrial applications.

To complete the panorama of the enormous impact that advanced polymeric materials can have in the everyday life, new-generation nanostructured polymers should be mentioned. Various techniques have been developed to produce nanosized polymer materials (polymer nanoparticles, nanocapsules, nanorods, nanobrushes) such as solvent evaporation, supercritical fluid technology, salting out, dialysis, micro-emulsion, mini-emulsion, surfactant-free emulsion, and interfacial polymerization. A recent review [10] covers the general description of the preparation of polymer nanostructures and the detailed description of the crucial parameters involved in techniques designed to obtain the desired properties.

Polymers properties can be substantially modified and strategically tuned by the addition of the so-called additives. These substances are categorized on the basis of their function as, for example, plasticizers, nucleating agents, flame retardant, crosslinking agents, slip agents, lubricants, reinforcing fillers, odorants. They can be completely miscible or completely immiscible to polymer bulk and are often added in a combination. On considering the extent of additives classes, it appears immediately clear as the wise use of different formulations of additives can remarkably extend the variety of polymers of technological interest.

To make an example, the classical polymer processing techniques normally involve extrusion, film casting, film blowing, blow molding, and injection molding. The processing rate in most of these techniques is limited by the appearance of flow instabilities, collectively known as the melt fracture phenomenon [11, 12]. Linear viscoelastic rheology and nonlinear rheology are dominant factors for good polymer processing. Additives aimed to better processing of polymers are proficiently exploited; they can be classified as acid neutralizers, antiblocks, pigments, antioxidants, compatibilizers, etc.

The improvement of many physical and mechanical properties of polymers (such as process viscosity, modulus, fracture toughness, and coefficient of thermal expansion) is efficiently carried out by the introduction of new types of additives, mainly intended for reinforcement of polymer hosts. This reinforcement can be comprised of the following:

1. Low-molecular-weight crystalline compounds, which simultaneously improve the impact or fracture toughness, increase or do not otherwise detrimentally decrease the modulus, and enhance the processability (through reduced viscosity).
2. Thermotropic liquid crystalline polymers, which are able to reduce the melt viscosity of the thermoplastics and can also improve their mechanical properties [13–15].
3. Carbon, glass, basalt, aramid, or natural fiber reinforcement [16]. Fiber-reinforced polymers are currently used in almost every type of advanced engineering

structures: aircraft, helicopters, automobiles, sports goods, civil infrastructures such as bridges and buildings, and so on.

4. Nanoparticle-, nanofiber-, or nanotubes-enriched polymers. Nanotechnology represents the emerging research area in innovative plastic-based devices. Nanofibers, in particular, have been used for a wide range of applications such as tissue engineering [17], filter media [18], reinforcement in composites [19] and micro/nano-electro-mechanical systems [20]. Such fibers can be made from polymers, carbon, and semiconductors into the form of continuous nanofibers, nanofibrous networks, or short nanowires and nanotubes. Moreover, reinforcement can also occur using particles of spherical shape as well as plate-like and fibrous nanofillers. As a result, nanofibers, because of their excellent mechanical properties coupled with very high aspect ratio, are theoretically the ideal candidates for reinforcing a polymer matrix.

In all these cases, a strong interface between the reinforcing phase and the host polymer matrix is always desirable to achieve desired properties.

1.2 Why to investigate a polymer surface?

The efforts during the last 20 years to quantify the chemical and physical properties of synthetic or naturally derived soft matter systems are mainly linked to the need to engineer highly sophisticated polymers. Today, synthetic chemistry can allow for excellent control over the bulk polymer properties. Meanwhile, the importance of polymer dynamics occurring on the surface is self-evident: the nature of polymer surfaces allows to control or modulate properties such as wetting, adsorption, bonding, friction, surface energy, compatibility with other substrates, chemical inertness or reactivity, presence of contaminants, surface morphology, crystallinity or roughness, etc.

Polymers, due to their very unstable nature, can undergo dramatic structural and thermodynamic changes in the confinement area, which may be produced by relatively weak variations of external conditions, such as application of external fields, changes in temperature, moisture, pH, mechanical stresses, etc. Therefore, surface rearrangements of the polymer chains as a consequence of environmental or interfacial interactions are often unpredictable. In this respect, the dynamics of polymer chains near surfaces has attracted growing attention and is analyzed in Section 1.2.1. Many analytical techniques have been devoted to the recognition and systematic investigation of the chemistry, morphology, and evolution of polymers surfaces.

In this book, the peculiarities, differences, and the recent progresses of the analytical tools adopted by polymer scientists to investigate polymer surfaces will be described and critically discussed.

1.2.1 Nature and dynamics of polymer surfaces

Soft materials such as polymers, especially amorphous or semicrystalline ones, cannot be simply considered as solids. They must be seen as semi-liquids or very viscous liquids. Moreover, they are characterized by weak aggregation linkages between macromolecules, if compared with conventional engineering materials such as inorganic systems, metals, ceramics, or semiconductors.

In this respect, the molecular motions occurring inside a polymer result in a wide range of cases:

1. the translation motion of entire macromolecules, which leads to a mutual flowing of polymer chains or, on the contrary, to a closed and blocked chains packing;
2. the cooperative wriggling and jumping of segments of molecules, which determines flexibility and elasticity;
3. the motion of a few atoms along the main chain or side groups, which allows the three-dimensional reorganization of the chains;
4. finally, the vibrational motions of the bonds between atoms of the polymer, common to all the molecules, from the simplest (i.e. low-molecular-weight compounds) to the most complicated one (i.e. synthetic polymers or copolymers, natural macromolecules, etc.).

These motions have a direct effect on many of properties in bulk, melt, and solution of polymers.

As far as the surfaces are concerned, it appears clear that the reorganization of polymer surfaces is strongly related to the characteristics of the interfacing medium (air, solvents, vacuum, other polymers or totally different materials, etc.). In this respect, dynamics on surfaces or interfaces must be treated as a different object of study when relevant to soft matter.

In detail, we will focus our attention on vibrational dynamics, changes of thermodynamic properties, rotation of functional groups in response to external stimuli, and inter- or out-diffusion of segments or chains from the bulk toward the surface, especially in polymer blends and in copolymers.

1.2.1.1 Vibrational dynamics of macromolecules

When we discuss about the different vibrational processes involving macromolecules, we must consider that, as far as polymers are concerned, these processes could cover an extremely broad range of characteristic time scales, spanning from about 100 fs up to years.

Theoretically speaking, the basic concepts discussed for molecular vibrations of finite molecules can be applied to polymer chains. The model of finite molecules consists of particles endowed with mass and held together by electrical forces such as covalent, dipole-dipole, van der Waals interactions, etc. Atoms in molecules are treated as particles in which all the mass is concentrated at a point. In particular, Wilson's GF matrix method has been developed in the past [21, 22]. Briefly, it uses internal coordinates that are bond stretches, in-plane angle bending, torsions, and waggings. For a molecule having n atoms, there are ($3n$-6) vibrational degrees of freedom, or ($3n$-5) if the molecule is linear. Using any set of generalized coordinates, the problem can be reduced to writing the secular equations to be solved to get eigenfrequencies and eigenvectors. However, as the polymer molecules are of infinite length, the order of the matrices becomes infinite. Polymers are subject to complex intramolecular and intermolecular interactions combined with many intramolecular degrees of motional freedom. To reduce the problem to one of workable dimensions, the screw symmetry of the polymer chain is exploited. In view of these facts, the number of mathematical and theoretical models aimed to describe polymers vibrational dynamics is very high and cannot be fully covered herein [23–34].

Today, computer simulations constitute an excellent tool for the investigation of the dynamic properties of complex systems [35–38]. These methods are usually *ab initio* molecular dynamics (MD) schemes [39, 40]. MD simulations can provide, within statistical averages, results that can be used to test the quality of the predictions of an approximate theory applied to the same model. If theory and simulations disagree, theory must be improved.

To obtain experimental results, several techniques have been used to characterize the conformation of polymeric systems. These include infrared (IR) absorption, Raman spectra, inelastic neutron scattering, X-ray diffraction (XRD), nuclear magnetic resonance, circular dichroism, and optical rotatory dispersion. In particular, IR spectroscopy is a very powerful analytical technique used to obtain information about molecular structure of polymers by measuring the frequency of IR radiation needed to excite vibrations in molecular bonds. IR spectroscopy in attenuated total reflection (ATR/FTIR) couples the analytical method of IR spectroscopy with the physical phenomena of total internal reflection (i.e. reflection and refraction of electromagnetic radiation at an interface of two media having different indices of refraction) to restrict the analyzed volume on the surface region of the sample. A deeper investigation of these spectroscopic techniques is discussed in Chapter 4.

1.2.1.2 Changes in thermodynamic properties on the surface

Based on the classical physicochemical theories typically developed for metals or ceramics, the thermodynamic properties of solids are also applicable to rigid polymers. However, the mobility of the constituents of the soft matter represented a considerable complication in this extension, due to the greater degrees of freedom of such liquid-like systems. Indeed, at room temperature, polymers are much more mobile than inorganic materials.

The assessment of the classical thermodynamic properties of the most common polymer systems is reported by Li *et al.* [41]. The polymer macromolecules can be thought of as segmental objects joined by covalent bonds. These segments continuously move to reach the configurational entropy, which determines the equilibrium state having the minimum free energy. A flexible polymer can adopt a large number of configurations of equal energy. Because of the opposite signs of the energy and entropy, maximizing the entropy is equivalent to minimizing the free energy. The effects of thermal fluctuations are then accounted for by minimizing the free energy with respect to all possible configurations.

From a statistical point of view, a long polymer chain can adopt a large number of microscopic conformation states; therefore, the long-chain conformational entropy, S, is described by the following formula:

$$S = K_b \cdot \ln \Omega,$$

where k_b is the Boltzmann constant and Ω is the number of long-chain conformation states. If a polymer long chain is deformed, whether compressed or elongated, its conformational entropy will be changed, and accordingly, its free energy.

Based on these considerations, we can argue that the thermodynamics of soft matter, such as polymers, is a very complicated field of investigation. In addition, when we focus our attention on the confinement area of polymers as well as for all the other atomic or molecular solids, other complications arise: in general, the surface represents a break of the symmetry experienced in the bulk, significantly changing the thermodynamic properties of the material (for example, the melting point or the surface energy). The statistical thermodynamics of surfaces is often treated by first determining the surface configuration that minimizes the internal energy [42].

Although much of the surface science of hard matter focuses on the atomic-scale structure, the study of soft surfaces involves the large-scale nature of the interface and especially its thermal fluctuations.

Polymer surfaces, in general, solid-phase surfaces, are intrinsically less energetically favorable than the bulk of the material: surface molecules have more energy compared with the molecules in the bulk because the latter experience balanced forces of interaction with all the neighbor molecules, whereas surface ones interact attractively with the underlying layer molecules; otherwise, there would be a driving force for surfaces to be created, removing the bulk of the material (i.e., sublimation).

The surface energy may therefore be defined as the excess energy at the surface of a material compared with the bulk. In the bulk – in the absence of strong enthalpic interactions – polymer chains typically take on a so-called random coil conformation, where the physical orientation of each repeat unit along the chain's backbone is essentially uncorrelated with the orientations of the chain's other repeat units. The presence of a surface, however, places tight constraints upon the conformations of chains near the surface such that the random coil conformation is perturbed. To alleviate this entropically unfavorable situation and lower the overall free energy of the system, it has been predicted [43] that polymer chains near the surface will segregate their chain ends to the boundary, thereby avoiding the required "reflection" at the material boundary and thus minimizing the loss in conformational entropy.

An outstanding question regarding the surface tension is whether this property depends solely on properties of species localized within a few angstroms of the surface or whether the contributions are more delocalized because of mobility in the fluid. Much of the complexity arises from surface orientation effects, which are especially pronounced in some copolymers and other inherently ordered systems such as liquid crystalline oligomers. Studies of surface free energies of solid monolayers are relevant for the understanding of fluids, and these and other data can be applied to the interpretation of polymer and copolymer melts.

The surface energy of polymers is a very important property that determines their wettability, adhesion, adsorption especially for applications such as coating, printing, gluing, biocompatible interfacing, self-cleaning, patterning, microfluidics, etc. The breaking of symmetry experimented in the confinement area strongly influences the melting point of polymers near the surface [44–46], and in the case of amorphous polymers, the mobility of the polymer chains also changes. Therefore, a fundamental question is whether the glass transition temperature, Tg, or the polymer mobility at a free surface is significantly different from that in the bulk.

Liu *et al.* [47] largely discussed about this question, reporting previous studies of different authors, with different conclusions. In particular, Keddie *et al.* [48] studied the thermal expansion of PS, showing that the Tg of a polymer surface is much less than that of the bulk, whereas Wallace *et al.* [49] found no indication of an increased mobility at the surface and, in fact, indicated that the mobility at the free surface may be suppressed. Xie *et al.* [50] initially concluded that there was no evidence of increased mobility, but subsequent experiments suggested the presence of a surface layer with enhanced mobility [51]. Kajiyama *et al.* [52] concluded that the surface mobility of PS was enhanced only for low-molecular-weight polymers where the preferential segregation of chain ends to the surface was sufficiently large to enhance the mobility. For higher-molecular-weight polymers, no enhanced mobility at the free surface was found. In response to these unclear conclusions, Liu *et al.* investigated the relaxation of PS segments at the free surface as a function of temperature, and they showed that the relaxation of the polymer within the first nanometer from the free surface is parallel to the surface, or planar, whereas in the first 10 nm the dominant relaxation occurs normal to the free surface. The full relaxation of the surface was found for temperatures higher than the bulk Tg. These studies have quantitatively shown that the segmental mobility of high-molecular-weight, amorphous polymer chains at a free surface is not significantly different from that in the bulk. Chain ends can enhance the surface mobility; however, to have high enough chain-end concentrations, the molecular weight of the polymer must be sufficiently low to have an effect.

As far as glassy polymers are concerned, Priestley *et al.* [53] demonstrated that the rate of structural relaxation of poly(methyl methacrylate) near surfaces and interfaces with respect to that of bulk was reduced by a factor of 2 at a free surface and by a factor of 15 at a silica substrate interface. Many studies have attempted to provide a physical understanding of changes in Tg induced by spatial confinement. Several factors have been invoked as being responsible for differences in structural and dynamic behavior of confined polymers compared with the bulk. These include geometry (e.g. pore size, shape) as well as specific interfacial interactions (e.g. hydrogen bonding, electrostatic forces) that can affect in a significant manner both chain conformation and segmental mobility. For weakly adsorbing surfaces, both positive and negative shifts in Tg are reported, yet these results still remain quite controversial [54]. Several methods can be used to obtain surface parameters related to the interfacial free energy and other thermodynamic interaction measures such as the enthalpy of adsorption or displacement [55].

A deep knowledge of the thermodynamic properties of polymer surfaces is the basis for the development of smart surfaces. It has been recognized that characterizing the surface thermodynamic properties (for example, hydrophobicity-hydrophilicity balance, electrokinetic properties, adhesion, thermal behavior) are vitally important for defining suitable applications and developing new markets. Awaja *et al.* [56] reviewed the main research efforts relevant to the study of polymer adhesion with a special focus on chemical, mechanical, and thermodynamic mechanisms linked to this property. Köstler *et al.* [57] focused their attention on the study of surface thermodynamic properties and wetting behavior of polyelectrolyte multilayers, which can be applied in biomedical or electronic devices, whereas Luzinov *et al.* [58] demonstrated the importance of understanding the relationships between the bulk properties of pristine polymeric materials and their surface characteristics in order to design materials with smart surface behavior. The dependence of surface free energy on molecular orientation in polycarbonate films, as determined by contact angle (CA) measurements, was demonstrated by Stachewicz *et al.* [59]. Geoghegan and Krausch [60] summarized the main theoretical and experimental aspects of wetting and dewetting phenomena relevant to polymer films. Wetting or measurements of the CA of a liquid test droplet on a solid surface represent a straightforward technique revealing surface energetic information inaccessible by the surface spectroscopies [61] CA measurements at polymer surfaces can be carried out by several different methods [62]. Polymer surface energetics and CA measurements will be described in detail in Chapter 7.

1.2.1.3 Rotation of functional groups on polymer backbones in response to different environmental conditions

In the past, the surface energetic properties of solids were assumed to be independent of the environment (polar and nonpolar) in which they are immersed. Successively,

Holly and Refojo [63], taking CA hysteresis experiments into consideration, concluded that hydrogel surfaces changed their polarity according to the nature of the environment. Since then, such surface modification phenomena have been investigated using X-ray photorelectron spectroscopy (XPS) analysis [64, 65] and dynamic CA measurements [66, 67].

Today, it is well accepted that polymer chains are not "frozen" or immobile entities, and environmentally induced changes in the surface structure have been observed on a variety of polymers, not only on those exposing relatively mobile surface layers or containing polar moieties. The polymer chains' mobility is an inherent property of the soft matter: the polymer configuration instantaneously changes. This enhanced mobility is also achieved through the rotation of functional groups along the polymer chain backbone rather than the long-chain segmental motion of macromolecules. From a mathematical point of view, adjacent bonds can be described by two angles in spherical coordinates: the zenith angle θ and the azimuth φ. The zenith angle represents the angle between adjacent bonds, and the azimuth represents the rotations about the zenith.

Here, we provide only a brief presentation of the major models of polymer rotation modes:

- Rotational isomeric model: In this model, the bond angle with respect to the adjacent one is allowed to take distinct rotational positions (usually about three).
- Freely rotating chain: The rotational restriction is removed, and the azimuth can take any value, but the zenith angle is fixed.
- Worm-like chain: A continuous version of the freely rotating chain, with vanishing bond lengths.
- Freely jointed chain: A sequence of freely rotating chains looses its bond angle correlation after one Kuhn length (i.e. in polymer melts or solutions, the ratio of mean square end-to-end distance, R_0^2, and the fully extended size, R_{max}). When the bond sequence is replaced by a sequence of Kuhn lengths, two adjacent Kuhn lengths have uncorrelated bond angles. They are like rods joined freely without any angular restrictions (both in φ and θ). In these models, the rod represents a Kuhn length only in the distance. The rods are assumed to be of fixed length and massless. The mass of the local collection of monomers (equivalent to one Kuhn length) is replaced by a spherical bead at the joints.
- Bead spring chain: In this model, as Kuhn length represents a collection of bonds, a collection of Kuhn lengths is represented by a spring. However, the spring is not a replacement of a fixed distance (i.e. a Kuhn length) but represents a variable distance.

Considering the polymer surfaces, the surface chains can respond to different surrounding conditions [68]. Given sufficient mobility, polymer surfaces will reorient or restructure in response to their local microenvironment to minimize their interfacial free energy with the surrounding phase. In particular, side chains of polymers are the most inclined to reorganization. The specific spatial arrangement of functional groups interfaced with a contacting medium is called the surface configuration. A typical rearrangement takes place when the polymer surfaces experience a change in polarity of solvents or interfacing media (Fig. 1.4). In this case, for example, when a polymer having polar functional groups experiences a polar medium, a rearrangement of these groups on surface is thermodynamically advantageous. On the contrary, when in contact with an apolar medium, these groups tend to minimize the interaction with this opposing environment.

An example of the influence of reorientation of functional groups on the features of the polymer surfaces is the configuration of atoms in a macromolecule of gelatin gel. Gelatin is known to contain hydrophilic groups; however, gelatin gels are surprisingly hydrophobic. This is because all the hydrophilic groups at the gelatin-air interface are oriented toward the bulk. Thus, it is not the configuration of hydrophilic groups with respect to the polymer backbone but rather the surface configuration of those groups that dictates wettability. The reorientation of surface-state functional groups in response to a change in the interfacing environment was called surface configuration change [69–72].

One of the most useful tools used to investigate the orientation of molecular chain was the reflection-absorption Fourier transform infrared spectroscopy (RA-FTIR) [73]. In principle, if there is a change accompanied by a reorientation of the dipole moment in the conformation of polymer chain, this difference can be detected through the RA-FTIR spectroscopy. For example, when a p-polarized IR beam is incident at a grazing angle to the substrate, the absorption for those dipole moments oriented normally to the substrate will be significantly enhanced. An increase or decrease in the intensity of a particular IR absorption peak is therefore indicative of a change in the molecular chain orientation. This feature

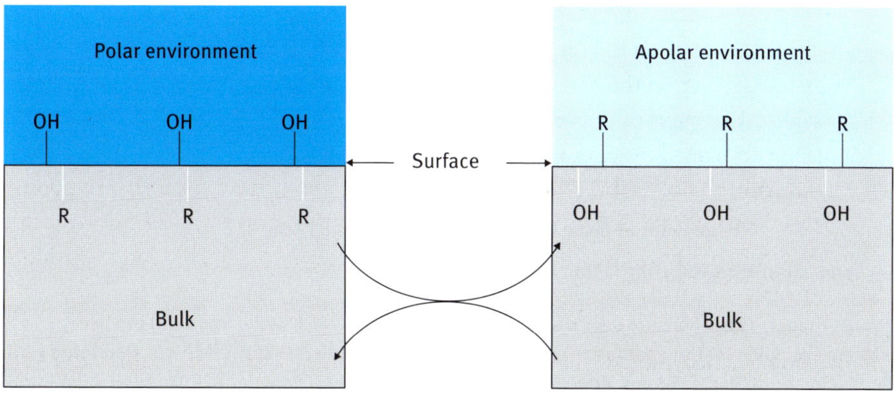

Fig. 1.4: Surface arrangement of functional groups of a polymer interfaced with different contact media.

is of great benefit when examining the orientation of a specific group within the bulk polymer film. For example, Guo and Yang [74] studied the polymer light-emitting diodes based on poly(2-methoxy-5-(2′-ethylhexyloxy)-1,4-phenylene-vinylene) (coded as MEH-PPV) under stress test to monitor the relation between the device performances and the conformational transformation of a conjugated polymer. Their experimental results indicated that the plane of the conjugated π-electron cloud in MEH-PPV tends to align parallel to the substrate. This rearrangement enhances the π-π-electron coupling and lowers the device operating voltage under high current densities. In Fig. 1.5, the two extreme cases for the orientation of conjugated chains inside the MEH-PPV film have been represented: in case 1, the plane of π-electron cloud is oriented parallel to the substrate; hence, the vibrating dipoles of C-H wagging mode are perpendicular to the surface plane. In case 2, the orientation is turned 90°.

The difference between these two conformations can be distinguished through the RA-FTIR because for the grazing-angle incident π-polarized IR beam, the absorption of this kind of vibrating mode is enhanced in Case 1 and lowered in Case 2. Interestingly, results indicate that a small change in the average orientation of the PPV unit of MEH-PPV causes a great deviation in the electrical and optical properties of the polymer film.

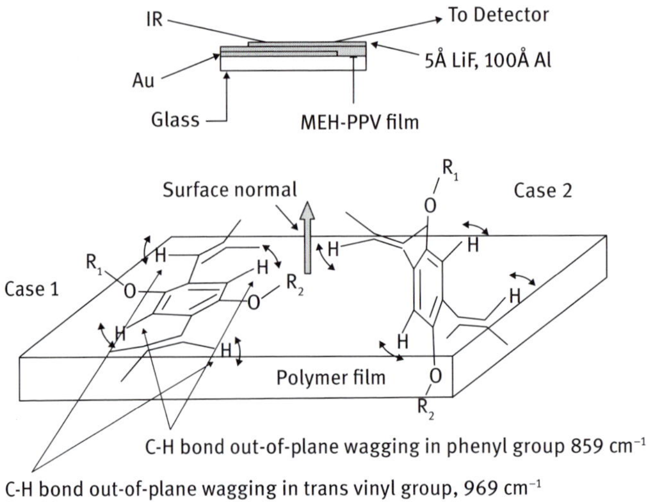

Fig. 1.5: (upper) The schematic device structure of the *in situ* RA-FTIR setup. (bottom) the two extreme examples for the orientation of conjugated chains inside the MEH-PPV films. Reprinted with permission from Guo TF, Yang Y, *Appl Phys Lett*, 80, 148–50, 2002. Copyright 2002, AIP Publishing LLC [74].

1.2.1.4 Surface interdiffusion or segregation of copolymers or polymer blends

In several applications, copolymers or polymer blends (i.e. physical mixtures of two or multicomponent polymers) are required to obtain enhanced properties and functions. Moreover, as described in Section 1.1.4, additives are also commonly added into polymeric materials used in our daily life to tune their properties: plasticizers, fillers, compatibilizers, antistatics, reinforcement or aesthetic enhancers, pigments, antioxidizing agents, etc.

Because many polymers are non-miscible, the constitutive components of block copolymers or blends tend to segregate into different domains, giving rise to the so-called phase separation. In copolymers, the covalent bonds of these components prevent the total phase separation from extending over a large length scale. Instead, only small domains with typical sizes in the range of tens of nanometers are formed. Phase separation or segregation phenomena lead to a different and often unpredictable surface polymer composition. Moreover, when polymer additives are added into the polymer host, they sometimes migrate toward the surface and on other times stay in bulk, according to different requirements. All these phenomena are particularly important for commercial polymer films because these systems normally comprise multiple polymer species, solvents, and additives, leading to numerous complex phase separation morphologies, such as bi-continuous structure, islands, or holes. Possible sources of phase segregation can be involved in polymer synthesis or formulation – composition, molecular weight, film thickness, solvents, additives – or can be linked to processing successive to their synthesis – changes in the external environment, pressure, temperature, interaction with other interfaces, etc. Recently, Xue *et al.* [75] summarized the fundamental theory and factors influencing phase separation in polymer thin films to develop new strategies for the utilization of phase separation to generate ordered pattern in polymer thin film systems.

As far as the surface migration of polymer additives is concerned, this feature can be considered both a problem and a desired phenomenon: in the latter case, surface active additives are intentionally added to the polymer host, so the enhancement of additive migration towards surface must be reached. However, for both the cases, the control and the establishment of principles for the selection of additives and for the optimized processing conditions to enhance or reduce the surface localization of these molecules play a primary role in the complex polymer systems manufacture.

In polymer systems, the surface concentration of additives is normally different from their bulk concentration. There are a number of factors that influence the surface migration of additives in polymers: Wu and Fredrickson [76] and Minnikanti and Archer [77, 78] proposed a linear response theory that explains this effect in terms of a so-called entropic attraction of chain ends to surfaces and thermodynamic features of the blend near a surface and in bulk. The surface segregation phenomenon in segmented or random copolymers, polymer blends, and polymer hybrid composites have been intensively studied by surface-sensitive spectroscopies by CA measurements and more recently by scanning probe techniques and mass spectrometry. Some significant examples of case studies will be illustrated in the following chapters.

1.2.2 Surface modification of polymers

Polymeric materials are designed for a given application primarily considering their favorable bulk properties (including thermal stability, mechanical strength, solvent resistance, cost, etc.). In addition to these fundamental characteristics, the selected polymers must also offer surface characteristics that could be modulated to the intended application. Today, to achieve this, the strategy of surface modification represents a common approach to obtain the ultimately desired properties for the myriads of polymer applications. Polymer surfaces can thereby be changed to reach a variety of goals, i.e. increasing adhesion, improving wettability, enhancing processability, reducing friction, reducing susceptibility to harsh chemicals or environmental agents, increasing dye absorption, etc. A brief excursus of the most important surface modifications of polymers is reported below (see Sections 1.2.2.1 – 1.2.2.5).

1.2.2.1 Improvement of wettability

The wettability, i.e. the ability of a liquid to maintain contact with a solid surface, is a fundamental property of solid surfaces both in the natural and technological world. The investigation of the polymer wetting phenomenon has been the subject of intensive studies both in the past (i.e. the spraying of a paint, the penetration of ink in paper, the liquid absorbency or repellency of cloths, and so on) and in the recent research (including, for example, microfluidic technologies, regulation of protein adsorption, biomaterials design). In particular, as far as the biomaterials research is concerned, surface wettability represents a property of primary importance [79, 80]. Polymers can be easily processed due to their good plasticity or fluidity after being heated, emulsified, mixed, or dissolved in an appropriate solvent. Indeed, various methods for modification of wetting behavior of polymer surfaces have been proposed. Super-hydrophilic surfaces with a WCA of about 0° generated by UV irradiation has been successfully used as a transparent coating with antifogging and self-cleaning properties [81]. Meanwhile, super-hydrophobic surfaces with a CA larger than 150° have been proposed to inhibit phenomena such as contamination, snow sticking, erosion, and current conduction [82–85].

The surface chemical composition of a polymeric system, determining its surface free energy, has great influence on wettability [86, 87]. In the case of segmented or random copolymers as well as polymer blends, as reported in a previous paragraph, these materials spontaneously rearrange to have low-energy non-wettable segments on the surface. The wettability of polymer surfaces could be tuned over a wide range by selecting polymer brushes with appropriate functional groups. A deeper discussion of surface grafting of macromolecules by post-polymerization modification (PPM) techniques was exposed by Galvin and Genzel [88]. The wettability modification of a surface may be also achieved by different chemical and/or physical treatments [89–92], including plasma treatment with the appropriate gases to render a surface hydrophilic or hydrophobic [93, 94]. Plasma is a highly reactive mixture of gases, such as argon, oxygen, hydrogen, or a mixture of these gases, consisting of a great concentration of ions, electrons, radicals, and other neutral species [95]. Plasma treatments provide an efficient, economic, environmentally friendly, and versatile technique for modifying the surface properties of materials, in particular polymers.

The surface topographic structure is also an important factor that influences the wettability. Based on these principles, intensive studies have been made to obtain super-hydrophobicity by improving the surface roughness [96–100]. It has been recognized that the cooperation between the surface chemical compositions and the topographic structures is crucial to construct special wettability. In this respect, a promising strategy to regulate wetting behavior is the combination of surface patterning and chemical surface modification. Therefore, polymer substrates can be patterned using physical methods to selectively degrade hydrophobic regions to result in surface portions of enriched hydrophobicity. Many elegant methods have been developed to construct microstructures and nanostructures, such as laser ablation and photolithography-based microfabrication [101–103], microwave plasma-enhanced CVD of trimethoxylmethoxysilane [104, 105], domain-selective oxygen plasma treatment [106], sol-gel method [107], etc.

To cite some representative examples of the most recent surface wettability modifications, IBM researchers developed a method to produce patterns with modified wettability *via* the deprotection of poly(*tert*-butyl acrylate) films to poly(acrylic acid) (PAA) using acids generated during lithographic patterning of the resist [108]. Wettability patterning could be achieved by masking only defined regions during a first modification, thus introducing hydrophobic pendant groups, and then backfilling in a second modification step with hydrophilic groups [109]. Moreover, Huck and colleagues [110] modified the wetting behavior of polyelectrolyte brushes simply by exchanging different ions into the brush.

Worth mentioning in the control of the wetting phenomenon is the ability of some "smart polymers" to switch between super-hydrophilicity and super-hydrophobicity. For example, Sun *et al.* [111] demonstrated this property for poly(*N*-isopropylacrylamide) (PNIPAAm), a thermal responsive polymer, and more interestingly, they demonstrated that this property can be greatly influenced by the surface roughness of the substrate (see Fig. 1.6).

1.2.2.2 Improvement of porosity or roughening

Porous polymer surfaces are required in applications such as separation processes, filters, foams, membranes, chromatography, biomaterials, paper coatings, superconductors, etc. Capability for mass transfer, liquid retention, improved opacity, lighter

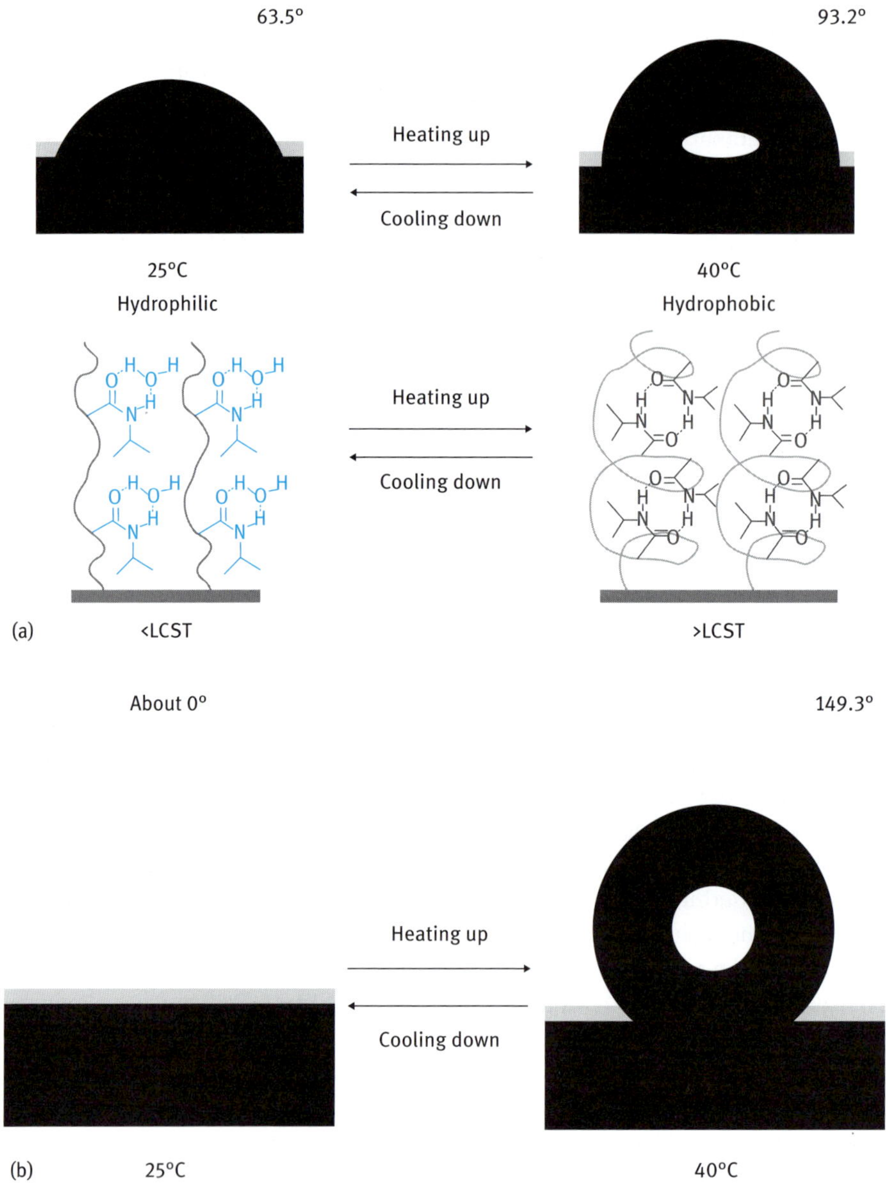

Fig. 1.6: Schematic diagram of reversible switching between super-hydrophilicity and super-hydrophobicity of PNIPAAm film: water drop profiles on a flat surface (a) and on a rough surface (b). The lower critical solution temperature (LCST) of PNIPAAm is about 32°C. Reproduced by Sun T, Wang G, Feng L, Liu B, Ma Y, Jiang L, Zhu D, *Angew Chem Int Ed* 43, 357–360, 2004, with permission. Copyright 2004, Elsevier [111].

weight, enhanced gloss, controlled spread and imbibition, control of heat conduction are properties strictly related to porous substrates. In all these contests, the pore size, distribution, geometry, stability could be critical features. Porous resins are widely used in chromatography [112]. Moreover, the presence of a variety of functional groups on polymer surface, which enable a specific sorption of various chemical species, allows the ability to operate as chromatographic separation systems [113]. Porous membranes are used in dialysis, desalination, filtration, gas separation, reverse osmosis, microfiltration, etc. Porous materials are also used as drug delivery systems [114]. Polymers such as poly(urethanes), poly(siloxanes), or poly(vinylic) systems have been used to control the release of pharmacologically active agents. Various typologies of pores (i.e. open, closed, transport, and blind pores) allow the polymer to adsorb and release drugs in a more reproducible and predictable manner. Pharmaceutically exploited porous adsorbents include mesoporous, macroporous, microporous, and nanoporous scales. Nanoporous systems represent the new and current challenge in pharmaceutical research [115]. When a porous polymeric drug delivery system is placed in contact with an appropriate medium, the release of drug to medium must be preceded by the drug dissolution from surface and successively by diffusion through the liquid filled channels or pores.

A controlled porosity is also imperative in superabsorbent polymers, i.e. polymers able to absorb very large amount of liquid, mostly used in the market of personal disposable hygiene absorbent products (baby diapers, adult protective underwear, and sanitary napkins) [116]. In all these fields, hydrogels, i.e. polymers or copolymers opportunely cross-linked and able to absorb great amounts of water without dissolving in this liquid, became the main class of polymers used. The pore structure in these materials is fundamental: a connected microporosity in the dry hydrogel is required because it increases the hydrogel sorption capacity for capillary retention effects. The hydrogel-interconnected, molecular-sized pores are also of great interest for absorption, separation, and catalytic applications.

As far as tissue engineering, and regenerative medicine research are concerned, the design, production, and characterization of porous networks result in demanding tasks: large pores and highly porous structures are required to promote *in vivo* direct osteogenesis and to allow vascularization and proper oxygen supply to the cells, whereas smaller pores result in osteochondral ossification. The interconnectivity of such pores is also significant because phenomena such as cell migration, ingrowth, vascularization, and nutrient diffusion for cell survival are inhibited if pores are not interconnected, even if the structures possess an elevated porosity.

Different procedures have been developed to obtain porous polymeric systems:

- Solvent casting/particle leaching: This technique is based on the dispersion of a porogen, which is usually an inorganic salt, into a polymer solution. The solvent evaporation produces a polymer-porogen network. The solute particles are subsequently leached or dissolved away by immersing the material in a selective solvent, resulting in the formation of a porous network [117]. The surface of these porous systems is particularly important because the solvent evaporation starts at the interface polymer/air and the first step of pore formation in the solvent for salt removal is governed by surface rearrangement of polymer surface in the selected liquid.

- Freeze drying: It consists of freezing the material and then reducing the surrounding pressure to allow the frozen water in the polymer to pass directly from the solid to the gas phase; thus, the porosity is generated from the direct sublimation of water.
- Gas foaming: The use of a blowing agent, in gaseous state, i.e. a gaseous porogen, generates a highly porous system. The porosity depends upon the amount of gas dissolved in the polymer. This fabrication method requires no leaching step and uses no harsh chemical solvents but usually requires high pressures and/or temperatures [118].
- Electrospinning: It is a simple, versatile, and widely accepted technique to produce fibers ranging from nanometers to micrometers, mostly in solvent solution and some in melt form. In this process, a high voltage is used to create an electrically charged jet of polymer solution or melt, which forms polymer fibers after drying or solidification [119]. Today, more than 200 polymers are used for electrospinning [120–123]. Scanning probe and electron microscopies (SPM and SEM) represent the most used techniques to ascertain the surface size, orientation, and uniformity of the electrospun nanofibers [124, 125].
- Phase inversion: In this technique, the principal pore-forming process is the use of a ternary system, i.e. polymer, solvent, and non-solvent. Pore size and distribution can be controlled for each specific application depending on the choice of ternary system and preparation parameters [126].
- Microfabrication processes or computer-aided additive biofabrication: These represent the newest and nonconventional scaffold manufacturing processes. The computer-aided fabrication processes include direct writing, fused deposition modeling, selective laser sintering, stereolithography, dimensional printing, etc. The pore size and geometry of the scaffold is entirely computer controlled. A complete presentation of these techniques was recently summarized by Bàrtolo *et al.* [127]. All these innovative processes should require a deep investigation on the surface properties of the processed polymers.

Meanwhile, the surface roughness represents another important property of polymer surfaces. For instance, in the biomaterial field, adhesion, proliferation, differentiation, and overall cell viability on polymer surfaces can be significantly affected by roughness [128, 129]. It could be increased by mechanical treatments, wet-chemical treatments with strong acids or bases, or by exposure to flames or corona discharge. At the nanometer scale, roughness could be obtained by grafting spacers having different molecular weight or different lengths to the polymer surface [130].

For other practical purposes, surface finish is crucial for both aesthetics and functional features of commercial polymers. In the case of reinforced thermoplastic materials, i.e. thermoplastic polymers reinforced with special additives during the molding step, surface finish variables must be considered in choosing the optimum material, in terms of additives and processing techniques. Aesthetics is important in some appearance products (glossy, shiny, lustrous): specular or metallic-like surfaces may be often desired; therefore, surface roughness must be controlled in order to conform to standard and variable molding parameters. Not only aesthetic but also functional features are linked to the surface finish when parameters such as wear resistance, coefficient of friction, or adhesion to other materials (plastics-on-plastics, plastics-on-metals, etc.) are affected by the surface smoothness or roughness.

1.2.2.3 Improvement of adhesion

Adhesion can be defined as the chemical (i.e. bond) or physical (i.e. adsorption) anchoring between two adjacent materials. Adhesion strength is measured as the cohesive force involved when separation occurs within two materials.

A typical embodiment where adhesion is a critical property is the coating of a substrate. The factors contributing to such property are essentially chemical or mechanical.

The durability of a coating onto the substrate is of prime importance in many applications: the coating must sustain mechanical, and thermal stresses, as well as elastoplastic distortions. If adhesion is poor, the localized break of the coating-substrate interface takes place, leading to blistering or even complete spalling off of the coating. This is particularly true if the coating is subject to corrosion or moisture: indeed, in an oxidative or humid environment, the tendency for the film to crack or peel from the substrate may be well aggravated.

Often, much more than two layers of different materials are required to develop a complex technological device. An example is supplied by solar cells, where usually a multilayer structure, such as that reported in Fig. 1.7, is proposed [131].

Usually, improvement of adhesion could be carried out by surface preparation of the substrate, for example, through contamination removal by different cleaning procedures (aimed to remove protective layers, organic contaminants, adsorbed layers, dust, and fingerprints) or by surface activation [132].

Different approaches can be adopted to alter the substrate surface and thus enhance adhesion:

- Mechanical: Roughening surfaces by grinding, brushing, blasting, or similar methods creates enough sites where glue or paint can physically grab on. The cutting action on thc surface during such a treatment may also cut polymer chains, creating open or pendant bonds exposed at the surface. This could allow for improved chemical bonding on the surface.
- Chemical: Treating surfaces with chemicals can alter the surface chemistry to improve the surface energy by providing active chemical bonds or groups on the polymer surface.
- Plasma treatments: Surface preparation of substrates can be carried out by plasma processes. Surface activation and contamination removal can be attained without creating hazardous by-products and without changing bulk properties. Surface activation by plasma treatment is performed using the right mixture of

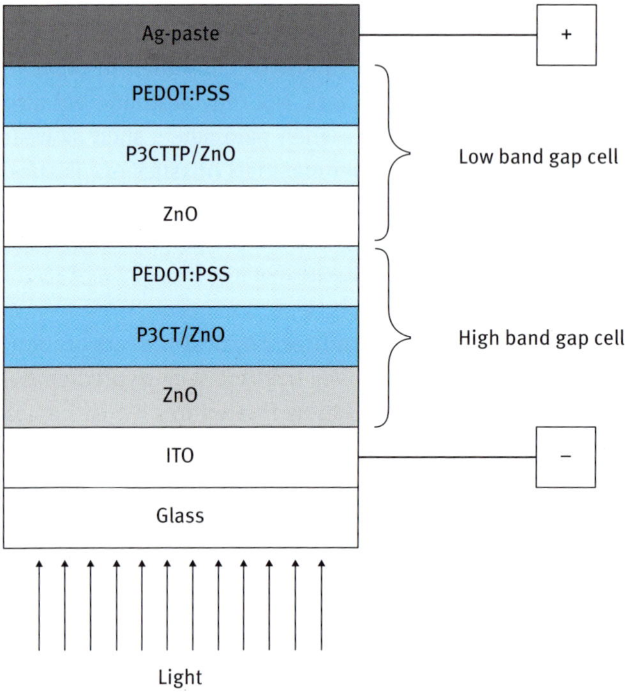

Fig. 1.7: An example of tandem polymer solar cell comprising eight layers. PEDOT:PSS, poly-(ethylenedioxythiophene):polystyrenesulfonic acid; P3CTTP/ZnO, poly(carboxyterthiophene-*co*-diphenylthienopyrazine) with ZnO nanoparticles; P3CT/ZnO, poly-(3-carboxydithiophene) with ZnO nanoparticles; ITO, indium tin oxide. Reproduced from Krebs FC, *Sol Energy Mater Sol Cells* 93, 394–412, 2009, with permission. Copyright 2009, Elsevier [131].

gases generating excited species that react with the polymer surface to create functional groups, such as hydroxyl (-OH), carbonyl (-C=O), carboxyl (-COOH), or amino (NH_x), that exhibit high polarity and change the base-acid interaction at the polymer surface. Different researchers highlighted the usefulness of plasma treatments for adhesion improvement [133, 134]. The plasma processes are used not only for adhesion promotion but also in a very wide variety of fields (microelectronics and optoelectronics, biomedical field, etc.) and with different aims (contamination removal, surface activation, polymerization, etching, cross-linking, etc.). In this respect, this chapter is not comprehensive on this matter, and in-depth dissertations on plasma processes are reported elsewhere [95].

Both mechanical and chemical improvements of adhesion can be monitored and evaluated by a number of analytical techniques, such as atomic force microscopy (AFM) and SEM, for morphological evaluation of surface modification, and FTIR, secondary ions mass spectrometry (SIMS), and XPS, for detection of new functional groups originating on the surface.

1.2.2.4 Interaction of polymers with biological environment: Biocompatibility

The biomaterial field probably represents the research field where the boundaries among traditional sciences, such as chemistry, physics, mathematics, and biology, are increasingly superimposing and connecting with those of engineering and medicine in order to find new and more sophisticated solutions to medical problems and finally to develop products to facilitate the recovery of the human body.

Biomaterials can be considered as all the materials that interface with a biological system such as catheters, prostheses, fracture fixation devices, intraocular lenses or soft contact lenses, pacemakers, artificial hearts, cardiac assist devices, and, more recently, drug delivery systems, microfluidics, biochips, robotics, etc. It is difficult to find a device that is not fabricated, totally or partially, using polymeric materials. Traditionally, a biomaterial, as defined by Park *et al.* [135], was considered a material used to replace a part of a living system or a material to be placed in intimate contact with living tissue. More recently, biomaterials are thought as materials that could deliberately elicit a particular response from the body to achieve a therapeutic goal. The effectiveness of a biomaterial placed in contact with human fluids or cells is strongly dependent on its interactions with this biological environment, which determine how it is accepted or rejected.

In general, a controlled biological response is primarily governed by the chemistry and topography of the outermost layers of biomaterials. The properties of interest for biomaterial surfaces include chemical composition, hydrophilicity/hydrophobicity balance, morphology (i.e. the domain structure), and topography (i.e. the surface roughness, planarity, and feature dimensions).

No common rules exist in selecting surface biomaterial properties: in the case of stents, for instance, no cell adhesion is desirable, whereas in tissue engineering applications, the main effort is to promote cell adhesion and proliferation on polymer surfaces [136]. For polymeric biomaterials, surface engineering is often required: the polymer surface, effectively, can be physically, chemically, or biologically modified by different techniques and for different purposes without affecting the key bulk physical properties of polymers. Surface engineering approaches to develop improved biomaterial performances can be generally divided into physical-chemical and biological methods.

Examples of physical-chemical methods are the following:

1. Selective etching: This method increases surface roughness, changes surface chemical constitution, degrades low-molecular-weight compounds that migrate to the surface, and mitigates residual surface stresses [137]. It can be carried out by acids or bases as well as by irradiation treatments exposed below.

2. Ionizing irradiation treatments such as plasma [138], ion [139], electron beam [140], laser treatments [141], and UV irradiation [142] are used to develop micro- and nano-patterning of biomedical surfaces or to create new surface functional groups.
3. Film deposition, i.e. self-assembled monolayer deposition [143] or Langmuir-Blodgett films [144].
4. Surface grafting of functional groups by wet chemical reactions, such as fluorination, organosilanization [145], or PPM [146]. Post-polymerization modification incorporates functionalities into polymer chains by decorating already-synthesized polymers with an appropriate chemical and/or biological species.

Meanwhile, examples of biological treatments, which are probably at the forefront of surface modification techniques of polymeric biomaterials, can be divided into these main classes:

- Covalent attachment of bioactive compounds to biologically functionalize polymer surfaces. It is possible to produce complex brush systems with position-dependent distributions of the chemical modifiers. A number of different types of biologically active molecules have been covalently incorporated into polymer systems and/or grafted on their surface: amino acids [147, 148], short sequences of peptides [149] or specific proteins and enzymes [150–152], antibodies [153], lipids [154], nucleotides or polynucleotides [155], etc.
- Surface absorption, embedding, or non-covalent immobilization of biologically active molecules using hydrogels [156–158], foams or scaffolds [159], microparticles, nanoparticles, or micelles [160, 161], biodegradable/bioerodible polymer matrices [162–164], etc.
- Cell pre-seeding on polymer biomaterials [165].

In any case, it is clear that an in-depth characterization of biomaterial surfaces is crucial to relate their modifications with changes in biological response: all the information gained from surface analysis may allow for tailoring future improvements in surface modification.

1.2.2.5 Improvement of conductivity

Plastics are usually conceived as insulating materials. Nevertheless, the improvement of polymer conductivity is one of the most representative examples of the possibility to act on specific physical properties, tailoring for a potential integration of polymers in electronic devices.

The use of polymers in electronics technology has several advantages over conventional inorganic (i.e. silicon-based) materials: lower costs and ease of manufacturing, large flexibility and variety of substrates and modifications. Indeed, the polymer structural order can be varied from purely amorphous to highly crystalline and its shape can range from fibers, films, roads, to spheres or pellets.

The majority of polymers does not possess an inherent conductivity; however, we have seen in a previous paragraph that polymers characterized by a conjugated backbone can exhibit inherent conductivity. In double bonds, the electrons are localized in one σ and one π bond; in the last one, the electrons are more mobile and can flow over the whole system, giving rise to an electron delocalization [166]. Although metals are able to conduct electricity *via* mobile electrons in the conduction band that travels across atoms in the lattice, an equivalent situation in organic materials is provided by the delocalized π-electrons. Only a few polymers are inherently conductive (see Fig. 1.8).

However, electron conjugation is not enough to make a polymer conductive. Therefore, a polymer material needs to be doped for electron flow to occur. The main strategy to improve the conductivity of a polymer is the oxidation or reduction by electron acceptors or donors, resulting in p- or n-type doped materials, respectively. The formation of a complex between the polymer and the dopant produces a considerable increase in electron mobility. Electrical conductivities can be varied by as much as 15 orders of magnitude by changing the dopant concentrations so that the electronic property control is feasible over the entire range from the insulator to the semiconductor. Although, typically, doping conductive polymers involves oxidation or reduction, chemical, electrochemical, photochemical, and ion implantation methods are also available [167, 168]. Moreover, different additives can be used to improve the conductivity of polymers, developing the so-called nanocomposite conductive materials: carbon fibers, carbon black, carbon nanotubes or nanorods, graphene oxide, layered titanates, rutiles, phosphates or phosphonates, metallic-based nanoparticles, etc. [169].

Modern applications of conductive polymers are in biosensors [170], electronic noses [171, 172], lab-on-a-chip [173], biomaterials [174], diagnostic devices [175], solar cells [176], etc.

A variety of techniques have been widely used to delineate the physical properties of the conjugated polymers. For example, electrochemical characterization provides

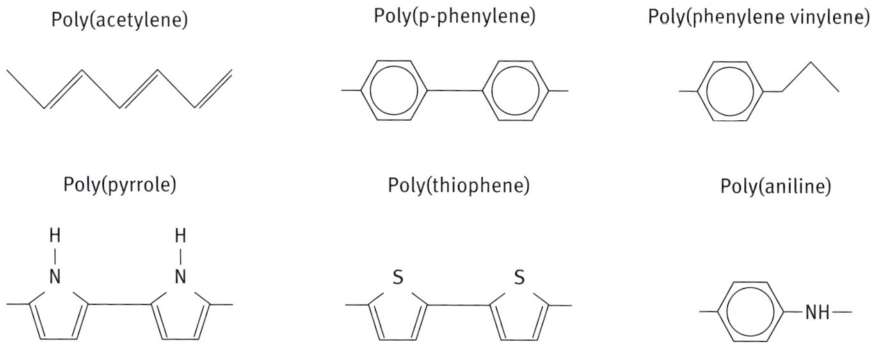

Fig. 1.8: Main examples of electrically conductive polymers.

information regarding redox behavior, number of electrons in the redox reaction, and diffusion coefficient estimation of conducting polymers [177].

Because, often, the interfacing of different polymers is carried out in these integrated devices and surface doping or surface modification of polymers is necessary, the outmost layer properties and features become an important issue. A typical surface technique currently used to analyze the nanoscale properties of thin films of conducting polymers is AFM, which considers information gained by its extension called phase imaging. For example, in the case of poly(bithiophene), which is deposited under potentiostatic and potentiodynamic conditions, the resulting morphologies of the films prepared using the two techniques are quite similar when analyzed by conventional AFM, whereas the phase contrast measurements revealed profound differences in the mechanisms of potentiostatic and potentiodynamic electropolymerization [178]. In addition, spectroscopic techniques have been successfully used to characterize conductive polymers: Meana-Esteban *et al.* [179] pointed out an experimental approach to correlate the Raman and the ATR spectra from azulene-based polymers according to the effective conjugation coordinate theory, observing that the IR spectra of the conducting state showed new doping-induced IR active vibrations.

Considering that the above-mentioned techniques are only two of the copious examples of surface investigations performed on conductive polymers, it must be emphasized that surface characterization plays a significant role for the correct application of polymers to electronic devices.

1.2.3 Possibility of predicting polymer performances by surface characterization techniques

Today, there is a huge demand for new functional properties of polymer interfaces in many academic and industrial branches, such as biotechnology, nanotechnology, biomaterials, electronics, coating applications, etc. Even if an almost complete description of methods for the characterization of bulk and surface properties of homopolymers is supplied by several books or databases, increasingly more complex structures (such as hybrid organic-inorganic materials, multilayer films, nanomaterials, copolymers or polymer blends) require more specialized characterization for end-use performance evaluation. Obviously, it is difficult to accurately predict end-use performance characteristics of the final product using tabulated data of individual components. As a result, an accurate characterization of these new systems is necessary, rather than using model polymers or components. In particular, knowing the surface properties and evolution of polymer materials when in contact with different environments can allow to understand the response of the material to external stimuli and ultimately to choose between the myriads of polymers and polymer combinations. If a rational design approach can be used to plan the synthetic conditions of a polymer-based material, feedback data obtainable from surface characterization in conjunction with molecular modeling programs yield a prediction of the desired properties.

Predicting the performance of a polymer-based device is clearly linked to the knowledge of molecular processes occurring on surface because those can drive the molecular, microscopic, and macroscopic structure and the resulting properties of the material.

Different studies have been devoted to develop a theoretical understanding of behavior of surfaces and interfaces in multi-constituent polymers, starting from the examination of pure component surfaces. These data can be used to either validate or construct predictive thermodynamic theories of polymeric interphase properties. An example is supplied by Penn and Bowler [180], who exploited CA data to predict the adhesive performances of different interfaces. Moreover, in nanoclay-reinforced polymers for high-technology applications, the surface characterization of the hybrid material, performed by XRD and SEM, was used to evaluate the impact of nanoclay reinforcement on hardness and dynamic-mechanical properties. Both mathematical regression models and measured values have been used for theoretical prediction [181]. Surface properties are also important in predicting the polymer service lifetime: to monitor the polymer changes with environmental exposures from the early stages of degradation, surface characterization techniques are required. Many analytical tools provide actual solutions in corrosion, hydrothermal, photothermal, and thermal degradation studies, giving information about the changes of functional groups by degradation processes. An interesting study of polymer oxo-biodegradation is reported in literature, evidencing the need of analytical tools to monitor the degradation process under simulated test conditions [182].

In industrial polymer research, the ability to predict the polymer performances is pressing: the main goal is to develop novel products able to meet specific end-use performance requirements while differentiating them from competitors and maintaining their production cost-effective. In this respect, surface characterization allows to select the right polymer for the right application, optimize the technological parameters of production, solve processing failures (such as degradation, incorrect mixture of additives, phase separation, aging, contamination, presence of reaction by-products, etc.), meet regulatory requirements, envisage suitable treatments for polymer modification, validate cleanliness, assess leachability/toxicity, predict shelf-life and finally collect all the obtained information to aid in the rational design of surface properties to accumulate a very-large-scale data sets and reference data, for interlaboratory comparisons. Meanwhile, in the academic field, the main goal is to find new materials to be successively transferred to the industrial field, supplying more technologically advanced properties: examples are self-cleaning ability, switchability, superior tribological/mechanical properties, biodegradability, long-term stability, corrosion protection, superhydrophobicity or super-hydrophilicity, etc.

Recently, a pioneering tendency is to study reactions occurring on surface under conditions similar to the real ones because a surface analysis may not be often able to simulate the real conditions of polymeric materials daily applications due to the operative restrictions required by many techniques. In this respect, new operating conditions are provided, for example, using ambient-pressure X-ray photoelectron spectroscopy (AP-XPS or APPES), ultra-low-energy secondary ion mass spectrometry, environmental scanning electron spectroscopy (ESEM), etc. A deep dissertation on the operative conditions, applications, possibilities, and limiting features of the main surface techniques used in polymer science is supplied in the course of this book.

References

[1] Baekeland LH. The chemical constitution of resinous phenolic condensation products. Ind Eng Chem 1913;5:506–11.

[2] Natta G, Pino P, Corradini P, et al. Crystalline high polymers of α-olefins. J Am Chem Soc 1955;77:1708–10.

[3] Ziegler K, Holzkamp E, Breil H, Martin H. Das Mulheimer Normaldruck Polyathylen-Verfahren. Angew Chem 1955;67:541–7.

[4] IUPAC. Basic classification and definitions of polymerization reactions. Pure Appl Chem 1994;66:2483–6.

[5] Metanomski WV. IUPAC. Compendium of macromolecular nomenclature. Oxford (UK): Blackwell Science; 1991.

[6] Young RJ, Lovell PA. Introduction to polymers. 3rd edn. Boca Raton (FL): CRC Press; 2011.

[7] Hiemenz PC, Lodge TP. Polymer chemistry. 2nd edn. Boca Raton (FL): CRC Press; 2007.

[8] Billmeyer FW Jr. Textbook of polymer science. 3rd edn. Manhattan (NY): Wiley; 1984.

[9] Stevens MP. Polymer chemistry: an introduction. 3rd edn. Oxford (UK): Oxford University Press; 1999.

[10] Prasad Raoa J, Geckelera KE. Polymer nanoparticles: preparation techniques and size-control parameters. Prog Polym Sci 2011;36:887–913.

[11] Ramamurthy AV. Wall slip in viscous fluids and influence of materials of construction. J Rheol 1986;30:337–57.

[12] Larson RG. Instabilities in viscoelastic flows. Rheol Acta 1992;31:213–63.

[13] Dutta D, Fruitwala H, Kohli A, Weiss RA. Polymer blends containing liquid crystals: a review. Polym Eng Sci 1990;30:1005–18.

[14] Pawlikowski GT, Dutta D, Weiss RA. Molecular composites and self-reinforced liquid crystalline polymer blends. Annu Rev Mater Sci 1991;21:159–84.

[15] LaMantia FP, Valenza A. Shear flow characterization of blends containing liquid crystal polymers. Makromol Chem Makromol Symp 1992;56:151–9.

[16] Rosato DV, Rosato DV, Reinforced plastics handbook. Oxford (UK): Elsevier Advanced Technology; 2005.

[17] Bhattarai S, Bhattarai R, Yi HN, et al. Novel biodegradable electrospun membrane: scaffold for tissue engineering. Biomaterials 2004;25:2595–602.

[18] Suthar A, Chase G. Nanofibres in filter media. Chem Eng 2001;726:26–8.

[19] Chatterjee A, Deopura BL. Carbon nanotubes and nanofibre: an overview. Fiber Polym 2002;3:134–9.

[20] Sundararajan S, Bhushan B, Namazu T, Isono Y. Mechanical property measurements of nanoscale structures using an atomic forcemicroscope. Ultramicroscopy 2002;91:111–8.

[21] Wilson EB, Wells AJ. Infra-red and raman spectra of polyatomic molecules XIV. Propylene. J Chem Phys 1941;9:319–22.
[22] Wilson EB, Decius JC, Cross PC. Molecular vibrations. New York (NY): McGraw-Hill; 1955.
[23] Volkenstein MV. Configurational statistics of polymer chains. New York (NY): Interscience; 1963.
[24] Birshtein TM, Pticin OB. Conformation of macromolecules. New York (NY): Interscience; 1966.
[25] Flory PG. Statistical mechanics of chain molecules. New York (NY): Interscience; 1969.
[26] Yamakava H. Modern theory of polymer solutions. New York (NY): Harper & Row; 1971.
[27] de Gennes PG. Scaling concepts in polymer physics. Ithaca (NY): Cornell University Press; 1979.
[28] Grosberg AYu, Khokhlov AR. Statistical physics of macromolecules. New York (NY): AIP; 1994.
[29] Gotlib YuYa, Darinskii AA, Svetlov YuE. Physical kinetics of macromolecules. Saint Petersburg (Russia): Khimia; 1986.
[30] Grosberg AYu, Khokhlov AR. Giant molecules: here, there, and everywhere. San Diego (CA): Academic Press; 1997.
[31] Doi M, Edwards SF. The theory of polymer dynamics. New York (NY): Clarendon Press, Oxford University Press; 1986.
[32] Fatkullin N, Kimmich R. Theory of field-gradient NMR diffusometry of polymer segment displacements in the tube-reptation model. Phys Rev E 1995;52:3273–6.
[33] Fischer E, Kimmich R, Fatkullin N, Yatsenko G. Segment diffusion and flip-flop spin diffusion in entangled polyethyleneoxide melts: a field-gradient NMR diffusometry study. Phys Rev E 2000;62:775–82.
[34] Dutcher JR, Marangoni AG. Soft materials: structure and dynamics. New York (NY): CRC Press; 2004.
[35] Car R, Parrinello M. Unified approach for molecular dynamics and density functional theory. Phys Rev Lett 1985;55:2471–4.
[36] Teraoka I. Polymer solutions: an introduction to physical properties. New York (NY): Wiley; 2002.
[37] Paul W, Grant DS. Structure and dynamics of amorphous polymers: computer simulations compared to experiment and theory. Rep Prog Phys 2004;67:1117–85.
[38] Reith D, Milchev A, Virnau P, Binder K. Computer simulation studies of chain dynamics in polymer brushes. Macromolecules 2012;45:4381–93.
[39] Grest GS, Kremer K. Molecular dynamics simulation for polymers in the presence of a heat bath. Phys Rev A 1986;33:R3628–31.
[40] Grest GS, Kremer K. Dynamics of entangled linear polymer melts: a molecular-dynamic simulation. J Chem Phys 1990;92:5057–86.
[41] Li YC, Wang CP, Liu XJ. Assessment of thermodynamic properties in pure polymers. Calphad 2008;32:217–26.
[42] Safran SA. Statistical thermodynamics of soft surfaces. Surf Sci 2002;500:127–46.
[43] de Gennes P-G. Mechanical properties of polymer interfaces. In: Sanchez IC, ed. Physics of polymer surfaces and interfaces. Boston (MA): Butterworth-Heinemann; 1992:55–71.
[44] Fuoss PH, Norton LJ, Brennan S. X-ray scattering studies of the melting of lead surfaces. Phys Rev Lett 1988;60:2046–9.
[45] Mailaender L, Dosch H, Peisl J, Johnson RL. Near-surface critical x-ray scattering from Fe_3Al. Phys Rev Lett 1990;64:2527–30.
[46] Kumar SK, Vacatello M, Yoon DY. Off-lattice Monte Carlo simulations of polymer melts confined between two plates. J Chem Phys 1988;89:5206–15.
[47] Liu Y, Russell TP, Samant MG, et al. Surface relaxations in polymers. Macromolecules 1997;30:7768–71.
[48] Keddie J, Jones RAL, Cory RA. Interface and surface effects on glass transition temperature in thin polymer films. Faraday Discuss 1994;98:219–30.
[49] Wallace WE, van Zanten JH, Wu WL. Influence of an impenetrable interface on a polymer glass-transition temperature. Phys Rev E 1995;52:R3329–32.

[50] Xie L, DeMaggio GB, Frieze WE, et al. Positronium formation as a probe of polymer surfaces and thin films. Phys Rev Lett 1996;74:4947–50.
[51] DeMaggio GB, Frieze WE, Gildey DW, Zhu M, Hristov HA, Yee AF. Interface and surface effects on the glass transition in thin polystyrene films. Phys Rev Lett 1997;78:1524–7.
[52] Kajiyama T, Tanaka K, Takahara A. Surface molecular motion of monodisperse polystyrene films. Macromolecules 1997;30:280–5.
[53] Priestley RD, Ellison CJ, Broadbelt LJ, Torkelson JM. Structural relaxation of polymer glasses at surfaces, interfaces, and in between. Science 2005;15:456–9.
[54] Carelli C, Young RN, Jones RAL, Sferrazza M. Chain length effects on confined polymer/polymer interfaces. Europhys Lett 2006;75:274–80.
[55] Werner C, Jacobasch HJ. Surface characterization of polymers for medical devices. Int J Artif Organs 1999;22:160–76.
[56] Awaja F, Gilbert M, Kelly G, Fox B, Pigram PJ. Adhesion of polymers. Prog Polym Sci 2009;34:948–68.
[57] Köstler S, Delgado AV, Ribitsch V. Surface thermodynamic properties of polyelectrolyte multilayers. J Colloid Interface Sci 2005;286:339–48.
[58] Luzinov I, Minko S, Tsukruk VV. Adaptive and responsive surfaces through controlled reorganization of interfacial polymer layers. Prog Polym Sci 2004;29:635–98.
[59] Stachewicz U, Li S, Bilotti E, Barber AH. Dependence of surface free energy on molecular orientation in polymer films. Appl Phys Lett 2012;100:094104-094104-4.
[60] Geoghegan M, Krausch G. Wetting at polymer surfaces and interfaces. Prog Polym Sci 2003;28:261–302.
[61] Johnston EE, Ratner BD. Surface characterization of plasma-deposited organic thin-films. J Electron Spectrosc Relat Phenom 1996;81:303–17.
[62] Klee D, Hocker H. Advances in polymer science: biomedical application/polymer blends. Biomed Appl Polym Blends 1999;149:1–55.
[63] Holly FJ, Refojo MF. Wettability of hydrogels. I. Poly (2-hydroxyethyl methacrylate). J Biomed Mater Res 1975;9:315–26.
[64] Ratner BD. Biomaterials: interfacial phenomena and applications. In: Cooper SL, Peppas NA, eds. ACS Adv Chem Ser. Washington (DC): American Chemical Society; 1982;199:9.
[65] Ratner BD. Surface structure of polymers for biomedical applications. Makromol Chem Makromol Symp 1988;19:163–78.
[66] Ruckenstein E, Gourisankar SV. Surface restructuring of polymeric solids and its effect on the stability of the polymer-water interface. J Colloid Interface Sci 1986;109:557–66.
[67] Lee SH, Ruckenstein E. Stability of polymeric surfaces subjected to ultraviolet irradiation. J. Colloid Interface Sci 1987;117:172–8.
[68] Andrade JD, Gregonis DE, Smith LM. Polymer surface dynamics. In: Andrade JD, ed. Surface and interfacial aspects of biomedical polymers. Vol I: surface chemistry and physics. New York (NY): Plenum Press; 1985;1:15–41.
[69] Yasuda H, Charlson EJ, Charlson EM, Yasuda T, Miyama M, Okuno T. Dynamics of surface property change in response to changes in environmental conditions. Langmuir 1991;7:2394–400.
[70] Drost-Hansen W. Structure of water near solid interfaces. Ind Eng Chem 1969;61:10–47.
[71] Yasuda T, Okuno T, Yasuda H. Contact angle of water on polymer surfaces. Langmuir 1994;10:2435–9.
[72] Yasuda H, Sharma AK, Yasuda T. Effect of orientation and mobility of polymer molecules at surfaces on contact angle and its hysteresis. J Polym Sci A 1981;19:1285–91.
[73] Suetaka W. Surface infrared and Raman spectroscopy methods and applications. New York (NY): Plenum Press; 1995.

[74] Guo TF, Yang Y. In situ study on the reorientation of polymer chains in operating polymer diodes. Appl Phys Lett 2002;80:148–50.
[75] Xue L, Zhang J, Han Y. Phase separation induced ordered patterns in thin polymer blend films. Progr Polym Sci 2012;37:564– 94.
[76] Wu DT, Fredrickson GH. Effect of architecture in the surface segregation of polymer blends. Macromolecules 1996;29:7919–30.
[77] Minnikanti VS, Archer LA. Surface enrichment of branched polymers in linear hosts: effect of asymmetry in intersegmental interactions and density gradients. J Chem Phys 2005;122:84904.
[78] Minnikanti VS, Archer LA. Surface migration of branched molecules: analysis of energetic and entropic factors J Chem Phys 2005;123:144902.
[79] Yul LY, Liu X, Vogler EA, Donahue HJ. Systematic variation in osteoblast adhesion and phenotype with substratum surface characteristics. J Biomed Mater Res A 2003;68:504–11.
[80] Xu LC, Siedlecki CA. Effects of surface wettability and contact time on protein adhesion to biomaterial surfaces. Biomaterials 2007;28:3273–83.
[81] Wang R, Hashimoto K, Fujishima A, et al. Light-induced amphiphilic surfaces. Nature 1997;388:431–2.
[82] Chen W, Fadeev AY, Hsieh MC, Öner D, Youngblood J, McCarthy TJ. Ultrahydrophobic and ultralyophobic surfaces: some comments and examples. Langmuir 1999;15:3395–99.
[83] Feng L, Li S, Li Y, et al. Super-hydrophobic surfaces: from natural to artificial. Adv Mater 2002;14:1857–60.
[84] Blossey R. Self-cleaning surfaces – virtual realities. Nat Mater 2003;2:301–6.
[85] Lafuma A, Quéré D. Superhydrophobic states. Nat Mater 2003;2:457–60.
[86] Woodward JT, Gwin H, Schwartz DK. Contact angles on surfaces with mesoscopic chemical heterogeneity. Langmuir 2000;16:2957–61.
[87] Sun T, Song W, Jiang L. Control over the responsive wettability of poly(N-isopropylacrylamide) film in a large extent by introducing an irresponsive molecule. Chem Commun (Camb) 2005:1723–5.
[88] Garvin CJ, Genzel J. Applications of surface-grafted macromolecules derived from post-polymerization modification reactions. Prog Polym Sci 2012;37:871– 906.
[89] Castell P, Wouters M, de With G, Fischer HR, Huijs F. Surface modification of poly(propylene) by photoinitiators: improvement of adhesion and wettability. J Appl Polym Sci 2004;92:2341–50.
[90] Maji D, Lahiri SK, Das S. Study of hydrophilicity and stability of chemically modified PDMS surface using piranha and KOH solution. Surf Interface Anal 2012;44:62–9.
[91] Arayanarakool R, Shui L, van den Berg A, Eijkel JCT. A new method of UV-patternable hydrophobization of micro-and nanofluidic networks. Lab Chip 2011;11:4260–6.
[92] Gang Z, Wang D, Gu ZZ, Möhwald H. Fabrication of superhydrophobic surfaces from binary colloidal assembly. Langmuir 2005;21:9143–8.
[93] Lawton RA, Price CR, Runge AF, Doherty WJ, Saavedra SS. Air plasma treatment of submicron thick PDMS polymer films: effect of oxidation time and storage conditions. Colloids Surf A 2005;253:213–5.
[94] Zhao LH, Lee J, Sen PN. Long-term retention of hydrophilic behavior of plasma treated polydimethylsiloxane (PDMS) surfaces stored under water and Luria-Bertani broth. Sens Actuators A 2012;181:33–42.
[95] Rossnagel SM, Cuomo JJ, Westwood WD. Handbook of plasma processing technology, fundamentals, etching, deposition, and surface interactions. Westwood (NJ): Noyes Publication; 1990.
[96] Onda T, Shibuichi S, Satoh N, Tsujii K. Super-water-repellent fractal surfaces. Langmuir 1996;12:2125–7.

[97] Erbil HY, Demirel AL, Avci Y, Mert O. Transformation of a simple plastic into a superhydrophobic surface. Science 2003;299:1377–80.
[98] Chaudhury MK, Whitesides GM. How to make water run uphill. Science 1992;256:1539–41.
[99] Guriyanova S, Semin B, Rodrigues TS, Butt HJ, Bonaccurso E. Hydrodynamic drainage force in a highly confined geometry: role of surface roughness on different length scales. Microfluid Nanofluid 2010;8:653–63.
[100] Kaggwa G.B, Froebe S, Huynh L, Ralston J, Bremmell K. Morphology of adsorbed polymers and solid surface wettability. Langmuir 2005;21:4695–704.
[101] He B, Patankar NA, Lee J. Multiple equilibrium droplet shake and design criterion for rough hydrophobic surfaces. Langmuir 2003;19:4999–5003.
[102] Extrand CW. Model for contact angles and hysteresis on rough and ultraphobic surfaces. Langmuir 2002;18:7991–9.
[103] Bico J, Marzolin C, Quéré D. Pearl drops. Europhys Lett 1999;47:220–6.
[104] Wu Y, Sugimura H, Inoue Y, Takai O. Thin films with nanotextures for transparent and ultra water-repellent coatings produced from trimethylmethoxysilane by microwave plasma CVD. Chem Vap Deposition 2002;8:47–50.
[105] Erbil HY, Demirel AL, Avci Y, Mert O. Transformation of a simple plastic into a superhydrophobic surface. Science 2003;299:1377–80.
[106] Teshima K, Sugimura H, Inoue Y, Takai O, Takano A. Ultrawater-repellent poly(ethyleneterephthalate) substrates. Langmuir 2003;19:10624–7.
[107] Shirtcliffe NJ, McHale G, Newton MI, Perry CC. Intrinsically superhydrophobic organosilica sol-gel foams. Langmuir 2003;19:5626–31.
[108] Husemann M, Morrison M, Benoit D, et al. Manipulation of surface properties by patterning of covalently bound polymer brushes. J Am Chem Soc 200;122:1844–5.
[109] Hensarling RM, Doughty VA, Chan JW, Patton DL. "Clicking" polymer brushes with thiol-yne chemistry: indoors and out. J Am Chem Soc 2009;131:14673–5.
[110] Azzaroni O, Brown AA, Huck WTS. Tunable wettability by clicking counterions into polyelectrolyte brushes. Adv Mater 2007;19:151–4.
[111] Sun T, Wang G, Feng L, et al. Reversible switching between superhydrophilicity and superhydrophobicity. Angew Chem Int Ed 2004;43:357–60.
[112] Xie S, Allington RW, Fréchet JMJ, Svec F. Porous polymer monoliths: an alternative to classical beads. Modern advances in chromatography. In: Freitag R, ed. Advances in biochemical engineering/biotechnology. Berlin: Springer; 2002;76:87–125.
[113] Eder K, Huber CG, Buchmeiser MR. Surface-functionalized, ring-opening metathesis polymerization-derived monoliths for anion-exchange chromatography. Macromol Rapid Commun 2007;28:2029–32.
[114] Jones D. Pharmaceutical applications of polymers for drug delivery. Rapra Technology; Shawbury (UK): 2004.
[115] Yan W, Hsiao VKS, Zheng YB, Shariff YM, Gao T, Huang TJ. Towards nanoporous polymer thin film-based drug delivery systems. Thin Solid Films 2009;517:1794–8.
[116] Buchholz FL, Graham AT. Modern superabsorbent polymer technology. New York (NY): Wiley-VCH; 1998.
[117] Ma PX, Langer R. Fabrication of biodegradable polymer foams for cell transplantation and tissue engineering. In: Morgan J, Yarmush M, eds. Tissue engineering methods and protocols. Totowa (NJ): Humana Press; 1999;18:47–56.
[118] Sachlos E, Czernuszka JT. Making tissue engineering scaffolds work. Review on the application of solid free from fabrication technology to the production of tissue engineering scaffolds. Eur Cell Mater 2003;5:29–40.
[119] Reneker DH, Chun I. Nanometer diameter fibres of polymer, produced by electrospinning. Nanotechnology 1996;7:216–23.

[120] Matthews JA, Wnek GE, Simpson DG, Bowlin GL. Electrospinning of collagen nanofibers. Biomacromolecules 2002;3:232–8.
[121] Zarkoob S, Eby RK, Reneker DH, Hudson SD, Ertley D, Adams WW. Structure and morphology of electrospun silk nanofibers. Polymer 2004;45:3973–7.
[122] Puppi P, Piras AM, Detta N, Dinucci D, Chiellini F. Poly (lactic-co-glycolic acid) electrospun fibrous meshes for the controlled release of retinoic acid. Acta Biomaterial 2010;6:1258–68.
[123] Kai D, Jin G, Prabhakaran MP, Ramakrishna S. Electrospun synthetic and natural nanofibers for regenerative medicine and stem cells. Biotechnol J 2013;8:59–72.
[124] Buchko CJ, Kozloff KM, Martin DC. Surface characterization of porous, biocompatible protein polymer thin films. Biomaterials 2001;22:1289–300.
[125] Theron A, Zussman E, Yarin AL. Electrostatic field-assisted alignment of electrospun nanofibers. Nanotechnology 2001;12:384–90.
[126] Young TH, Chert LW. A two step mechanism of diffusion-controlled ethylene vinyl alcohol membrane formation. J Membr Sci 1991;57:69–81.
[127] Bàrtolo PJ, Domingos M, Patrìcio T, Cometa S, Mironov V. Biofabrication strategies for tissue engineering. In: Fernandes P, Bártolo PJ, eds. Advances on modeling in tissue engineering series: computational methods in applied sciences. 1st edn. Berlin: Springer; 2011;20:137–76.
[128] Kieswetter K, Schwartz Z, Hummert TW, et al. Surface roughness modulated the local production of growth factors and cytokines by osteoblast-like MG-63 cells. J Biomed Mater Res 1996;32:55–63.
[129] Lampin M, Warocquier-Clerout R, Legris C, Degrange M, Sigot-Luizard MF. Correlation between substratum roughness and wettability, cell adhesion and cell migration. J Biomed Meter Res 1997;36:99–108.
[130] Chung TW, Liu DZ, Wang SY, Wang SS. Enhancement of the growth of human endothelial cells by surface roughness at nanometer scale. Biomaterials 2003;24:4655–661.
[131] Krebs FC. Fabrication and processing of polymer solar cells: a review of printing and coating techniques. Sol Energy Mater Sol Cells 2009;93:394–412.
[132] Bull SJ. Techniques for improving thin film adhesion. Vacuum Vol 1992;43:517–20.
[133] Liston EM, Martinu L, Wertheimer MR. Plasma surface modification of polymers for improved adhesion: a critical review. J Adhes Sci Technol 1993;7:1091–127.
[134] Hegemann D, Brunner H, Oehr C. Plasma treatment of polymers for surface and adhesion improvement. Nucl Instrum Methods Phys Res B 2003;208:281–6.
[135] Park JB, Lake RS. Biomaterials, an introduction. 2nd edn. New York (NY): Plenum Press; 1992.
[136] Guney A, Kara F, Ozgen O, et al. Surface modification of polymeric biomaterials. In: Taubert A, Mano JF, Rodríguez-Cabello JC, eds. Biomaterials surface science. Weinheim (Germany): Wiley-VCH Verlag; 2013:89–158.
[137] Mijovic JS, Koutsky JA. Etching of polymeric surfaces: a review. Polym Plast Technol Eng 1977;9:139–79.
[138] Chu PK, Chen JY, Wang LP, Huang N. Plasma-surface modification of biomaterials. Mater Sci Eng: R 2002;36:143–206.
[139] Satriano C, Carnazza S, Guglielmino S, Marletta G. Differential cultured fibroblast behavior on plasma and ion-beam-modified polysiloxane surfaces. Langmuir 2002;18:9469–75.
[140] Afrany A, Deelder A. Surface modification by electron beam irradiation for improved immunoassay. Radiat Phys Chem 1999;55:681–4.
[141] Hunt JA, Williams RL, Tavakoli SM. Laser surface modification of polymers to improve biocompatibility. J Mater Sci Mater Med 1995;6:813–17.
[142] Ahad AU, Bartnik A, Fiedorowicz H, et al. Surface modification of polymers for biocompatibility via exposure to extreme ultraviolet (EUV) radiation. J Biomed Mater Res A 2013; published online DOI: 10.1002/jbm.a.34958.

[143] Zhang S. Fabrication of novel biomaterials through molecular self-assembly. Nat Biotechnol 2003;21:1171–8.
[144] Kado Y, Aoki A, Miyashita T. Surface modification using polymer Langmuir-Blodgett films. J Mater Sci 2002;37:4839–43.
[145] Goddard JM, Hotchkiss JH. Polymer surface modification for the attachment of bioactive compounds. Prog Polym Sci 2007;32:698–725.
[146] Galvin CJ, Genzer J. Applications of surface-grafted macromolecules derived from post-polymerization modification reactions. Prog Polym Sci 2012;37:871– 906.
[147] De Giglio E, Cometa S, Calvano CD, et al. A new titanium biofunctionalized interface based on poly(pyrrole-3-acetic acid) coating: proliferation of osteoblast-like cells and future perspectives. J Mater Sci Mater Med 2007;18:1781–9.
[148] De Giglio E, Cometa S, Cioffi N, Torsi L, Sabbatini L. Analytical investigations on poly(acrylic acid) coatings electrodeposited on titanium based implants: a versatile approach to biocompatibility enhancement. Anal Bioanal Chem 2007;389:2055–63.
[149] De Giglio E, Sabbatini L, Colucci S, Zambonin G. Synthesis, analytical characterization, and osteoblast adhesion properties on RGD-grafted polypyrrole coatings on titanium substrates. J Biomater Sci Polym Ed 2000;11:1073–83.
[150] Denizli A, Piskin E, Dixit V, Arthur M, Gitnick G. Collagen and fibronectin immobilization of PHEMA microcarriers for hepatocyte attachment. Int J Artif Organs 1995;18:90–5.
[151] Sun L, Dai J, Baker GL, Bruening ML. High-capacity, protein-binding membranes based on polymer brushes grown in porous substrates. Chem Mater 2006;18:4033–9.
[152] Cullen SP, Liu X, Mandel IC, Himpsel FJ, Gopalan P. Polymeric brushes as functional templates for immobilizing ribonuclease A: study of binding kinetics and activity. Langmuir 2008;24:913–20.
[153] Iwata R, Satoh R, Iwasaki Y, Akiyoshi K. Covalent immobilization of antibody fragments on well-defined polymer brushes via sitedirected method. Colloids Surf B 2008;62:288–98.
[154] Salvador-Morales C, Zhang L, Langer R, Farokhzad OC. Immunocompatibility properties of lipid-polymer hybrid nanoparticles with heterogeneous surface functional groups. Biomaterials 2009;30:2231–40.
[155] Taira S, Yokoyama K. Self-assembly DNA-conjugated polymer for detection of single nucleotide polymorphism. Biotechnol Bioeng 2004;88:35–41.
[156] De Giglio E, Cometa S, Ricci MA, et al. Development and characterization of rhVEGF-loaded poly(HEMA-MOEP) coatings electrosynthesised on titanium to enhance bone mineralization and angiogenesis. Acta Biomater 2010;6:282–90.
[157] De Giglio E, Cometa S, Ricci MA, et al. Ciprofloxacin-modified electrosynthesised hydrogel coatings to prevent titanium implant-associated infections. Acta Biomater 2011;7:882–91.
[158] Halliday AJ, Moulton SE, Wallace GG, Cook MJ. Novel methods of antiepileptic drug delivery – polymer-based implants. Adv Drug Deliv Rev 2012;64:953–94.
[159] Rahman CV, Ben-David D, Dhillon A, et al. Controlled release of BMP-2 from a sintered polymer scaffold enhances bone repair in a mouse calvarial defect model. J Tissue Eng Regen Med 2012. doi: 10.1002/term.1497.
[160] Mundargi R, Babu V, Rangaswamy V, Patel P, Aminabhavi T. Nano/micro technologies for delivering macromolecular therapeutics using poly(D,L-lactide-co-glycolide) and its derivatives. J Control Release 2008;125:193–209.
[161] Yoo HS, Park TG. Folate receptor targeted biodegradable polymeric doxorubicin micelles. J Control Release 2004;96:273–83.
[162] Jain RA. The manufacturing techniques of various drug loaded biodegradable poly (lactide-coglycolide) (PLGA) devices. Biomaterials 2000;21:2475–90.

[163] Grayson ACR, Choi IS, Tyler BM, Wang PP, Brem H, Cima MJ, Langer R. Multi-pulse drug delivery from a resorbable polymeric microchip device. Nat Mater 2003;2:767–72.
[164] Sivakumar PM, Cometa S, Alderighi M, Prabhawathi V, Doble M, Chiellini F. Chalcone. Carbohydr Polym 2012;87:353– 60.
[165] Vunjak-Novakovic G, Radisic M. Cell seeding of polymer scaffolds. Biopolym Methods Tissue Eng Methods Mol Biol 2004;238:131–45.
[166] Naarmann H. Polymers, electrically conducting. Ullmann's encyclopedia of industrial chemistry. Weinheim (Germany): Wiley-VCH Verlag; 2002.
[167] Shirakawa H, Louis EJ, MacDiarmid AG, Chiang CK, Heeger AJ. Synthesis of electrically conducting organic polymers: halogen derivatives of polyacetylene, $(CH)_x$. J Chem Soc Chem Commun 1977:578–80.
[168] Diaz AF, Kanazawa KK, Gardini GP. Electrochemical polymerization of pyrrole. J Chem Soc Chem Commun 1979:635–6.
[169] Eftekhari A. Nanostructured conductive polymers. Wiley-Blackwell, ed. Chicester (UK): Wiley; 2011.
[170] Cotrone S, Cafagna D, Cometa S, De Giglio E, Magliulo M, Torsi L, Sabbatini L. Microcantilevers and organic transistors: two promising classes of label-free biosensing devices which can be integrated in electronic circuits. Anal Bioanal Chem 2012;402:1799–811.
[171] Hatfield JV, Neaves P, Hicks PJ, Persaud KC, Tavers P. Toward an integrated electronic nose using conducting polymer sensors. Sens Actuat 1994;18:221–8.
[172] Torsi L, Tanese MC, Cioffi N, Gallazzi MC, Sabbatini L, Zambonin PG, Raos G, Meille SV, Giangregorio MM. Side-chain role in chemically sensing conducting polymer field-effect transistors. J Phys Chem B 2003;107:7589–94.
[173] Henderson RD, Guijt RM, Andrewartha L, Lewis TW, Rodemann T, Henderson A, Hilder EF, Haddad PR, Breadmore MC. Lab-on-a-Chip device with laser-patterned polymer electrodes for high voltage application and contactless conductivity detection. Chem Commun (Camb) 2012;48:9287–9.
[174] Owens RM, Malliaras GG. Organic electronics at the interface with biology. MRS Bull 2010;35:449–56.
[175] Jones IL, Livi P, Lewandowska MK, Fiscella M, Roscic B, Hierlemann A. The potential of microelectrode arrays and microelectronics for biomedical research and diagnostics. Anal Bioanal Chem 2011;399:2313–29.
[176] Li G, Zhu R, Yang Y. Polymer solar cells. Nat Photon 2012;6:153–61.
[177] Diaz AF, Martinez A, Kanazawa KK, Salmon M. Electrochemistry of some substituted pyrrole. J Electroanal Chem 1980;130:181–7.
[178] O'Neil KD, Semenikhin OA. AFM phase imaging of thin films of electronically conducting polymer polybithiophene prepared by electrochemical potentiodynamic deposition. Russ J Electrochem 2010;46:1345–52.
[179] Meana-Esteban B, Lete C, Kvarnström C, Ivaska A. Raman and in situ FTIR-ATR characterization of polyazulene films and its derivate. J Phys Chem B 2006;110:23343–50.
[180] Penn LS, Bowler ER. A new approach to surface energy characterization for adhesive performance prediction. Surf Interface Anal 1981;3:161–4.
[181] Lingaraju D, Ramji K, Mohan Rao NBR, Rajya lakshmi U. Characterization and prediction of some engineering properties of polymer – clay/silica hybrid nanocomposites through ANN and regression models. Procedia Eng 2011;10:9–18.
[182] Chiellini E, Corti A, D'Antone S, Baciu R. Oxo-biodegradable carbon backbone polymers – oxidative degradation of polyethylene under accelerated test conditions. Polym Degrad Stab 2006;27:2739–47.

Beat A. Keller

2 Investigation of polymer surfaces by time-of-flight secondary ion mass spectrometry

2.1 Introduction

2.1.1 Analysis of surfaces

Many surface-sensitive analytical techniques are known today that provide the scientist who is working in the field of polymer surface characterization with a set of tools to address the basic questions of analytical science. Generally, one seeks answers about which compounds are found on the surface, how they are laterally distributed, and how much of the material is found. Certainly, there are numerous additional questions to pose, but these are the most frequent. It is common knowledge that no single technique provides high-quality answers to all three of the basic analytical challenges. At the same time, it means mastering the field by choosing the most powerful analytical tool to address the problem at hand.

This chapter will provide basic insight about the possibilities and limitation of time-of-flight secondary ion mass spectrometry (TOF-SIMS) to obtain useful information about the chemistry of polymer surfaces [1–7]. Experiments concerning measurements about polymer characterization, end-group chemistry, chemically or physically modified polymers, molecular weight determination, polymer blends and additive chemistry have been reported. In recent years, TOF-SIMS has become important for the characterization and utilization of biopolymers [8]. However, this book chapter does not cover this topic.

In principle, the technique is straightforward. A high energy focused and pulsed primary ion beam is directed at the location of interest on the surface, inducing a more or less controlled sputter process. As a consequence, atomic and molecular secondary ions that are characteristic of the surface chemistry are released, collected, and directed through a mass analyzer system to a counting detector. Mass and identity can then be estimated according to simple kinetic velocity-to-mass relations given by classical charged particle mechanics: $t_{TOF} \propto U_{acc}^{-1} \propto (m/z)^{1/2}$ [9, 10]. The detection of co-emitted secondary neutral species and secondary electrons requires additional detection schemes [11].

The first observation of the ion impact-induced emission of positive and neutral ions from a metal plate target was observed in 1910 by J. J. Thomson [12] during his famous "Kanalstrahlen" experiment. After World War II and during the moon exploration period, the first commercial instruments were constructed for the isotopic composition analysis of space-borne materials. At the same time, the analysis of semiconductor and thin film samples has become more important. In fact, the semiconductor industry today is still one of the most important users of applications for SIMS analysis. The generic outline of a TOF-SIMS experiment is schematically shown in

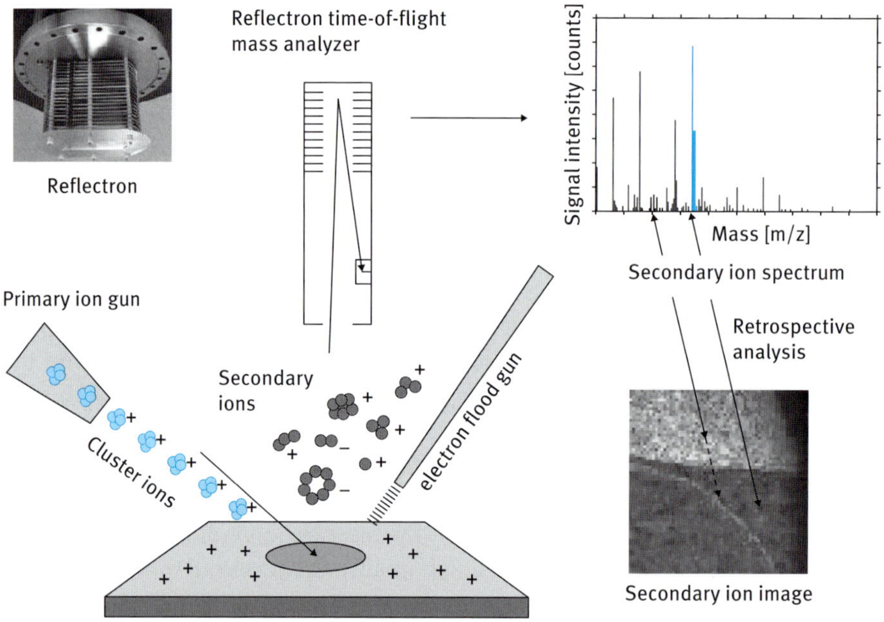

Fig. 2.1: Schematic drawing of a generic TOF-SIMS experiment. Gold or bismuth primary ion sources have replaced classical gallium, indium, or noble gas guns almost completely. The use of a reflectron ion energy-focusing element is standard equipment for high-mass-resolution TOF-SIMS applications. Data can be represented as mass spectrum or mass-resolved ion image. The electron flood gun operates between measurement cycles to compensate for excessive surface charging. Copyright 2012, Empa.

Fig. 2.1. Data can be represented as a spectrum or image. Access to subsurface or three-dimensional (3D) analysis via sputtering is also important but is not within the scope of this work.

Raw data obtained from a two-dimensional (2D) imaging experiment can be stored and reloaded for retrospective data analysis. This analytical method is based on selecting areas of interest on the image and subsequently reconstructing a TOF-SIMS spectrum of it. Inversely, signals from a spectrum can be used to reconstruct an image in which the intensity scale of colors assigned to individual mass fragments is proportional to the number of counts in each pixel and therefore represents the 2D distribution of peaks found in the TOF-SIMS spectrum. No preselection of signals for TOF-SIMS mapping is necessary because all data are available for inspection long after finishing the experiment.

2.1.2 The SIMS process: A detailed approach of theory and instruments

True surface sensitivity of the SIMS technique was achieved when Benninghoven and colleagues [13–16] introduced static SIMS (SSIMS) using a high-resolution TOF mass analyzer that allowed limiting the ion current from the primary source to a value of less than approximately 10^{-10} to 10^{-9} A/cm^2. A fraction as little of 1% of the outermost surface layers of a material is typically consumed to obtain a meaningful TOF-SIMS spectrum. The mode of operation was called SSIMS technology. Of course, the quality of information depends on signal strength and particle yield. From a number of experiments, the conclusion was drawn, and now, it is widely accepted that through SSIMS experiments, a good description of polymer surface chemistry is possible [17, 18].

The impact of energetic charged projectiles is always locally destructive. The degree of disruption of the affected area around the site of collision is dependent on the nature of the primary ion used for analysis. Most of the particles that are released as a consequence of primary ion impact are neutral (90%). These secondary particles are accessible *via* post-ionization using a temporally correlated pulsed laser system [11, 19]. The surface sensitivity of the SIMS process is explained by the fact that the secondary species emitted from the surface are originating from the topmost monolayers (1 nm) of a material.

2.1.2.1 Sputter process and SSIMS approach

Early calculations indicated that the energy and momentum transfer of the primary ion to the solid matrix is taking place in many cases *via* a collision cascade introduced by Sigmund [20]. The interaction of energetic ions with a solid can be described by the following mechanism (stopping power describes the energy loss inside the solid as function of distance from the surface):

- Nuclear stopping power: Direct interaction between ion and target nuclei (elastic Coulomb interaction). This process is dominant for primary energies $E \leq 100$ keV/z.
- Electronic stopping power: Inelastic coupling between ion and target atom electrons (primary ions in the megaelectron-volt range).
- Exchange stopping power: Exchange of electric charges between ion and target (small).

The typical diameter of a collision cascade is about 10 nm around the ion impact site. The linear cascade theory has been successful in describing sputtering in many materials. However, it does not include extreme thermal processes and therefore situations with high damage. Amorphization, site damage, or ion-induced reaction are taking place within a time span of a few femtoseconds after impact. In complex materials such as polymers or biological samples, the cascade theory, which is based on elastic collisions between point charge particles fails. In cluster ion experiments, the average energy of individual particles is low, and interactions take place *via* large distance and directional energy transfer. It has been realized by Benninghoven and colleagues [13, 14] that an ion current density of 10^{-10} up to 10^{-9} A/cm^2 (ion dose $\leq 10^{13}$ ions/cm^2) is low enough that each impinging projectile finds pristine surface conditions in most materials, and therefore, only *ca* 1–10% of the surface is influenced during measurement. The kinetic energy of the secondary particles is low (≤100 eV) and shows a Thompson distribution for atomic particles [21] and a Maxwell-Boltzmann distribution for molecular fragments. The former is obviously broader than the latter. The angular distribution of atomic secondary ion particles exhibits roughly a cosine dependence of the form [21]

$$\frac{\partial N}{\partial \Omega} = \pi^{-1} \cos\theta. \tag{2.1}$$

However, in particular, when ions such as sodium or silver as contaminants (or by intention) are present at the surface, ion-induced cationization can give rise to different, nonclassical situations. The most important factors determining the efficiency of the sputtering process are the mass of the primary ion, its angle of incidence, and the energy. After ion impact, all subsurface atoms within the collision zone are moved (species redistribution). This is called “collisional mixing” and limits the depth resolution of the eroding process for monoatomic primary ions (e.g. in depth profiling studies).

For Cs^+ ions, the following formula can be applied [22]:

$$\lambda = 1.84\, E^{0.68} \cos(\theta), \tag{2.2}$$

where λ is the penetration depth (nm), E is the impact energy (keV), and θ is the angle of incidence.

Numerous other material-related effects influencing the emission of secondary particles have been discussed in the literature [3, 22].

In reality, the situation is often more complex because materials exhibit individual character in terms of composition, contamination, crystal structure, and density. This individualism is frequently called matrix effect and limits the possibility of SIMS investigations requiring precise quantification. In addition, insulating materials such as ceramics or polymers accumulate positive electric charges deposited during a TOF-SIMS experiment. Unless properly compensated by a low-energy electron gun (as shown in Fig. 2.1), surface charging will change the acceptance properties of the analyzing system and degrade mass resolution as well as signal intensity (SI). Because of matrix effects, TOF-SIMS is usually not suitable for precise quantitative analysis. Although studies in this direction have been published [23], X-ray photoelectron spectroscopy (XPS), which provides much more reliable quantitative data, is often used as complementary analytical technique to TOF-SIMS. Quantification is one of the most important limitations of the technique. However, when compared with TOF-SIMS, XPS is less surface sensitive and has a lower molecular specifity [7, 24–25].

Sputtered particles of molecular fragments generally show vibrational and rotational excitation that sometimes can lead to dissociation if a metastable doorway state is excited. If this takes place in the field free drift tube of a TOF analyzer, such particles can only be detected using a straight-line detection geometry.

The high sensitivity of SIMS also allows for trace analysis of many atomic species in the parts-per-billion range (10^{-6}), which is particularly important in the investigation of electronic materials. In this case, the mass resolution is more important than the mass range. However, the rather low-duty cycle can yield to long measurements, e.g. for deeply located quantum wells. When investigating surfaces of polymeric samples, the analytical accessibility of high mass species or molecular fragments provides the experimenter with detailed information about differences in composition and end-group chemistry.

2.1.2.2 Mass analyzer systems

Three standard types of mass analyzers are used in SIMS technology:
- quadrupole analyzer;
- magnetic sector analyzer;
- TOF analyzer.

The theory of TOF analyzer systems has been discussed in detail in the literature [9, 10]. Compared with quadrupole and magnetic sector analyzers, TOF systems have up to 10 times better transmission efficiency. Combined with the parallel (multiplex) mass transfer, the overall sensitivity can outnumber quadrupole as well as magnetic sector analyzers by a factor 10^3 to 10^4. The attainable mass range limit depends on the length of the field free drift tube, the data rate, and also the performance of the detector and counting electronics. Detected secondary particles can be assigned to a specific mass with parts-per-million (ppm) accurateness with respect to the exact mass calculated from isotopic masses of the fragment constituents. Parts per million is defined as

$$\text{ppm} = \frac{M_{\text{obs}} - M_{\text{exact}}}{M_{\text{exact}}} \times 10^6,$$

where M_{obs} is the observed m/z mass and M_{exact} is the exact m/z mass.

2.1.2.3 Ion sources and primary ions

Inorganic and metal oxide materials show restricted breakup into fragment secondary ions and generate most of the signals in the low-mass range ("fingerprint domain"), which is restricted to signals with mass $m/z \leq 250$. Many organic species do not yield useful characteristic signals at masses $m/z \leq 150$, particularly in positive secondary ion spectra. Since the generation of high-mass secondary ions is strongly dependent on the type, mass, energy, and impact angle of the primary ion beam, fragmentation is not uniform from all regions of the area influenced by the impinging ion. Local heating is highest in the immediate vicinity of the impact site, and therefore, fragmentation is excessive at this location. With increasing distance, quasi-molecules remain intact. Obviously, signals collected from the periphery of the impact crater of the primary ion reflect the initial composition and structural features more realistically. Gentle SIMS (G-SIMS) (Section 2.2.2.7) is based on this concept.

The first commercially available ion sources were operating with Ga^+, Cs^+, or In^+, which all provided relatively small focused beams (*ca* 100 μm) with sufficiently high ion densities. However, in the high mass range, signal intensity is decreasing rapidly for monoatomic projectiles because of excessive fragmentation and low yield of intact material. Property-specific molecular fragments are of low abundance. Therefore, the majority of molecular fragments induced by monoatomic ion sources are detected in the fingerprint range of the SIMS spectrum, which leaves the experimenter with complex spectra and unspecific mass signals. When dealing with organic materials such as polymers, the impact of a high-energy-charged projectile not only induces the emission of secondary particles but also disrupts the molecular identity of the solid. The consequence is a crater of damage extending over several nanometers around the impact site. With the introduction of cluster ion sources to TOF-SIMS spectrometry in the mid to late 1980s, this effect could be minimized at least for surface spectrometry. Access to subsurface domains by using depth-profiling experiments became realistic with the use of gas cluster ion beams (GCIBs) as sputter source.

Routine use of large data libraries, the application of the G-SIMS theory and the introduction of principal component analysis (PCA) to TOF-SIMS were all developed to facilitate the interpretation of SSIMS experiments. Some of these concepts are discussed in Section 2.2.2.7.

2.1.2.4 Surface ionization ion guns

Surface ionization ion guns have first been developed for use in ion implant technology [26]. For some time, monoatomic alkali metal sources such as Cs^+ or In^+ were in use to acquire surface spectra. With the development of cluster primary ion guns, the application of monoatomic ion sources for surface analytical investigations was replaced completely, and this type of primary ions are almost entirely used to perform depth profiling studies of electronegative species in semiconductor samples. Ion emission is stimulated by heating an adsorbed layer of Cs (the most commonly used element) on a low-work-function material. The ionization potential of cesium is low, and therefore highly mobile electrons can move to the substrate (e.g. a porous tungsten or iridium plug). Under vacuum conditions, using a suitable extraction potential, ion pulses with sufficient intensity and energy spread can be generated. Although beam brightness often succeeds 5×10^{-5} A/cm^2 sr, the focus diameter of Cs^+ beams is limited to values of *ca* 100 μm and therefore the gun is not suitable for high-lateral-resolution applications (Section 2.1.2.8).

2.1.2.5 Duoplasmatron and gas ion sources

Gaseous elements are ionized in duoplasmatron ion sources by electron impact. High and stable beams of Ar^+, Xe^+, O^-, and SF_5^+ can be generated. Oxygen and noble gas primary ion beams are used to obtain depth profile data from nonorganic materials.

Gillen and Roberson [27] demonstrated, for the first time, true depth profiling through poly(methyl methacrylate) (PMMA) using an SF_5^+ cluster beam source for sputtering. SF_5^+ clusters are produced by leaking SF_6 gas into an electron impact chamber. Other smaller clusters are formed in parallel inside the ionization plume, but SF_5^+ is the most abundant species. Later, depth profiling of polymers and biological samples [28] became a vital field using cluster ion beams. For some time, C_{60}^+ ion beams were popular [29, 30] due to the high mass of the primary ion and energy splitting among a large number of carbon atoms. It is a thermal effusive source, in which a powder of C_{60} (or other carbon compounds, e.g. C_{70} or coronene) is heated up to 700 K. The carbon gas is then ionized by electron impact. Although providing a sufficiently high ion current (up to 500 nA, continuous), the source suffers from lack of long-term stability and ease of operation. GCIBs (Section 2.1.2.7) show superior performance for this application and are nowadays the best choice for depth profiling of organic compounds.

2.1.2.6 Liquid metal ion guns

Liquid metal ion guns (LMIGs) [31] have become the most popular type for day-to-day operation of TOF-SIMS instruments. They are easy to operate, reliable, and provide a stable ion current. For a long period, gallium has been used as source material because of its low melting point (302.9 K) and excellent focusing properties, which proved ideal for imaging applications. Ga beams can routinely be focused down to 60 nm and are therefore also widely used in focused ion beam (FIB) ion microscopy. However, secondary ion efficiency is low, and high-mass fragments usually are not accessible with sufficient intensity for small-area mapping. Because gallium does not occur as a pristine element in nature, it has to be purified to avoid split signals in high-resolution SIMS.

As sources in SSIMS technology and surface analysis, gold and bismuth cluster ion sources provide up to typically 1,000 times higher signal levels above the fingerprint mass region and most often also show less fragmentation because of primary ion energy distribution over several beam particles [32, 33].

Walker and Winograd [34] described the use of a LMIG ion source generating Au^+, Au^{2+}, and Au_2^+ primary ion particles. An eutectic alloy with Si, Be, and Ge was experimentally tested. The melting point of a SiBeAu tip lies around 693 K. For the materials investigated, Au^{2+} projectiles induced the least damage to the surface. However, all gold particles produced considerably stronger secondary ion yields than achieved by Ge-LMIG sources. Commercially available gold sources were introduced in 2003 [35] and produced routinely Au_n^+ clusters ($n = 1–5$).

The use of Bi_n^{m+} clusters (n denoting the cluster size and m the number of charge) has replaced most other types of primary ions for surface characterization experiments because of the high secondary ion yield at high mass range. The cluster size can vary between 1 and 7, with Bi_1^+, Bi_3^+ and Bi_3^{2+} the most often used species. The metal or metal alloy compound deposited on a fine needle with a reservoir is liquid at *ca* 540 K. The most recent designs [36] can be focused almost down to the ultimate physical limit (*ca* 15–30 nm) as defined by Orloff *et al.* [37]. Cluster ions break up after impact into component atoms, which carry only a small fraction of the initial kinetic energy and therefore induce less damage on the surface and subsurface layers. The penetration depth of primary ions is strongly dependent on impact energy and angle [38].

Numerous other, more exotic, clusters consisting of ionic species that were produced by ^{252}Cf fission fragment-induced desorption of source material deposited as thin layer on an aluminum-Mylar foil (CsI^+, Cs_2I^+, and $Cs_3I_2^+$ [39, 40] or $(Bi_2O_3)_nBiO^+$ [40]) were experimentally tested.

LMIG sources are capable of profiling through thin polymer films but show increased damage accumulation as compared with gas cluster and C_{60}^+ sources [34].

2.1.2.7 Gas cluster ion sources

GCIBs are presently the source of choice for organic depth profiling, including investigations of polymers. Although many different gases have been studied, the most often used system is argon.

In a first step, the noble gas is condensed into a neutral cluster by supersonic cooling. This effect dates back to early experiments performed for gas phase reaction studies [41]. The still neutral clusters of different sizes are ionized by electron impact and afterwards mass selected. Modern commercially available ion sources produce argon clusters with component numbers ranging from 50 up to more than 10,000.

A study comparing the emission of secondary ions from PMMA generated by a single atom and by large, size selected argon clusters (m_{av} = 700, 1,000 and 1,350) was published by Hashinokuchi *et al.* [42] in 2008. The SI for fragment ions with masses m/z = 100, 125, 300, 500, and 800, respectively, was found to be strongly enhanced with increasing cluster size for a PMMA film deposited on silver when compared with single atom experiments. Later experiments on PMMA, polystyrene (PS), polycarbonate (PC), and indium tin oxide (ITO)/glass samples bombarded with 5.5-keV Ar_{700}^+ ions under SSIMS conditions were presented by Ninomiya *et al.* [43]. The signals detected were compared with the literature and found in good agreement for PMMA [88] (m/z = 69, $C_4H_5O^+$, and m/z = 115, $C_6H_{11}O^+$), PS [44] (m/z = 91, $C_7H_7^+$, and m/z = 115, $C_9H_7^+$), and PC [45] (m/z = 77, $C_6H_5^+$, and m/z = 135, $C_9H_{11}O^+$), respectively.

However, to make the equipment useful as pulsed beam in commercially available high-mass resolution TOF-SIMS instruments, a small fraction of the cluster beam is selected by special 90° beam geometry. This allows for the generation of short, well-defined ion pulses enabling the acquisition of high-resolution surface mass spectra.

Kayser *et al.* [46] investigated the differences in negative ion TOF-SIMS spectra of PC using a Bi_3^+ and an Ar_{7500}^+ primary ion beam. While the former exhibited average particle energy of 6.6 keV/atom, the GCIB source only deposited 2.6 eV/atom onto the surface. The bismuth beam generated a considerable amount of fragmentation. In contrast, the spectrum obtained from the argon cluster source was dominated by characteristic molecular fragments.

As proven in earlier investigations for other cluster guns [47, 48], the degree of surface charging was considerably reduced in several TOF-SIMS experiments using GCIB cluster guns.

2.1.2.8 High-resolution ion images obtained from cluster LMIG sources

Highly focused ion beams are also a prerequisite for high-resolution imaging TOF-SIMS by raster scanning the beam over a selected surface area. Collecting mass spectra at each pixel of the image as a function of position (usually 128 × 128 or 256 × 256) and retrospectively analyzing the signal intensities provides the experimentalist with a powerful tool for two-dimensional analysis of surface chemistry. All ions are detected in parallel. Secondary electrons co-produced after each primary ion impact yield secondary electron images of the analyzed area [49]. Charles Evans and Associates developed in the early 1990s, a TOF-SIMS design based on coherent ion transfer through a TRIFT mass analyzer and subsequent position-sensitive ion detection (resistive anode encoder, or RAE) capable of mass-resolved imaging. Modes of detection for TOF-SIMS imaging are represented in Fig. 2.2.

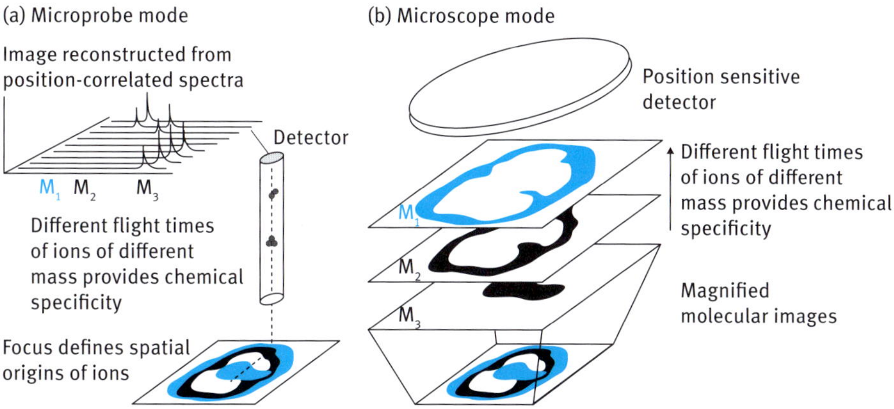

Fig. 2.2: Two different modes of detection are possible for TOF-SIMS imaging. (a) In the microprobe configuration, the mass spectra are collected from a preselected pixel array by rastering the primary ion beam and images are rebuilt retrospectively from an extended raw-data set. (b) In the microscope mode, images are transferred coherently through a TOF analyzer system and image distribution is magnified before being detected at a position-sensitive RAE detector. Retrospective analysis is also possible. Reprinted with permission from Luxembourg SL, Mize TH, McDonnell LA, Heeren RMA, *Anal Chem* 76, 5339–44, 2004. Copyright 2004, American Chemical Society [49].

The lateral resolution of a TOF-SIMS image is instrumentally limited by the degree of beam focus and the detection efficiency of a specific characteristic fragment ion from the area under investigation. The number of secondary ions that are emitted from a defined area is limited by the damage cross section of the material. Therefore, the achievable lateral resolution is not only limited by the primary ion focus but rather by the material-dependent number of ions that can be detected from a small area in an SSIMS experiment.

The efficiency, E, is defined as

$$E(M\pm) = \frac{Y(M\pm)}{\sigma(M\pm)},$$

where Y is the secondary ion yield and σ is the disappearance cross section.

A sample-dependent useful lateral resolution Δ_l from an area A_D can be defined [50] as $\Delta_l = (A_D)^{1/2} = (N/E(M\pm)^{1/2}$. The useful lateral resolution is related to the secondary ion emission (E) and is sample dependent. It should not be confused with lateral resolution, which is dependent on the spot size of the primary ion beam. N is the number of secondary ions that is detected from the area A_D before the uppermost monolayer of the material is removed by the primary ion beam:

$$N(t) = N(0)\exp(\sigma It/Ae),$$

where $N(t)$ is dependent from the disappearance cross section, I is the primary ion current, t is the time, A is the bombarded area, and e is the electric charge, respectively [50]. The issue of lateral resolution in SSIMS experiments is of increasing importance for detecting small, nanosized features in imaging mode. Touboul *et al.* [50] estimated values for Δ_l for the cholesterol $[M+H]^+$ signal at $m/z = 385$ after Bi_n^{q+} bombardment (Tab. 2.1).

A full imaging data set is composed of the complete set of time (i.e. mass)-dependent counts arising from each pixel. However, in cases where very small

Tab. 2.1: Ion bombardment efficiency and useful lateral resolution for Bi_n^{q+} primary ions measured for the cholesterol $[M+H]^+$ signal at $m/z = 385$ in rat brain sections

Primary ion	Ion bombardment efficiency (cm^{-2})	Useful lateral resolution Δ_l (nm)
Bi_1^+	3.40×10^8	1,148
Bi_3^+	1.71×10^9	483
Bi_3^{2+}	1.29×10^9	557
Bi_5^+	7.14×10^8	749
Bi_5^{2+}	2.81×10^9	377
Bi_7^+	2.12×10^8	1,374

Data from Touboul D, Kollmer F, Niehuis E, Brunelle A, Laprévote O, *J Am Soc Mass Spectrosc* 16, 1608–18, 2005, with permission from the American Society of Mass Spectrometry [50].

features are investigated, the dimension can approach the minimum size of a pixel. In such cases, it is more realistic to compare neighboring pixel intensities than observe average signal intensities over a number of pixels (binning). Several authors defined a number $N = 4$ as adequate pixel counts. However, it remains speculative to which extent this number really describes small-scale features on a sample surface. Counting statistics of TOF-SIMS imaging of complex surfaces is not yet fully described in all details.

2.2 TOF-SIMS investigations of polymer materials

2.2.1 General remarks

The physical and chemical characterization of materials surfaces requires, in most cases, the application of several complementary techniques. The most important are discussed in other chapters of this textbook. However, the use of mass spectrometric methods was identified as powerful analytical approach to obtain information about the chemical composition of surfaces and subsurface domains.

Several very elaborate and detailed review articles on the application of mass spectrometry and TOF-SIMS spectrometry to the analysis of polymeric materials and their surface have been published [5, 51–57].

2.2.2 Polymers

Bombardment of polymers with energetic charged particles can induce different phenomena such as cross-linking, chain scission, loss of chemical groups, depolymerization, loss of hydrogen or oxygen, monomer units, and even the formation of new single or double bonds.

Polymers that tend to favor cross-linking [e.g. PS, PC, polypropylene (PP), polyethylene (PE)] show excessive signal decay with increasing fluence. On the other hand, materials that are known to degrade *via* a random main-chain scission process (e.g. polymethacrylates, polyesters) tend to retain characteristic molecular signals with increasing beam intensity [58].

Depending on the type of information that is required, the experimental setup and sample handling have to be chosen or adapted. Thick-layer or bulk polymer surfaces need little or no sample preparation and are usually analyzed by inserting a small piece of the material in the instrument. Certainly, in the case of highly insoluble materials, this is a good experimental approach. Because of the sometimes rather high surface charging [59], which can be difficult to fully compensate using a low-energy electron gun, the mass spectra of bulk polymers are often restricted to the fingerprint domain ($m/z \leq 150$). This is most evident in experiments using a monoatomic primary ion source. Surface charging has been described as less distinct in cluster ion TOF-SIMS experiments. However, data generated with monoatomic ions provide information about the repeat unit of the polymer, possible surface contaminants, and also the end-group chemistry. Leggett [60] divided the formation of molecular fragments in SSIMS spectra of poly(styrene) into three distinct processes:

- Violent fragmentation, close to the impact site, characterized by high-energy transfer to the polymer. This leads to the formation of low-mass fragments, often containing unspecific chemical information.
- Unzipping region at a more distant area surrounding the impact site. Energy density is decreasing, and larger molecular fragments are formed. Often, information about monomer interactions, cross-linking behavior, and structural rearrangements is found. Laser post-ionization can be used to obtain further information of the neutral particle rich domain. Mahoney [58] recently published a compilation of polymeric structures according to their cross-linking behavior as type I or type II under irradiation of charged particles.
- Simple low-energy fragmentation characterized by formation of large fragments. The energy (in the range of electron volts) available for fragmentation is low, and therefore, bond breaking in certain chemical structures is impossible. The ejection of monomer fragments is dominating. The information about subtle molecular rearrangements is often very useful to obtain structural information.

As outlined earlier, some of these results are questioning the application of the simple linear Sigmund cascade theory [20] for the description of fragment ejection from polymers.

Gilmore and Seah [61] developed a model based on the dependence of SI upon ion dose to characterize the probability of bond breaking and fragmentation pattern of linear polymers at the end and in the middle. This effect was confirmed by investigating the fragmentation of poly(terahydrofluoroethylene) (PTFE) and poly(ethyleneterephthalate) (PET). Whereas PTFE is characterized by multiple bond breaking decay, PET shows a fast exponential decay for the molecular fragments at masses $m/z = 149$ and $m/z = 193$, but slow decay was observed at $m/z = 93$. These early investigations later became the foundations of G-SIMS (Section 2.2.2.7).

However, if dealing with structure determination and molecular weight assignment, elaborate sample preparation techniques are sometimes required. The most common is the dissolution of the polymer in a suitable solvent and subsequent spreading on a solid support in a concentration of about 1 mg/ml. Thoroughly cleaned silicon wafers or acid etched thin silver sheets (dilute HNO_3) are most often used for this purpose. The formation of secondary ions in this case is not generated *via* the collision cascade mechanism but by cationization of the parent ion or molecular fragments. Etched silver provides characteristic $^{107}Ag/^{109}Ag$ doublets after cationization. In many cases, the cation is added to the polymer solution (e.g. Na^+) in advance. Keller and Hug [62] investigated the use of ion-exchanged Nafion117 as source for selective silver [63] or rubidium cationization. In this way, TOF-SIMS has many similarities with matrix-assisted laser desorption ionization mass spectrometry (MALDI-MS). Cationization becomes more probable with increasing polarizability of a molecule. Therefore, some uncertainty remains if independence from end-group chemistry is a realistic approach of measured SI of cationized species [64].

Molecular weight information from TOF-SIMS experiments is obtained by investigation of cationized fragment pattern. Lee *et al.* [65] reported a reasonable good correlation with gel permeation chromatography and MALDI-TOF for data from the derivatives poly (α-methylstyrene, poly(4-methylstyrene), poly(4-*tert*-butyl styrene) investigated as thin films of different average mass on silver substrates in the mass range 0–5,000. For the aromatic residue-substituted poly(4-vinyl biphenyl) and poly(vinylnaphthalene), however, experimental values were slightly lower than the manufacturer specifications. Galuska [66] analyzed the low-mass fingerprint region of polyisoprene, 1,4-poly(butadiene), polyisobutylene, PE, PS, PP, and poly(1-butene) (P1-B) with respect to possible correlation between the average molecular weight and apparent monomer signal intensities in the TOF-SIMS spectra. In this experiment, the protonated ($m + H^+$) fragment was observed. Signal intensities correlated well with molecular weight values higher than 20,000 but lacked a general linear relationship at lower molecular masses. A general linear dependence was reported for a wide variety of polymers:

$$\text{Relative (M+H) SI} = M(\text{MW}/1{,}000)E + B, \tag{2.3}$$

where B describes the ion ratio intercept and E is a polymer type-dependent exponent with values between −0.5 and −0.6 [66]. The same type of molecular weight estimation was applied by Coullerez *et al.* [67] to study a series of 2,2-bis (hydroxymethyl) propionic acid with different molecular weights. Signals in the fingerprint range of positive and negative TOF-SIMS mode were assigned to the fragmentation of the (bis-MPA) hyperbranched aliphatic polyester and used to calibrate the molecular weight.

2.2.2.1 Polydimethylsiloxane

Polydimethylsiloxane (PDMS) is one of the most extensively studied polymers from the polysiloxane (often termed silicones) family. The polymer has found widespread application in the industry as anti-foaming component in shampoos and household cleaners, as lubricating agent, water repellant, and anti-fouling protective coat. PDMS is frequently found on surfaces showing adhesive or printing failure and is also found as contaminant on extruded parts because of its use as low-temperature mold-release agent (e.g. baking paper or chocolate process casting mold). Its low toxicity, clear appearance, and safe handling make PDMS an ideal food-processing additive.

The use of PDMS as a component for microfluidic devices has been reviewed by Ng *et al.* [68]. The authors demonstrate the versatility of the material for fabrication of pneumatically actuated switches, passive structures, and chaotic mixers. Because of the straightforward surface modification by oxidative processing [69], PDMS can be easily made hydrophilic by exposure to an oxidizing plasma (e.g. air, oxygen). This process destroys the methyl (Si-CH_3) groups and introduces silanol functional groups (Si-OH) to the surface.

However, as most siloxanes, PDMS is migrating to materials surfaces, and due to its high mobility, it can form a dense overlayer that completely masks chemical entities of interest. In such cases, the fingerprint regions of TOF-SIMS spectra are dominated by the characteristic signals of PDMS [53] identified as Si^+ (m/z = 29.79), CH_3Si^+ (m/z = 42.99), CH_3SiO^+ (m/z = 58.98), $C_3H_7O^+$ (m/z = 59.03), and $(CH_3)_3Si^+$ (m/z = 73.06). $Si_2OC_5H_{15}{}^+$ (m/z = 147), $Si_3O_3C_5H_{15}{}^+$ (m/z = 207), respectively. Cyclic fragments are observed at mass m/z = 221 ($Si_3O_2C_7H_{21}{}^+$) [70]. Figure 2.3 shows the typical positive-ion PDMS TOF-SIMS spectrum in the fingerprint range below m/z = 300. Molecular weight distributions were obtained by Dong *et al.* [71, 72] and later Inoue *et al.* [73] by analyzing the $(M+Ag)^+$ fragments of different silver deposited PDMS modifications. Molecular weight distribution with C_2H_6SiO monomer repeat units (m/z = 74) was detected in the high-mass range ($m/z \geq 1{,}000$) of the TOF-SIMS spectra.

2.2.2.2 Polystyrene

PS rapidly degrades under ion irradiation. In contrast to polyacrylates, no difference in fragmentation was observed after irradiation of PS with different cluster ions. The inability to sputter intact polymer chain fragments because of entanglement of the

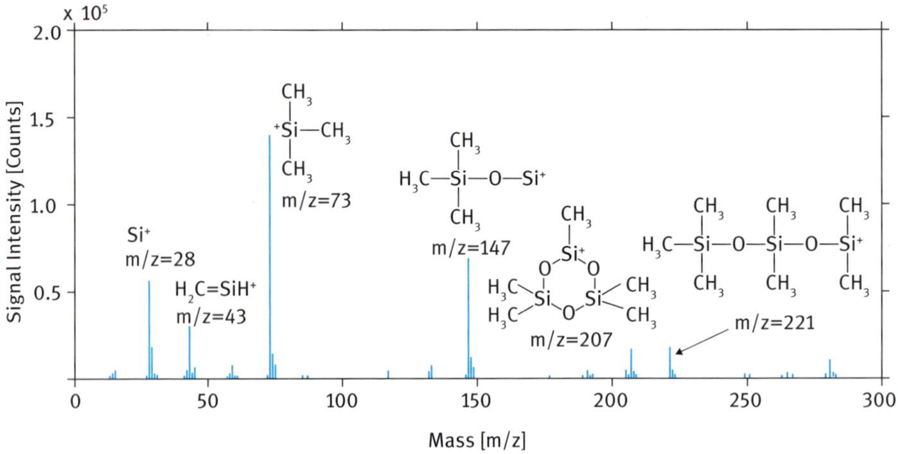

Fig. 2.3: Static TOF-SIMS spectrum of PDMS with characteristic fragmentation signals in the fingerprint spectrum region. The formation of hexacyclic fragment ions is energetically favored over linear species. Adapted from Keller BA, Hug P, *Anal Chim Acta* 393, 201–12, 1999. Copyright 1999, Elsevier [53].

rather bulky benzyl group might be responsible for this effect. Chain scission is the dominant degradation process with the formation of multiple aromatic fragment ions.

2.2.2.3 Polyacrylates

The application of TOF-SIMS to the surface analysis of polymethacrylates was described by Belu *et al.* [74]. The authors investigated a series of well-defined polyacrylate polymers, which were identical in their structural identity but had incremental additions of C_2H_4 groups to the backbone and side chain. Characteristic signals were identified for all investigated polyacrylates. In addition, the specific backbone and side-chain chemistry revealed unique mass signals in the TOF-SIMS spectra, which could be assigned to individual well-controlled polymer materials. Leeson *et al.* [75] showed that molecular weight had a large effect on signal intensities obtained from solvent cast films produced of monodispersed PMMA standards with M_w = 2,965–1,200,000. Negative secondary ion intensities were found to be very sensitive to trace amounts of solvent.

The degradation of methacrylate polymers under irradiation of a 5-keV SF_5^+ primary ion beam and an ion dose of approximately 2×10^{14} ions/cm^2 was described by Wagner [76–78] in a series of publications. Clearly, such extended bombardment exceeds the SSIMS limit considerably, but certain trends of degradation behavior also occur in SSIMS fragmentation. Modifications in the main-chain and end-group chemistry of poly(acrylates) revealed improved degradation of poly(methyl acrylate) and poly(methacryl acrylate) as compared with PMMA. These results were suggesting that main-chain or pendant methyl groups clearly reduce depolymerization and increase the rate of intermolecular or intramolecular cross-linking in the polymer matrix.

Mahoney [79] recently published data on PMMA thin films with variable tacticity. Damage cross-sections of the stereospecific polymer were estimated in a temperature-dependent study between –198 and –123 K under SF_5^+ bombardment. The isotactic samples exhibited lower damage accumulation compared with atactic and syndiotactic PMMA, which was explained by increased main-chain scission due to increased strain along the polymer chain.

2.2.2.4 Fluorinated polymers

Positive and negative secondary ion spectra of neat Teflon surfaces were published by Léonard *et al.* [80]. Characteristic signals in the fingerprint mass range included C^+ (m/z = 12), CF^+ (m/z = 31), CF_3^+ (m/z = 69), $C_2F_5^+$ (m/z = 119), $C_3F_5^+$ (m/z = 131), and $C_3F_7^+$ (m/z = 169), respectively. The negative ion spectra were dominated by F^- (m/z = 19), CF^- (m/z = 31), F_2^- (m/z = 38), CF_3^- (m/z = 69), and $C_2F_5^-$ (m/z = 119).

In an early study, Fowler *et al.* [64] analyzed perfluorinated polyether (PFPE) on Ag and Si substrates. The use of PFPE (Krytox™, DuPont) as surface lubricant stimulated the investigation. Characteristic molecular fragments and the molecular weight distribution were detected by TOF-SIMS. The complete chemical structure could be reconstructed using integrated intensities of a series of molecular ion signals extracted from investigated compounds. Information of end-group chemistry was also obtained from the spectra, however, with an uncertainty of fragment assignment.

Nafion® (DuPont) is a block copolymer consisting of a fluoropolymer backbone and several other fluoro-homopolymer units. The polymer is often poorly defined, but the most popular representative, Nafion117®, with only one extended fluoropolymer group and otherwise unity numbered subunits, has the well-defined structure shown in Fig. 2.4a. The super acidic end-group proton of the sulfonic acid in the α-position is very mobile, and the polymer has found widespread application as membrane material in fuel cell technology [81]. The polymer is structurally very cavernous and exchange reactions with water and small cations take place with almost 100% yield [62, 63]. As in most ionomers, proton exchange with various cations can be induced in suitable media [82].

Liu *et al.* [83] studied Aquivion (Hyflon), which is a perfluorosulfonate polymer built of a Teflon backbone and a shorter sulfonic acid side chain than that found

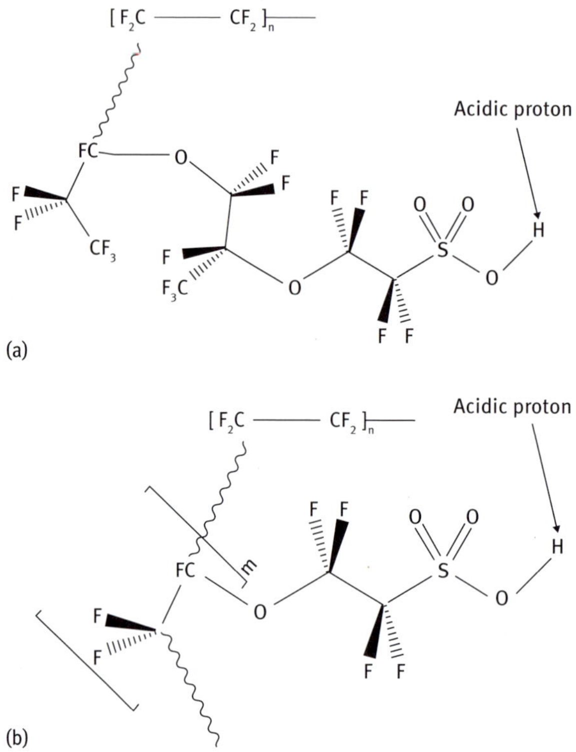

Fig. 2.4: Chemical structures of (a) Nafion and (b) Aquivion (Hyflon). The shorter side chain of Hyflon promotes a more pronounced formation of subsurface-located ionic clusters because of increased chain mobility. Copyright 2011, Empa.

in Nafion. The chemical structure of Hyflon is shown in Fig. 2.4b. An electro-active actuator of Hyflon with 1-butyl-2,3-dimethylimidazolium chloride (BMMI-Cl) was analyzed using a C_{60}^+ ion gun. Ions originating from BMMI (BMMI+, $m/z = 153$) and chlorine ($^{35}Cl^-$ and $^{37}Cl^-$) were detected with high intensity. Fragments characteristic to the backbone included CF^+ ($m/z = 31$) and F^- ($m/z = 19$). A signal assigned to SO_3^- was observed at mass $m/z = 80$. Depth profiles were recorded with these signals as markers for the individual components ≈ 7 μm inside the polymer.

2.2.2.5 Poly(ethyleneterephthalate)

Because its often very inhomogeneous surface, the analysis of PET has attracted much interest to use TOF-SIMS for characterizing the chemical groups present. Nysten *et al.* [84] studied surface inhomogeneity induced into a sample obtained by compression molding of the two incompatible polymers PET and PP. TOF-SIMS data in combination with scanning electron microscopy (SEM) and lateral force microscopy (LFM) revealed a surface structure consisting of a continuous phase with inclusions of nodules whose size were between 1 and 100 μm. The continuous phase could be assigned to PET and the nodules to PP.

2.2.2.6 Polyethylene glycol

Keller and Hug [53] published a positive-ion TOF-SIMS spectrum of polyethylene glycol (PEG) in the mass range of $m/z = 0–800$, recorded with a 8-kV Cs^+ primary ion beam. As shown in Fig. 2.5, the fingerprint range of the spectrum is dominated by the monomer fragment signal $(CH_2\text{-}CH_2\text{-}OH)^+$ at $m/z = 45$. Protonated fragments with the general chemical composition $(CH_2\text{-}CH_2\text{-}O)_nH^+$ were also detected in the mass range $\leq m/z = 150$. For larger oligomers (index $n \geq 5$), a second series with repeat units $(CH_2\text{-}CH_2\text{-}O)_nOH_3^+$ showing an intensity maximum around $m/z = 415$ $(CH_2\text{-}CH_2\text{-}O)_9OH_3^+$ was reported. The insert in Fig. 2.5 indicates higher signal intensities for oxygen terminated oligomer series.

2.2.2.7 Spectra libraries and G-SIMS approach

In the early days of TOF-SIMS analytical experiments, the interpretation of spectra has largely benefited from the compilation of spectra libraries, mainly covering the fingerprint region. Characteristic masses from often-analyzed materials were compared with stored signals from the library. This approach has a long tradition in analytical chemistry and dates back to the developing of mass spectrometry as tool for the organic chemist. However, in contrast to traditional electron impact ionization, which can easily be standardized, the impact of charged projectiles onto a polymer surface with several kiloelectron-volt energy induces fragmentation and, most often, also matrix-dependent structural changes in secondary ions. Of these, only a very small

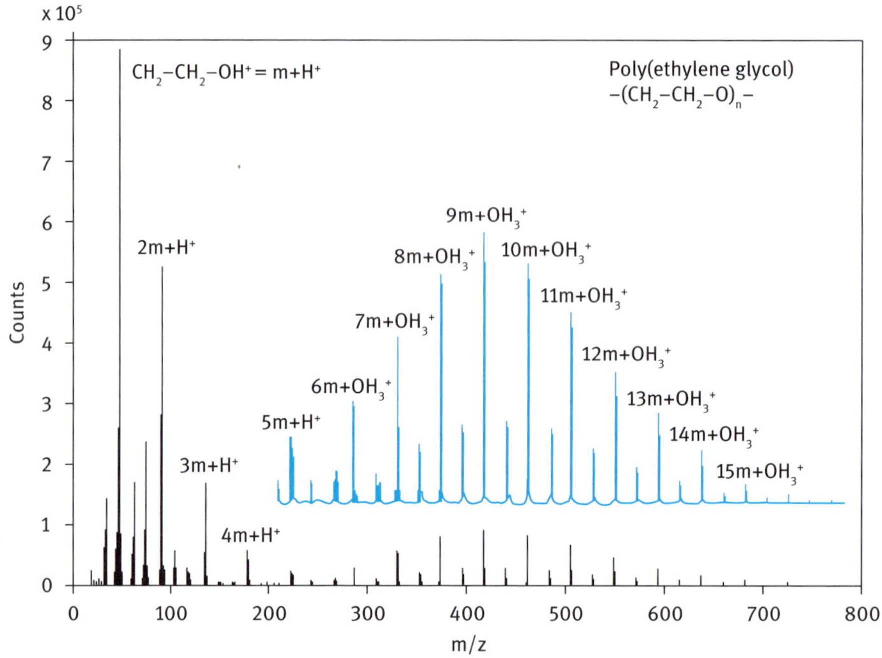

Fig. 2.5: Positive-ion TOF-SIMS spectrum of PEG obtained after bombardment with an 8-keV Cs^+ primary ion beam. The sample was prepared as thin film deposited from a 1-mg/ml ethanol solution. With permission from Keller BA, Hug P, *Anal Chim Acta* 393, 201–12, 1999. Copyright 1999, Elsevier [53].

fraction is usually related to the original chemical structure, and therefore the spectra of complex surfaces are very difficult and time consuming to interpret. Although the concept of library matching works rather well in some cases for monoatomic primary ions (e.g. Ga^+, Cs^+, In^+, or Ar^+) and serves often as starting point for more advanced interpretation, it completely fails for the use of cluster ion guns. Green *et al.* [85] estimated that the contents of the popular public chemical database PubChem has grown by more than two orders of magnitude containing a very large number of complex molecules, all accessible within TOF-SIMS mass range.

To obtain a more detailed picture of fragmentation mechanism in polymers, Gilmore and colleagues [86–88] introduced the concept of G-SIMS. Although a complete description of G-SIMS is not within the scope of this book, a short outline is given. Briefly, the concept is based on the fact that secondary ions released after ion impact from areas close to the impinging point exhibit more fragments than those more distant from it due to different energy density. Therefore, a set of secondary ions is generated with signals arising from structure-related ions together with fragmented secondary particles. If intelligently steering the energy of impact, the amount of the former species can be enhanced and the latter decreased.

A G-SIMS theory [88] was developed based on a partition function for a population of intact molecules with a characteristic temperature T_p. The theory requires two spectra with signal intensities S_{1x} and S_{2x} and masses M_x acquired under different conditions and therefore with different surface temperatures T_1 and T_2 ($T_1 < T_2$). A G-SIMS spectrum can be computed as

$$G_x = M_x S_{1x} (S_{1x}/S_{2x})^g, \tag{2.4}$$

where g is the so-called g-index. For atomic primary ions, this number is typically about 13, but it has a considerably lower value for cluster primary ions, e.g. Bi_n^{m+}. In this latter situation, the fragmentation is low, and the nonlinear intensity enhancement of fragments in the high-mass range largely simplifies peak specificity. Because the kinetic energy is distributed over several atoms in a cluster beam, modern LMIG sources operating with gold or bismuth reservoirs are providing very effective low-energy conditions [89]. To correlate spatial alignment of the two independent beams, a Bi/Mn-g-tip has been introduced recently [90] for use in commercial LMIG ion sources.

G-SIMS has been successfully applied to the analysis of polymers [89, 91].

2.2.2.8 Multivariate analysis and principal component analysis

Usually, materials collected from the real world or modified by advanced laboratory processing exhibit a complex surface chemistry [91]. Reducing the dimension of this complexity by multivariate statistical analysis (MVA) allows the extraction of meaningful information about individual compounds and their relationship. MVA is concerned with data acquired from multiple independent measurements of samples. A fundamental relationship between variables is obtained from measurements and subsequent data analysis. TOF-SIMS spectra contain signals from all components, such in high concentration and sometimes equally important low-concentration compounds as well. Yet knowing the information buried within the complexity of acquired mass spectra or TOF-SIMS images, it is possible to extract sample chemical composition, orientation of surface adsorptions, contaminants, or compound homogeneity. Technically, a list of fragments is selected from a series of 10–15 individual spectra. After preprocessing (e.g. peak normalization to the sum of intensities, mean centering), a matrix is generated with one row describing the samples and the other describing the signal intensities (normalized). A software is computing a scores matrix, describing the relationship between the samples (i.e. comparing the chemical identity), and the loadings matrix is illustrating the relationship between the original variables and the principal components. Timely computer equipment allows for running data processing algorithm to reduce complexity in SIMS spectra and identify important molecular signals. MVA methods such as PCA and partial least squares (PLS) have been successfully used for data reduction in analytical science and also in TOF-SIMS. However, it remains in many cases difficult to find the proper preprocessing and data selection to obtain useful information about a given analytical question [92].

Aoyagi *et al.* [91] recently applied G-SIMS and multivariate curve resolution (MCR) to the investigation of fragment ion patterns of PEG with several different molecular weight distributions (PEG600, PEG1000, mixtures). The authors prepared thin films of the neat materials and mixtures on silicon substrates, and were able to extract pure spectra of each deposited compound by MCR. The use of G-SIMS allowed for classification of the relationship between secondary ions and proposed fragmentation mechanism. Samples were measured with Mn^+, Bi^+, and Bi_3^+, respectively. A G-SIMS spectrum of PEG showing enhanced specific signals at mass $m/z = 113$, 157, 201, and 245 $(C_2H_4O^+)_n$ ($n = 2$–5) was discussed. Signals at $m/z = 113$ were assigned to $(C_2HC_2H_4O^+)_2$. As concluded from the experiments, G-SIMS spectra obtained with 25 keV Bi^+ and Mn^+ primary ions (g index = 13) showed the most prominent signal enhancement in the fingerprint mass range ($m/z \leq 300$) and MCR provided pure spectra from the investigated complex polymer mixtures.

2.2.3 Polymer additives

Additives are essential parts of synthetic technical polymers and enhance the application range of the product. Common types of additives are lubricants, plasticizers, antioxidants, UV absorbers, impact modifiers, and fillers. Release agents are usually not classified as typical additive compounds but are often found on technical polymer surfaces and notorious as harmful to desired properties of the later product. A broad range of additives are routinely used in polymer manufacturing, and their relative composition is specifically tailored to the later application. Organic materials include classes of alkyl phenols or hydroxybenzophenones, inorganic and organometallic additives, fillers comprised of different oxides, salts, metallocarboxylates, and finally accelerators (Ni, Zn). For an extensive discussion of the topic, the reader is referred to the monographs of Jan Bart [93, 94] and citations therein.

Materials properties such as brittleness, thermal and light stability, or impact resistance are achieved by additive concentrations ranging from the parts per million (ppm) to the percentage range. A combination of several additives is usually necessary to give the polymer its desired performance. It has been realized [91] that compounding, i.e. the incorporation of the additive into the polymer matrix, is often a prerequisite to avoid dynamic processes at the surface [94]. If not controlled properly, the dynamic behavior of near-surface- or subsurface-located additives can lead to phenomena such as blooming, color change or adhesion and printing failure [53].

The surface segregation of Irgafos 168 (Ciba), one the most used antioxidant additives in industry, was investigated by Médard *et al.* [95] on the surface of poly(ethylene terephthalate-ethylene isophthalate) (PETI). Thin films of PETI, containing various amounts of Irgafos 168, were spin casted from chloroform. The film thickness was estimated to be 1.5 nm at 3 wt% additive concentration. TOF-SIMS spectra obtained in the framework of a small round robin test at three independent laboratories and acquired using different instrument platforms were reported. To compare signals, a Ga^+ primary ion source was used for all investigations. Strong segregation of Irgafos 168 to the surface of the substrate accompanied by subsequent crystallization for samples containing 5 wt% additive concentration was observed. A uniform covering of the surface with additive was detected by imaging TOF-SIMS and further confirmed by angle-dependent XPS investigations. Surface segregation (blooming) of mobile polymer additives usually occurs when the concentration overpasses the solubility within the polymer matrix and decreases the intensity of characteristic substrate signals.

A surface-migration study of brominated flame retardants in high-impact PS (HIPS) was recently published by Holbrook *et al.* [96]. The authors detected blooming of polybrominated diphenyl ether (PBDE, BDE-209) on a HIPS surface using micro-X-ray fluorescence spectroscopy, FIB-SEM, energy dispersive X-ray spectroscopy, and TOF-SIMS. Experiments based on a Bi_3^+ primary ion beam revealed characteristic signals of BDE-209 ($C_{12}OBr_{10}$) on the HIPS surface at masses at $m/z = 485$, 487, 489, and 491, respectively. The effect was also confirmed by imaging TOF-SIMS investigations (Fig. 2.6).

2.2.4 Copolymers

Homopolymers are built of identical monomer subunits, whereas copolymers consist of two or more different species. Classes of copolymers are defined depending on the alternating arrangement of the subunits:

- periodic copolymers with a well-defined sequence such as ABAABBABAABB;
- alternating copolymers, which are a special case with sequence ABABABABAB;
- block copolymers where homopolymer subunits are linked by a covalent bond;
- statistical copolymers with a completely irregular order of monomer sequence.

Kita-Tokarczyk *et al.* [97] recently published a study discussing the interaction between 1,2-dipalmitoyl-*sn*-glycero-3-phosphocholine (DPPC) and 1,2-dioleoyl-*sn*-glycero-3-phosphocholine with amphiphilic poly(2-methyloxazoline)-block-poly(dimethylsiloxane)-block-poly(2-methyloxazoline) block copolymers. For simplicity, the polymer consisting of a symmetric structure of two terminally poly(2-methyloxazoline) $(PMOXA)_{15}$ units and one $PDMS_{110}$ polymer, located at the center ($PMOXA_{15}$-$PDMS_{110}$-$PMOXA_{15}$) was called $A_{15}B_{110}A_{15}$ (mass 10,700 g/mol). Among other characterizing techniques, TOF-SIMS imaging was used to observe the formation of phase irregularities occurring from fluorescent dye attachment to the lipids under investigation. TOF-SIMS spectra of a LB film (Langmuir-Blodgett) consisting of neat $A_{15}B_{110}A_{15}$ copolymer, and a similarly prepared film with co-deposited lipids were recorded to identify characteristic signals. Peaks

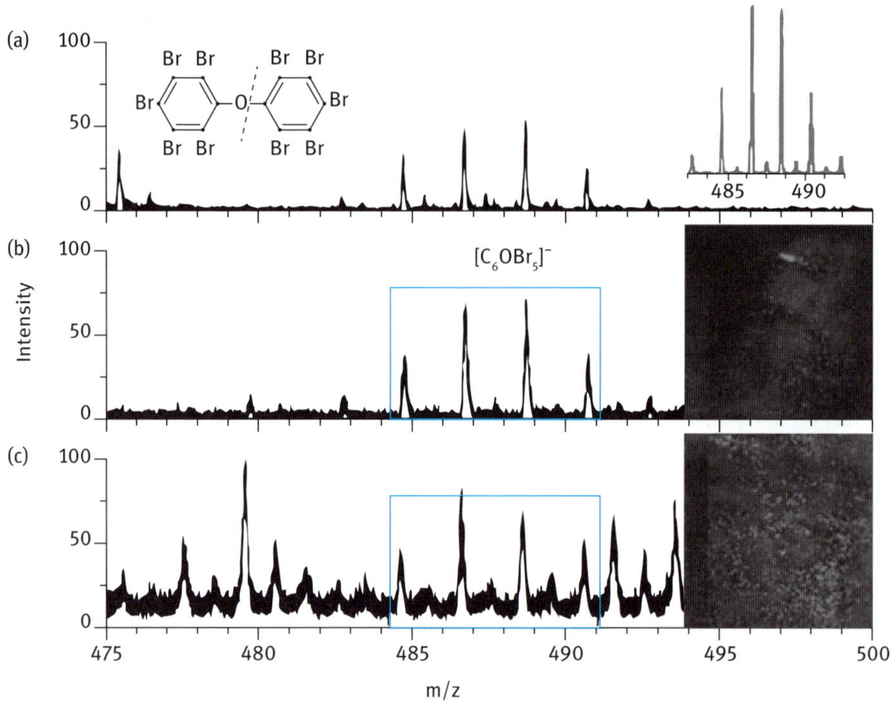

Fig. 2.6: Negative-ion secondary mass spectra and images of BDE-209 obtained from (a) standard (on silicon) and HIPS samples. The intense high-mass signal $C_6OBr_5^-$ is generated from symmetric bond breaking at the central ether group and reflects the calculated isotopic distribution of bromine at m/z = 485, 487, 489, and 491, respectively, as shown in the insert at the upper right-hand corner. TOF-SIMS spectra b and c were obtained from the (b) gray and (c) black sides of the HIPS surface. Overlay images (500 × 500 μm) show the green (BDE-209), blue (b, m/z = 183, HIPS fragment), and (c, m/z = 293 blue and m/z = 463, red, HIPS fragment). Reprinted from Holbrook RD, Davis JM, Scott KC, Szakal C, *J Microsc* 246, 143–52, 2012, with permission of John Wiley & Sons Ltd [96]. Original color figure in supplement material at http://dx.doi.org/10.1515/9783110288117.

assigned to the polymer included CN^- (m/z = 26) and CNO^- (m/z = 42), whereas DPPC was identified by PO_2^- (m/z = 62.97), PO_3^- (m/z = 78.92), and a peak attributed to the alkyl chain with composition $C_{16}H_{31}O_2^-$ (m/z = 255.25), respectively. As shown in Fig. 2.7, the formation of star-like structures consisting predominantly of DPPC was detected in the TOF-SIMS images of the lipid-covered LB films. This phenomenon was observed particularly in films with high surface pressure where part of the lipid is excluded from the polymer film (collapse pressure) and condensation into the star-like structure is energetically favored.

Keller and Mayerhofer [98] detected vesicles of an $A_xB_yA_x$ copolymer in rat liver samples. The spherical objects with an average diameter of 200 nm were identified by mass resolved imaging of $(CH_3)_3Si^+$ (m/z = 73.047) as characteristic signal of PDMS. Signal assignment, however required mass discrimination from conflicting fragments of $C_3H_5O_2^+$ (m/z = 73.029), $C_3H_7NO^+$ (m/z = 73.053), and $C_4H_{11}N^+$ (m/z = 73.089)

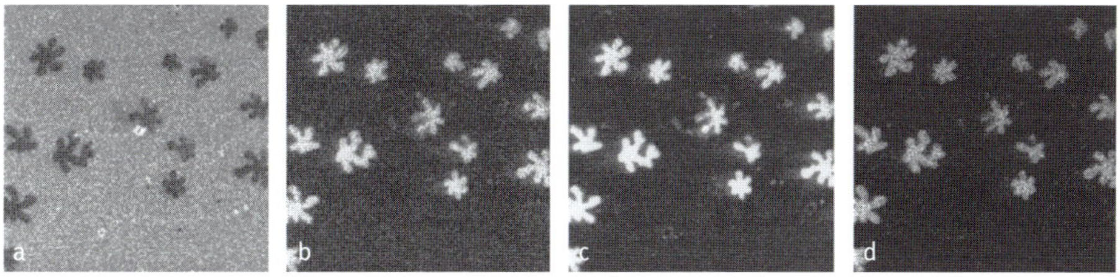

Fig. 2.7: TOF-SIMS images (100 × 100 µm) of Langmuir-Blodgett monolayers from $A_{15}B_{110}A_{15}$-DPPC with molar ratio of 0.3–0.7 transferred on a silicon surface at 25 mN/m. Brighter colors indicate higher signal intensity. The images were recorded using a 25-keV Bi_3^{2+} ion beam. Fragments were assigned to (a) CN^-, $A_{15}B_{110}A_{15}$ (m/z = 26.00), (b) PO_2^-, DPPC (m/z = 62.97), (c) PO_3^-, DPPC (m/z = 78.92), and (d) $C_{16}H_{31}O_2^-$, alkyl chain (m/z = 255.25). Reprinted with permission from Kita-Tokarczyk K, Itel F, Grzelakowski M, Egli S, Rossbach P, Meier W, *Langmuir* 25, 9847–56, 2009. Copyright 2009, American Chemical Society [97].

originating from surrounding tissue. Because the sample was measured below 148 K, condensation of water was one of the problems arising with transfer of the material into the UHV measuring chamber of the TOF-SIMS instrument. A water cluster with composition $H_9O_4^+$ (m/z = 73.050) was detected at the surface of the tissue sample. The possibility to perform imaging of secondary ions with high mass and high lateral resolution at the same time is therefore required to distinguish between fragments arising from different origin. Tight focusing of temporally narrow ion pulses is not possible in conventional ion guns. The development of primary ion beam technology has met this challenge by operating the LMIG source in a so-called burst mode. The temporally broad ion pulse is split into several sub-pulses (patented technology of IonTof GmbH, Münster, Germany), of which one can be selected for high-mass-resolution imaging. A mass resolution $m/\Delta m \geq 2{,}500$ is achieved for signals in the fingerprint region of the mass spectrum. However, the mass interference between $H_9O_4^+$ and $(CH_3)_3Si^+$ requires a mass resolution of $m/\Delta m > 24{,}000$ to be clearly resolved. This figure is beyond the performance of LMIG technology, and was only attainable because in reality the $(CH_3)_3Si^+$ signal was shifted by +200 ppm due to incomplete charge compensation during measurement.

An electric potential-dependent adsorption/desorption process of the triblock copolymer poly(propylene sulfide-*bl*-ethylene glycol) (PPS-PEG) from a conductive ITO substrate was studied by Tang *et al.* [99]. Characteristic signals of PEG were identified at masses out of a series with $C_aH_bO_c^+$, eg. $C_3H_7O^+$ (m/z = 59). As most intense fragments of the PPS backbone, the authors assigned $C_2H_3S^+$ (m/z = 58) and $C_3H_7S^+$ (m/z = 75). TOF-SIMS spectra from PPS-PEG adlayers contained signals at masses Na^+ (m/z = 22.992), CH_3O^+ (m/z = 31.019), S^+ (m/z = 32.98), $C_2H_5O^+$ (m/z = 45.037), $C_4H_9^+$ (m/z = 57.071), $C_2H_3S^+$ (m/z = 58.995), $C_3H_7O^+$ (m/z = 59.054), $C_3H_7S^+$ (m/z = 75.027), $C_6H_9^+$ (m/z = 81.067), In^+ (m/z = 114.905), and In-S^- (m/z = 146.866), respectively.

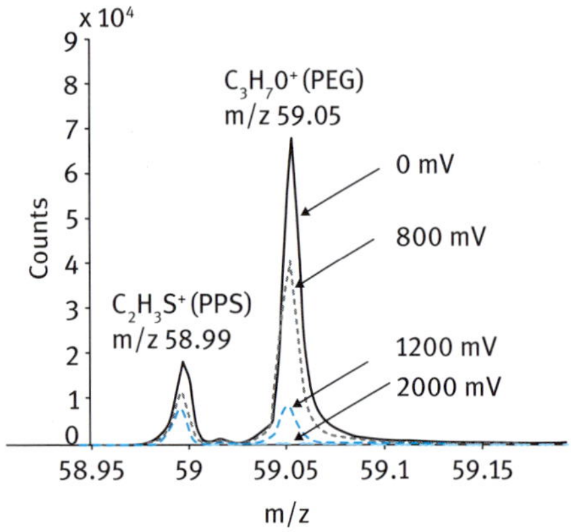

Fig. 2.8: Averaged positive-ion high-resolution TOF-SIMS spectra from three independent samples shown around the nominal mass $m/z = 59$. Spectra were recorded as function of applied anodic potential with fragments $C_2H_3S^+$ ($m/z = 58.995$) characteristic of PPS and $C_3H_7O^+$ ($m/z = 59.052$) originating from PEG. Reprinted from Tang C, Feller L, Rossbach P, *et al.*, *Surf Sci* 600, 1510–17, 2006, with permission from Elsevier [99].

Electric potential-dependent polymer desorption was observed by comparing signal intensities of $C_3H_7O^+$ (m/z = 59.05, PEG) and $C_2H_3S^+$ (m/z = 58.99, PPS) after applying potentials between 0 and 2,000 mV to the copolymer-coated ITO substrate. Maximum intensities were measured at 0 mV, which corresponds to an intact surface film. Spectra shown in Fig. 2.8 indicate a significant decrease for both signals at anodic potential ≥ 800 mV as consequence of increasing polymer desorption from the surface. TOF-SIMS spectra recorded from samples exposed to 2,000 mV showed similar peak distribution as the neat ITO surfaces. An intact desorption in molecular form was postulated rather than oxidation of sulfur or chain degradation by XPS investigations of the process.

2.2.5 Multicomponent polymers (polymer blends)

Blending of polymers is one of the technologically most important fields of polymer chemistry today because it allows the manufacturing of materials that combine properties of single components. This synthetic concept has been successfully applied to provide many consumer products or system components with new and superior qualities [100, 101].

However, one of the most remarkable and often cumbersome physico-chemical facts is the often found phase separation of its constituents due to minimal miscibility or partial immiscibility. Depending on the application of the material, the formation of micro-mechanical defect sites are more or less important for the performance of the polymer blend. Feng *et al.* [102] published a compatibility study of PC and poly(vinylidene fluoride) blends using TOF-SIMS and SEM. Yuan and Shoichet [103] investigated the enrichment of poly(trifluorovinyl ether) (PTFVE) in PS blends at low bulk concentration. PTFVE is chemically characterized by a fluorocarbon backbone with oligoether side groups. The surface of the blended material was characterized using TOF-SIMS and XPS and was found to be enriched by PTFVE at concentrations <1 wt%. Large aggregates with diameters around 200 nm were detected. It was suggested that PTFVE in PS/PTFVE blends forms core-corona structure elements at the surface with the oligoether groups as corona. The fingerprint mass range of positive-ion TOF-SIMS spectra was investigated with mass $m/z = 91$ ($C_7H_7^+$, tripylium ion) as dominant signal in pristine PS. In blends of PS with PTFVE, this signal decreased dramatically for only as low as 0.5 wt% of PTFVE, and strong peaks at masses $m/z = 29$ ($C_2H_5^+$), $m/z = 45$ ($C_2H_5O^+$), and $m/z = 73$ ($C_2H_5OC_2H_4^+$), respectively, were detected as characteristic fragments of the oligoether group of PTFVE. Spectra collected from polymer blends with 2.5 wt% PTFVE showed no tripylium ions at mass $m/z = 91$ anymore. Because poly(trifluorovinyl ether) is a viscous liquid at room temperature (Tg = 213 K), this effect is probably due to a surface energy-driven rearrangement process. Polymer films cast from toluene, which is a more potent solvent for PS than e.g. chloroform, required greater PTFVE concentration to show similar surface activity. A surface reconstruction process was also observed by Hug and Keller [82] in modified ethylene-methacrylic acid copolymers.

2.2.6 Plasma modification and deposition

Léonard *et al.* [80] studied the adhesion properties of metals following H_2-, O_2-, and N_2-plasma modification of Teflon PFA and Teflon AF1600 surfaces. TOF-SIMS measurements confirmed that ablation of fluorine from surfaces exposed to H_2-plasma treatment showed the strongest effect. However, surface incorporated hydrocarbons during the plasma process strongly decreased adhesion efficiency by formation of weak boundary layers. Signals assigned to CN^- on the surface were detected after N_2-plasma exposure and enhanced occurrence of surface oxygen functional groups was characteristic for O_2-plasma treatment.

The introduction of epoxy groups to biologically active surfaces has been recognized to form stable and reliable bonds to peptides. Silanization by solvent based (3-glycidyloxypropyl)trimethoxysilane has previously found widespread application. However, the practical use is restricted by excessive pre-polymerization in the liquid phase and incomplete formation of a surface layer. Hydrolysis of siloxal bonds is also often observed. Plasma deposition of surface films offers an alternative to solvent based epoxy functionalization. Pulsed plasma polymerization of glycidyl-methacrylate [104, 105] produced well-characterized epoxy-rich and almost pinhole-free surfaces. Thierry *et al.* [106] used a DC plasma deposition process of allyl-glycidylether (AGE) monomer onto various surfaces and demonstrated the homogeneity of the epoxy-carrying surface by XPS and TOF-SIMS experiments. Fingerprint spectra in the range of m/z = 0–150 of pulsed and DC-deposited films were compared. Most of the structural information of epoxy-functionalized surfaces was found within this relatively restricted mass span (Fig. 2.9).

The inspection of TOF-SIMS spectra under high-mass resolution conditions reveals a split-up of unit mass signals into several distinct peaks. For example, the insert in Fig. 2.9 at mass m/z = 57 shows fragment masses at m/z = 57.03 ($C_3H_5O^+$)

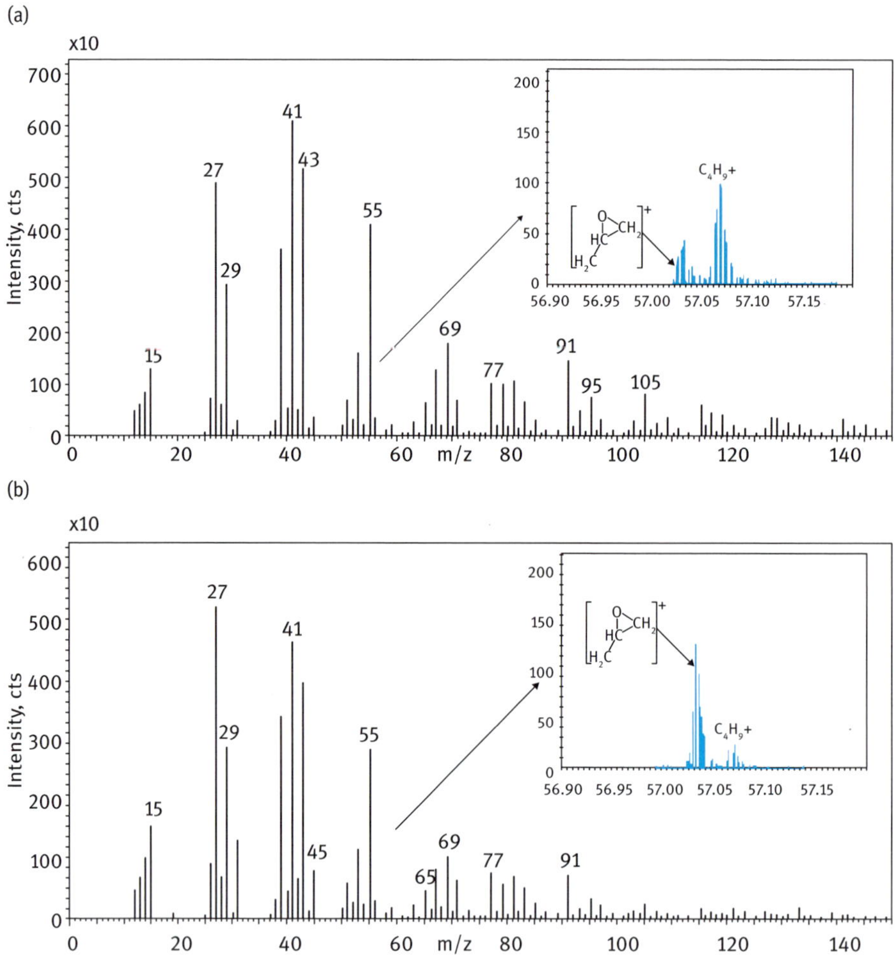

Fig. 2.9: TOF-SIMS spectra of (a) continuous and (b) plasma-deposited films of AGE. $C_3H_5^+$ at mass m/z = 41 is dominating spectrum *a*, whereas $C_2H_3^+$ at m/z = 27 is the most intense signal in spectrum b. The formation of enhanced low-mass fragments in spectrum *b* indicates a better retention of the monomer chemical structure under pulsed deposition conditions. Reprinted with permission from Thierry B, Jasieniak M, de Smet LCPM, Vasilev K, Griesser HJ, *Langmuir* 24, 10187–95, 2008. Copyright 2008, American Chemical Society [106].

and $m/z = 57.07$ ($C_4H_9^+$). Comparing the intensities of close-spaced signals of different origin allows an estimation of the influence of process parameters. Because $C_3H_5O^+$ is related to the epoxy functionality, the relative enhancement of this signal in spectrum b indicates the superior retention of this chemical group under pulsed plasma deposition conditions. PCA was used to detect subtle differences between films produced under distinct conditions. Clustering of related data was observed, which confirmed the chemical homogeneity of the processed surfaces. For details about the peak list and PCA parameters, the reader is referred to the literature [106].

Pulsed plasma polymerization is a very effective and straightforward experimental way to produce uniform surface polymer films of various chemical compositions. It was shown [107] that pulsed plasma polymerization of 2-hydroxyethyl-methacylate leads to the formation of a well-defined poly-HEMA film.

Kinmond *et al.* [108] detected a high structural retention of perfluoroalkyl groups after the plasma polymerization of 1*H*,1*H*,2*H*-perfluorododecene using NMR and TOF-SIMS. In contrast, continuous-wave conditions lead to a high level of fragmentation and considerable cross-linking in the film.

2.2.7 Other applications

The application of TOF-SIMS imaging to small, almost nano-dimensional structures has made great progress in recent years. The focal diameter of timely LMIG cluster ion sources has been decreased to close to theoretical limit [36, 37]. The number of secondary ions generated from small areas is now high enough to acquire images with good quality. However, restricted information can be obtained from the surface topography of the material at this lateral dimension. This has classically considered the domain of force microscopy (atomic force microscopy, AFM; scanning probe microscopy, SPM), where lateral resolution of atomic dimension is possible under suitable experimental conditions. Chemical information gained by SPM or AFM is poor and is usually restricted to phase contrast AFM or Kelvin probe force microscopy. A recently developed instrument combining a TOF-SIMS instrument equipped with a 10-nm-resolution gun with a high-quality, in-line long-range SPM assembly in the same vacuum chamber allowed for the first time true complementary access to high-resolution TOF-SIMS imaging combined with the ultra-high lateral surface morphology information of SPM (3D NanoChemiscope Project) [109]. The instrument was applied to the 3-D analysis of a PMMA/PS blend on silicon via GCIB sputtering.

Although the equipment is intended for use in a depth profiling mode and applications in metrology, the shallow profiles classify the instrument as surface-sensitive technique. A similar experimental set-up was presented by Wirtz *et al.* [110]. Combined SIMS-SPM 3D-mapping of an annealed PS/PMMA blend was measured using a high-precision sample stage mounted inside a NanoSIMS instrument. Secondary ions recorded included $^{12}C^-$, $^{12}C^1H^-$, $^{16}O^-$, and $^{24}C_2^-$, respectively. Images of $^{12}C^-$ and $^{16}O^-$ were presented with a field of view of $10.6 \times 9.8\ \mu m^2$. An area of 256×256 pixels was scanned with a 50-nm-diameter primary ion beam. Two separate phases were detected, from which PS was identified by the absence of oxygen. Topographical information was simultaneously obtained, revealing a distinct crater-like surface roughness as result of considerably lower sputter yield of PS compared with PMMA.

References

[1] Vickerman JC, Briggs D, ed. TOF-SIMS: surface analysis by mass spectrometry. 1st and 2nd edn. Chichester (UK): IM Publications and Surface Spectra; 2001 and 2013.

[2] Vickerman JC. Molecular surface mass spectrometry by SIMS. In: Vickerman JC, Gilmore IS, eds. Surface analysis: the principal techniques. 2nd ed. Chichester (UK): Wiley; 2009:113–205.

[3] Briggs D, Seah MP, ed. Practical surface analysis: ion and neutral spectroscopy. Vol 2. Chichester (UK): Wiley; 1983.

[4] Eccles AJ. SIMS instrumentation. In: Vickerman JC Brown A, Reed NM, eds. Secondary ion mass spectrometry: principles and applications. Oxford (UK): Oxford University Press; 1989:73–104.

[5] Grams J. New trends and potentialities of TOF-SIMS in surface studies. New York (NY): Nova Science Publishers; 2007.

[6] Smith R, Walls JM. Ion erosion in surface analysis. In: Walls JM. Methods of surface analysis: techniques and applications. Cambridge (UK): Cambridge University Press; 1989:20–56.

[7] Reed NM. Static SIMS for applied surface analysis. In: Vickerman JC Brown A, Reed NM. Secondary ion mass spectrometry: principles and applications. Oxford (UK): Oxford University Press; 1989:186–243.

[8] Thiel V, Sjövall P. Using time-of-flight secondary ion mass spectrometry to study biomarkers. Annu Rev Earth Planet Sci 2011;39:125–56.

[9] Beavis RC. The interpretation of reflectron time-of-flight mass spectra. In: Cotter RJ, ed. ACS Symp Ser. Washington (DC): American Chemical Society; 1994;549:49–60.

[10] Cotter RJ. Time-of-flight mass spectrometry: instrumentation and applications in biological research. Washington (DC): ACS Professional Reference Books; 1997.

[11] Wood M, Zhou Y, Brummel CL, Winograd N. Imaging with on beams and laser postionization. Anal Chem 1994;66:2425–32.

[12] Thomson JJ. Rays of positive electricity. Philos Mag 1910;20:752–67.

[13] Benninghoven A. The analysis of monomolecular layers of solids by secondary ion emission. Z Physik 1970;230:403–17.

[14] Benninghoven A, Jaspers D, Sichterman W. Secondary ion emission of amino acids. Appl Phys 1976;11:35–9.

[15] Hagenhoff B. Sekundärionenmassenspektrometrie molekularer Oberflächenstrukturen. Thesis. Münster (Germany): University Münster; 1993 [in German].

[16] Niehuis E, Heller T, Benninghoven A. A new high resolution time-of-flight secondary ion mass spectrometer. J Vac Sci Technol 1987;A5:1243–46.

[17] Van Vaeck L, Adriaens A, Gijbels R. Static secondary ion mass spectrometry: (S-SIMS) Part 1: methodology and structural interpretation. Mass Spectrom Rev 1999;18:1–47.

[18] Adriaens A, Van Vaeck L, Adams F. Static secondary ion mass spectrometry (S-SIMS) Part 2: material science applications. Mass Spectrom Rev 1999;18:48–81.

[19] Brummel CL, Willey KF, Vickerman JC, Winograd N. Ion beam induced desorption with postionization using high repetition femtosecond lasers. Int J Mass Spectrosc Ion Process 1995;143:257–70.

[20] Sigmund P. Theory of sputtering. I. Sputtering yield of amorphous and polycrystalline targets. Phys Rev 1969;184:383–416.
[21] Garrison BJ. Energy and angular distributions of atoms sputtered from polycrystalline surfaces: Thompson and beyond. Nucl Instrum Method Phys Res 1989;B40/41:313–16.
[22] Benninghoven A, Ruedenauer FG, Werner HW. Secondary ion mass spectrometry: basic concepts, instrumental aspects, applications and trends. New York (NY): Wiley; 1987.
[23] van den Eynde X. Quantitative characterization of polymer surfaces. In: Vickerman JC, Briggs D, eds. TOF-SIMS: surface analysis by mass spectrometry. 1st edn. Chichester (UK): IM Publications and Surface Spectra; 2001:543–73.
[24] Kovac J. Surface characterization of polymers by XPS and SIMS techniques. Materials and Technology 2011;45:191–97.
[25] Hauert R, Keller BA. Chemical surface analysis with nanometer depth resolution. Chimia 2006;60:800–4.
[26] Wilson RG. Surface ionization ion sources. IEEE Trans Nucl Sci 1967;14:72–74.
[27] Gillen G, Roberson S. Preliminary evaluation of an SF_5^+ polyatomic primary ion beam for analysis of organic thin films by secondary ion mass spectrometry. Rapid Commun Mass Spectrom 1998;12:1303–12.
[28] Fletcher JS, Lockyer NP, Vickerman JC. Developments in molecular SIMS depth profiling and 3D imaging of biological systems using polyatomic primary ions. Mass Spectrom Rev 2011;30:142–74.
[29] Wong SCC, Hill R, Blenkinsopp P, Lockyer NP, Weibel DE, Vickerman JC. Development of a C_{60}^+ ion gun for static SIMS and chemical imaging. Appl Surf Sci 2003;203–4:219–22.
[30] Cheng J, Wucher A, Winograd N. Molecular depth profiling with cluster ion beams. J Phys Chem B 2006;110:8329–36.
[31] Swanson LW. Liquid metal ion sources: mechanism and applications. Nucl Instrum Methods Phys Res 1983;218:347–53.
[32] Winograd N. The magic of cluster SIMS. Anal Chem 2005;77:143A–9A.
[33] Appelhans A, Delmore JE. Comparison of polyatomic and atomic primary beams for secondary ion mass spectrometry of organics. Anal Chem 1989;61:1087–93.
[34] Walker EV, Winograd N. Prospects for imaging with TOF-SIMS using gold liquid metal ion sources. Appl Surf Sci 2003;203–4:198–200.
[35] Davies N, Weibel P, Blenkinsopp N, Lockyer N, Hill R, Vickerman JC. Development and experimental application of a gold liquid metal ion source. Appl Surf Sci 2003;203–4:223–27.
[36] Kollmer F, Paul W, Krehl M, Niehuis E. Ultra high spatial resolution SIMS with cluster ions approaching the physical limits. Surf Interface Anal 2013;45:312–14.
[37] Orloff J, Utlaut M, Swanson L. High resolution focused ion beams: FIB and its application. New York (NY): Kluwer Academic; 2003.
[38] Kollmer F. Cluster ion bombardment of organic materials. Appl Surf Sci 2004;231–2:153–8.
[39] Blain MG, Della-Negra S, Joret H, Le Beyec Y, Schweikert EA. Secondary-ion yields from surfaces bombarded with keV molecular and cluster ions. Phys Rev Lett 1989;63:1625–28.
[40] Harris RD, Baker WS, Van Stipdonk MJ, Crooks RM, Schweikert EA. Secondary ion yields produced by keV atomic and polyatomic ion impacts on a self-assembled monolayer surface. Rapid Commun Mass Spectrom 1999;13:1374–80.
[41] Levine RD. Molecular reaction dynamics. Cambridge (UK): Cambridge University Press; 2005.
[42] Hashinokuchi M, Moritani K, Nakagawa J, Kashiwagi T, Toyoda N, Mochiji K. Secondary ion mass spectrometry using size-selected gas cluster ion beam. J Surf Anal 2008;14:387–90.
[43] Ninomiya S, Ichiki K, Yamada H, et al. Precise and fast secondary ion mass spectrometry depth profiling of polymer materials with large Ar cluster ion beams. Rapid Commun Mass Spectrom 2009;23:1601–06.

[44] Vanden Eynde X, Bertrand P. TOF-SIMS quantification of polystyrene spectra based on principal component analysis (PCA). Surf Interface Anal 1997;25:878–88.
[45] Henry M, Dupont-Gillain C, Bertrand P. Conformation change of albumin adsorbed on polycarbonate membranes as revealed by TOF-SIMS. Langmuir 2003;19:6271–76.
[46] Kayser S, Rading D, Moellers R, Kollmer F, Niehuis E. Surface spectrometry using large argon clusters. Surf Interface Anal 2013;45:131–33.
[47] Cheng J, Winograd N. Depth profiling of peptide films with TOF-SIMS and a C_{60} probe. Anal Chem 2005;77:3651–59.
[48] Cheng J, Winograd N. Molecular depth profiling of multi-layer systems with cluster ion sources. Appl Surf Sci 2006;252:6498–501.
[49] Luxembourg SL, Mize TH, McDonnell LA, Heeren RMA. High-spatial resolution mass spectrometric imaging of peptide and protein distributions on a surface. Anal Chem 2004;76:5339–44.
[50] Touboul D, Kollmer F, Niehuis E, Brunelle A, Laprévote O. Improvement of biological time-of-flight secondary ion mass spectrometry imaging with a bismuth cluster ion source. J Am Soc Mass Spectrom 2005;16:1608–18.
[51] Peacock PM, McEwen CN. Mass spectrometry of synthetic polymers. Anal Chem 2006;78:3417–28.
[52] Hanton SD. Mass spectrometry of polymers and polymer surfaces. Chem Rev 2001;101:527–69.
[53] Keller BA, Hug P. Time-of-flight secondary ion mass spectrometry of industrial materials. Anal Chim Acta 1999;393:201–12.
[54] Reed NM, Vickerman JC. The application of static secondary ion mass spectrometry (SIMS) to the surface analysis of polymeric materials. In: Sabbatini L, Zambonin PG, eds. Surface characterization of advanced polymers. Weinheim (Germany): VCH Verlagsgesellschaft; 1983:83–162.
[55] Peacock PM, McEwen CN. Mass spectrometry of synthetic polymers. Anal Chem 2002;74:2743–48.
[56] Peacock PM, McEwen CN. Mass spectrometry of synthetic polymers. Anal Chem 2004;76:3417–28.
[57] Hakkarainen M, ed. Mass spectrometry of polymers. New techniques. Adv Polym Sci. Heidelberg (Germany): Springer-Verlag; 2012;248.
[58] Mahoney CM. Cluster secondary ion mass spectrometry of polymers and related materials. Mass Spectrom Rev 2010;29:247–93.
[59] Legget GJ, Vickerman JC. Sample charging during static SIMS studies of polymers. Appl Surf Sci 1995;84:253–66.
[60] Leggett GJ. Mechanisms of ion production and fragmentation in static SIMS. In: Vickerman JC, Briggs D, Henderson A, eds. The static SIMS library. Manchester (UK): Surface Spectra; 1998:19–38.
[61] Gilmore IS, Seah MP. Static SIMS: a study of damage using polymers. Surf Interface Anal 1996;24:746–62.
[62] Keller BA, Hug P. Cationization in TOF-SIMS spectra of biologically active peptides using modified ion-containing polymers as support. In: Benninghoven A, Bertrand P, Migeon HN, Werner HW, eds. Secondary ion mass spectrometry, SIMS XII. Amsterdam (The Netherlands): Elsevier; 2000:885–88.
[63] Keller BA, Hug P. A new type of support based on ion-containing polymers for selective generation of cationized species in TOF-SIMS spectra. In: Benninghoven A, Bertrand P, Migeon HN, Werner HW, eds. Secondary ion mass spectrometry, SIMS XII. Amsterdam (The Netherlands): Elsevier; 2000:749–52.

[64] Fowler DE, Johnson RD, van Leyen D, Benninghoven A. Quantitative time-of-flight secondary ion mass spectrometry of a perfluorinated polyether. Surf Interface Anal 1991;17:125–36.
[65] Lee Y, Han S, Yoon J-H, Lim H, Cho J. Studies of polystyrenes by secondary ion mass spectrometry. J Surf Anal 1999;6:50–53.
[66] Galuska AA. TOF-SIMS determinations of molecular weights from polymeric surfaces and microscopic phases. Surf Interface Anal 1997;25:790–98.
[67] Coullerez G, Lundmark S, Malmström E, Hult A, Mathieu HJ. TOF-SIMS for the characterization of hyperbranched aliphatic polyesters: probing their molecular weight on surfaces based on principal component analysis (PCA). Surf Interface Anal 2003;35:693–708.
[68] Ng JMK, Gitlin I, Stroock AD, Whitesides GM. Components for integrated poly (dimethylsiloxane) microfluidic systems. Electrophoresis 2002;23:3461–73.
[69] Morra M, Occhiello E, Marola R, Garbassi F, Humphrey P, Johnson D. On the aging of oxygen plasma-treated polydimethylsiloxane surfaces. J Colloid Interface Sci 1990;137:11–24.
[70] Delcorte A, Befahy S, Poleunis C, Troosters M, Bertrand P. Improvement of metal adhesion to silicone films: a TOF-SIMS study. In: Mittal KL, ed. Adhesion aspects of thin films. Vol 2. Boca Raton (FL): CRC Press; 2005:155–66.
[71] Dong X, Proctor A, Hercules DM. Characterization of poly(dimethylsiloxane)s by time-of-flight secondary ion mass spectrometry. Macromolecules 1997;30:63–70.
[72] Dong X, Gusev A, Hercules DM. Characterization of polysiloxanes with different functional groups by time-of-flight secondary ion mass spectrometry. J Am Soc Mass Spectrom 1998;9:292–98.
[73] Inoue M, Murase A, Sugiura M. Molecular weight evaluation of poly(dimethylsiloxane) on solid surfaces using silver deposition TOF-SIMS. Anal Sci 2004;20:1623–28.
[74] Belu AM, Brocchini S, Kohn J, Ratner BD. Characterization of combinatorially designed polyacrylates by time-of-flight secondary ion mass spectrometry. Rapid Commun Mass Spectrom 2000;14:564–71.
[75] Leeson AM, Alexander MR, Short RD, Briggs D, Hearn MJ. Secondary ion mass spectrometry of polymers: a TOF-SIMS study of monodispersed PMMA standards. Surf Interface Anal 1997;25:261–74.
[76] Wagner MS. Degradation of poly(acrylates) under SF_5^+ primary ion bombardment studied using time-of-flight secondary ion mass spectrometry. 1. Effect of main chain and pendant methyl groups. Surf Interface Anal 2004;36:42–52.
[77] Wagner MS. Degradation of poly(acrylates) under SF_5^+ primary ion bombardment studied using time-of-flight secondary ion mass spectrometry. 2. Poly(n-alkyl methacrylates). Surf Interface Anal 2004;36:53–61.
[78] Wagner MS. Degradation of poly(acrylates) under SF_5^+ primary ion bombardment studied using time-of-flight secondary ion mass spectrometry. 3. Poly(hydroxymethyl methacrylate) with chemical derivatization. Surf Interface Anal 2004;36:62–70.
[79] Mahoney CM. Cluster SIMS depth profiling of stereo-specific PMMA thin films on Si. Surf Interface Anal 2010;42:1393–401.
[80] Léonard D, Bertrand P, Shi MK, Sacher E, Martinu L. Plasma surface modification of fluoropolymers studied by ToF-SIMS. Plasmas and Polymers 1999;4:97–111.
[81] Ma ZQ, Chen P, Zhao TS. A palladium-alloy deposited Nafion membrane for direct methanol fuel cells. J Membr Sci 2003;215:327–36.
[82] Hug P, Keller BA. Surface reconstruction in modified ethylene-methacrylic acid copolymers probed by TOF-SIMS. In: Olefjord I, Nyborg L, Briggs D, eds. ECASIA'97 (European Conference on Applications of Surface and Interface Analysis). Chichester (UK): Wiley; 1997:727–30.
[83] Liu Y, Lu C, Twigg S, et al. Direct observation of ion distributions near electrodes in ionic polymer actuators containing ionic liquids. Sci Rep 2013;3:973.

[84] Nysten B, Verfaillie G, Ferain E, et al. SEM, TOF-SIMS and LFM morphological study of an heterogeneous polymeric surface. Microscopy Microanal Microstruct 1994;5:373–80.
[85] Green FM, Gilmore IS, Seah MP. Mass Spectrometry and informatics: distribution of molecules in the PubChem database and general requirements for mass accuracy. Anal Chem 2011;83:3239–43.
[86] Gilmore IS, Seah MP. Static SIMS: towards unfragmented mass spectra-the G-SIMS procedure. Appl Surf Sci 2000;161:465–80.
[87] Aoyagi S, Gilmore IS, Mihara I, Seah MP, Fletcher IW. Identification and separation of protein, contaminant and substrate peaks using gentle-secondary ion mass spectrometry and the g-ogram. Rapid Commun Mass Spectrom 2012;26:2815–21.
[88] Gilmore IS, Seah MP. Organic molecule characterization-G-SIMS. Appl Surf Sci 2004;231:224–29.
[89] Straif CJ, Hutter H. Investigation of polymer thin films by use of Bi-cluster-ion-supported time-of-flight secondary ion mass spectrometry. Anal Bio Anal Chem 2009;393:1889–98.
[90] Green FM, Kollmer F, Niehuis E, Gilmore IS, Seah MP. Imaging G-SIMS: a novel bismuth-manganese source emitter. Rapid Commun Mass Spectrom 2008;22:2602–08.
[91] Aoyagi S, Mihara I, Kudo M. Mixed polymer sample evaluation using gentle secondary ion mass spectrometry and multivariate curve resolution. Surf Interface Anal 2012;44:789–92.
[92] Graham DJ, Wagner MS, Castner DG. Information from complexity: challenges of TOF-SIMS data interpretation. Appl Surf Sci 2006;252:6860–68.
[93] Bart JCJ. Additives in polymers: industrial analysis and applications. Chichester (UK): Wiley; 2005.
[94] Bart JCJ. Polymer additive analytics: industrial practice and case studies. Firenze (Italy): Firenze University Press; 2006.
[95] Médard N, Benninghoven A, Rading D, et al. Antioxidant segregation and crystallisation at polyester surfaces studied by ToF-SIMS. Appl Surf Sci 2003;203–4:571–4.
[96] Holbrook RD, Davis JM, Scott KC, Szakal C. Detection and speciation of brominated flame retardants in high-impact polystyrene (HIPS) polymers. J Microsc 2012;246:143–52.
[97] Kita-Tokarczyk K, Itel F, Grzelakowski M, Egli S, Rossbach P, Meier W. Monolayer interactions between lipids and amphiphilic block copolymers. Langmuir 2009;25:9847–56.
[98] Keller BA, Mayerhofer K. The role of TOF-SIMS in nanomedicine. European J Nanomedicine 2009;2:37–42.
[99] Tang C, Feller L, Rossbach P, et al. Adsorption and electrically stimulated desorption of the triblock copolymer poly(propylene sulfide-bl-ethylene glycol) (PPS-PEG) from indium tin oxide (ITO) surfaces. Surf Sci 2006;600:1510–17.
[100] Mayerhofer KE, Heier J, Maniglio Y, Keller BA. Three dimensional analysis of self-structuring organic films using time-of-flight secondary ion mass spectrometry. Thin Solid Films 2011;519:6183–89.
[101] Yu BY, Lin WB, Ilda SI, et al. Effect of fabrication parameters on three-dimensional nanostructures of bulk heterojunctions imaged by high-resolution scanning TOF-SIMS. Nano 2010;4:833–40.
[102] Feng J, Weng L-T, Li L, Chan C-M. Compatibilization of polycarbonate and poly(vinylidene fluoride) blends studied by time-of-flight secondary ion mass spectrometry and scanning electron microscopy. Surf Interface Anal 2000;29:168–74.
[103] Yuan Y, Shoichet MS. Surface enrichment of poly(trifluorovinyl ethers) in polystyrene blends. Macromolecles 2000;33:4926–31.
[104] Harris LG, Schofield WCE, Badyal JPS. Multi-functional molecular scratchcards. Chem Mater 2007;19:1546–51.

[105] Tarducci C, Kinmond EJ, Badyal JPS, Brewer SA, Willis C. Epoxide-functionalized solid surfaces. Chem Mater 2000;12:1884–89.

[106] Thierry B, Jasieniak M, de Smet LCPM, Vasilev K, Griesser HJ. Reactive epoxy-functionalized thin films by a pulsed plasma polymerization process. Langmuir 2008;24:10187–95.

[107] Tarducci C, Schofield WCE, Badyal JPS, Brewer SA, Willis C. Monomolecular functionalization of pulsed plasma deposited poly(2-hydroxyethyl methacrylate) surfaces. Chem Mater 2002;14:2541–45.

[108] Kinmond EJ, Coulson SR, Badyal JPS, Brewer SA, Willis C. High structural retention during plasma polymerization of 1H,1H;2H-perfluorododecene: an NMR and TOF-SIMS study. Polymer 2005;46:6829–35.

[109] Moellers R, Bernard L, Dianoux R, et al. Combined SIMS-SFM instrument for the 3-dimensional chemical analysis of nanostructures. Supported by FP7 of the European Commission, Münster (Germany); 2008. (Accessed September 6, 2013, at http://www.3dnanochemiscope.eu/3D_NanoChemiscope/Home.html.

[110] Wirtz T, Fleming Y, Gerard M, et al. Design and performance of a combined secondary ion mass spectrometry-scanning probe microscopy instrument for high sensitivity and high-resolution elemental three-dimensional analysis. Rev Sci Instrum 2012;83:063702.

Elvira De Giglio, Nicoletta Ditaranto, and Luigia Sabbatini

3 Polymer surface chemistry: Characterization by XPS

3.1 Introduction

We have seen in Chapter 1 that when a polymer interacts with another material or with its environment, the chemical and physical structures of the polymer surface determine the nature of the interaction and can undergo dramatic changes. Therefore, the understanding of the chemical behavior needs surface characterization with a high degree of chemical specificity in terms of composition and structure. This has been the driving force for the evolution of traditional analytical methods as well as for the development of new techniques. The application of X-ray photoelectron spectroscopy (XPS) to polymer surface analysis since its beginning largely contributed to its widespread use. Indeed, the requirements for polymer surface analysis have driven the development of peculiar instrumental facilities because of the importance of this materials field. The XPS perspectives in this sector are widely recognized, and there are plenty of literature that cover, in more or less details, the application of this technique to polymer-based materials. It is clear, however, that most polymer surface studies devoted to chemical characterization benefit from the combined use of XPS with other techniques such as secondary ions mass spectrometry (SIMS), ion scattering spectroscopy, and Fourier transform infrared spectroscopy in attenuated total reflectance mode (ATR/FTIR), which provide information from varying surface thicknesses, ranging from the top few angstroms (SIMS), to 2–10 nm (XPS), to a few micrometers (ATR/FTIR).

Some reviews describe many types of information available from XPS measurements of polymers [1–3]. XPS is now routinely used to obtain surface composition of homopolymers, copolymers, polymer blends, hybrid (i.e. organic/inorganic) systems, composites, and to follow processing steps, application-oriented surface modifications, such as chemical derivatization, grafting, or physical treatments (plasma, ion or electron beam, laser irradiation, etc.) as well as polymer degradation chemistry [4–6].

In this chapter, we will see how XPS is able to answer to important questions such as:

- Which elements are present on the surface? Which is their chemical state?
- Which is the atomic/mass percent surface composition?
- If the material is present as a thin film at the surface, how thick is the film? How uniform is the thickness and/or the chemical composition of the film?
- Which is the spatial distribution of elements (chemical mapping)?

Some fundamentals of the technique will be provided as well as basic information on the instrumental apparatus; applications to polymer surface analysis, both routine and specialized, will be illustrated.

3.2 Photoelectron spectroscopy: A brief history

The origin of XPS can be traced back to the end of the nineteenth century with the discovery of the photoelectric effect by Hertz in 1887. After that, several groups analyzed the energies of electrons emitted from metals bombarded by hard X-rays, the most active in the field being Rutherford's group in Manchester. By 1914, Rutherford provided the early basic XPS equation:

$$\mathrm{KE} = h\nu - \mathrm{BE}, \tag{3.1}$$

where KE is the photoelectron kinetic energy, $h\nu$ is the exciting photon energy, and BE is the electron binding energy in the solid.

Different groups performed research in the field, but the turning point was the construction by Kai Siegbahn [7] of an instrument capable of measuring electron kinetic energy with a resolving power of 10^5. This allowed KE, and hence BE, to be measured accurately for the first time. Siegbahn had full consciousness of the potential of XPS in chemical analysis and coined the acronym ESCA, for electron spectroscopy for chemical analysis, which is still largely used to address the technique [7]. A further spread was driven by the development of suitable commercial instruments that started to appear around 1969–1970 with the construction of the first UHV system in 1972.

3.2.1 Basic principles

XPS is a spectroscopic technique capable of providing atomic and molecular information about the surface of a solid material. In XPS, the sample surface is irradiated with X-rays providing surface atoms to emit electrons after the entire photon energy is transferred to them.

The relationship among the parameters involved in an XPS experiment is very similar to Eq. 3.1, implemented with a further term:

$$\mathrm{KE} = h\nu - \mathrm{BE} - W, \tag{3.2}$$

where W is the spectrometer work function.

The electron kinetic energy is an experimental quantity that can be measured by the spectrometer, but this is dependent on the photon energy of the X-rays used; the BE of the electron is the parameter that identifies the electron specifically, both in terms of its parent element and atomic energy level. Because no two elements possess the same set of BE, the measured kinetic energy provides for the elemental analysis (except for hydrogen and helium). From Eq. 3.2, it is a simple matter to calculate the BE of the electron because all the other three terms are known or, at least, measurable. In practice, this task will be performed by the control electronics or data system associated with the spectrometer and the operator merely selects a binding or kinetic energy scale, whichever is considered the more appropriate.

A scheme of the photoemission process is reported in Fig. 3.1.

The scheme represents an incident X-ray impinging the material and causing an electron from an internal shell to be ejected from the atom (Fig. 3.1a). The photoelectron spectrum will reproduce the electronic structure of an element quite accurately because all electrons with a BE less than the photon energy ($h\nu$) will feature in the spectrum. Because of the high energy of X-ray radiation, the photoelectrons lines coming from the core levels are the most intense in an XP spectrum; unresolved lines of low intensity also occur in the low-binding-energy region, produced by photoelectron emission from molecular orbitals (valence band). Those electrons that are excited and escape without energy loss contribute to the characteristic peaks in the spectrum; those that undergo inelastic scattering contribute to the background of the spectrum. The former comes from a depth that depends on the inelastic mean free path (IMFP, λ) of the electrons through the sample and on geometric factors (see Fig. 3.5 for details) and is called "sampling depth." This varies within the range of few nanometers and determines the surface-sensitive nature of XPS. After the photoemission, the ionized atom returns to its ground state through two main processes: emission of an X-ray photon, known as X-ray

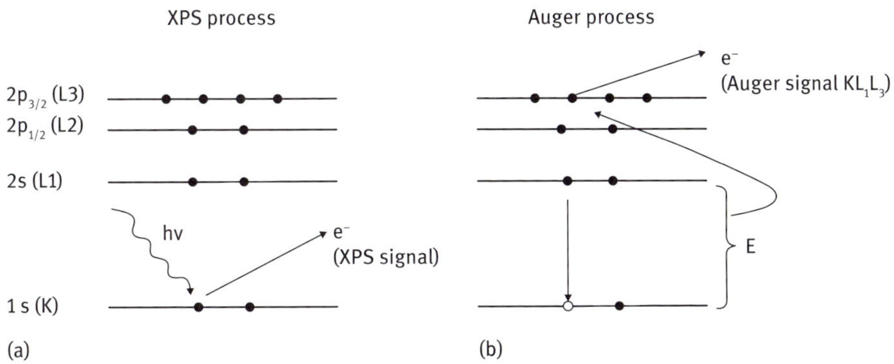

Fig. 3.1: (a) Schematic diagram of the XPS process and (b) of the Auger process.

fluorescence, or ejection of an Auger electron (Fig. 3.1b) with a KE given by the relationship:

$$\mathrm{KE} = E_K - E_{L_1} - E_{L_3} - W, \tag{3.3}$$

where suffixes K, L_1, and L_3 are related to the different levels depicted in Fig. 3.1. Thus, Auger electrons are produced as a consequence of the XPS process, and this kind of relaxation is not competitive at all with the photoemission process. Moreover, the energy of an Auger electron is independent of the incident $h\nu$, as evident from Eq. 3.3. Ultimately, Auger signals yield valuable chemical information usually complementary to that obtained from XPS signals. Other important features that can be observed in an XP spectrum are X-ray satellites and shake-up lines. The former comes from a nonmonochromatic source (see Section 3.3.2 for details) and are related to minor X-ray components at higher photon energies. The latter comes from a complex photoelectric process because there is a finite probability that an ion (after photoionization) will be left in an excited energy state a few electron volts above the ground state. When this happens, the kinetic energy of the emitted photoelectron is reduced, and this will result in a "shake-up" peak at a higher BE than the main line.

XP spectra will then appear with these features, deriving from both photoelectron and Auger emission. To better understand the "language" of the spectra, it can be useful to learn some nomenclature rules.

3.2.2 Spectroscopic and X-ray notations

The formalism used to describe which electrons are involved with each of the observed transitions refers to two different notations: XPS uses the spectroscopic notation, whereas Auger electrons are identified by the X-ray notation. They are substantially equivalent, and the differences arise from historical reasons. Basically, both are based on the *j-j* coupling scheme describing the electron motion around the nucleus and whose total angular momentum is the vectorial sum of electron spin and angular momenta. Referring to n, l, and s, quantum numbers, it is possible to derive the total angular momentum for each electron, which is given by $j = l + s$. The spectroscopic notation provides that states with $l = 0, 1, 2, 3, \ldots$ are, respectively, named $s, p, d, f, \ldots$, preceded by the number n. It should be noted that with $l = 0$, j can assume only one value, and the corresponding signals deriving from these orbitals are single signals. For each value of l different from zero, the quantum number j assumes two values; this is in accordance with doublet signals originating from orbitals with $l > 0$.

In X-ray notation, states with $n = 3, 4, \ldots$ are designated with the capital letters K, L, M, N, ... with a number as suffix, depending on the different j values. Table 3.1 better explains the notations and their relationship.

Tab. 3.1: The "language"

Quantum numbers				Spectroscopic notation	X-ray notation
n	l	s	$j = l + s$		
1	0	+1/2	1/2	1s	K
2	0	+1/2	1/2	2s	L1
2	1	−1/2	1/2	$2p_{1/2}$	L2
2	1	+1/2	3/2	$2p_{3/2}$	L3
3	0	+1/2	1/2	3s	M1
3	1	−1/2	1/2	$3p_{1/2}$	M2
3	1	+1/2	3/2	$3p_{3/2}$	M3
3	2	−1/2	3/2	$3d_{3/2}$	M4
3	2	+1/2	5/2	$3d_{5/2}$	M5
...	...	...	...	...	...

3.3 Instrumentation

An XPS apparatus substantially consists of a high-vacuum environment, a source of fixed energy radiation, an electron energy analyzer coupled with a lens system to disperse the emitted electrons according to their kinetic energy, an electron detector to measure the flux of emitted electrons of a particular energy, and a flood gun for charge compensation. Typically, the apparatus is then connected to a computer with a proper software interface, which registers the signal coming from the detector, and a spectrum of counts rate *vs* kinetic energy will be displayed in real time. After acquisition, data processing tools are generally provided to extract all the required information. Figure 3.2 depicts a block diagram of a typical XPS spectrometer.

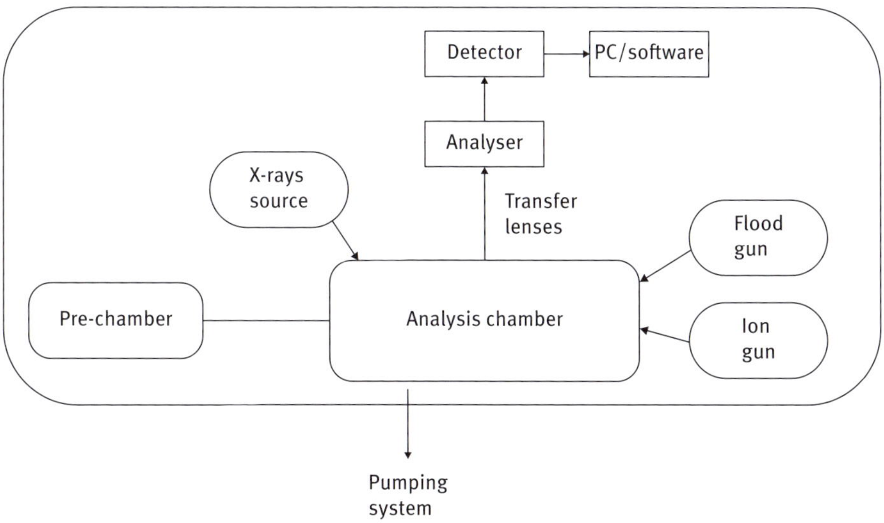

Fig. 3.2: Schematic diagram of an XPS spectrometer.

3.3.1 Vacuum system

Electron spectroscopic techniques require strict vacuum conditions, that is, ultra-high vacuum (UHV) range of 10^{-8} to 10^{-10} mbar. The reasons for this are itemized in the following:
- The analytical signal of low-energy electrons is easily scattered by the residual gas molecules in the environment, thus leading to the collapse of the signal.
- More importantly, the UHV environment is necessary because of the surface sensitivity of the techniques themselves. Indeed, at 10^{-6} mbar, it is possible for a monolayer of gas to be adsorbed onto a solid surface in about 1 s, too short compared with the time usually required for a typical spectral acquisition. Therefore, the need for a UHV environment during analysis is mandatory.

How to establish such a vacuum will depend on different manufacturer preferences. Typically, the chambers are made of stainless steel, and joints are made using flanges that are tightened onto copper gaskets. UHV conditions are then usually obtained and maintained by dynamic pumping down of the whole spectrometer through a proper pumping system. Turbomolecular pumps along with a titanium sublimation pump to assist the primary pumping are generally used to achieve the desired vacuum level. All UHV systems need baking from time to time to remove adsorbed layers from the chamber walls, abd the baking temperature is usually in the range of 100°C–160°C for routine use.

3.3.2 X-ray sources

3.3.2.1 Dual Mg/Al anode X-ray tube

The photon beam energy for XPS source is usually obtained from an X-ray tube with a hot filament at high voltage ranging from 10 to 15 kV and a current of 10–15 mA, which emits electrons. These electrons are accelerated toward a grounded potential anode. The maximum anode current is set by the efficiency with which the heat can be dissipated; therefore, X-ray anodes are usually water-cooled. The choice of anode material determines the energy of the X-ray transition generated, and it is usually chosen to be high enough to excite an intense photoelectron peak from all elements of the periodic table. It must also possess a natural X-ray line width, narrow enough to achieve a good resolution. Based on these considerations, the most popular anode materials that satisfy the requirements are aluminum and magnesium. These are usually supplied in a single X-ray gun with a twin-anode configuration providing AlKα or MgKα photons of nearly mono-energetic (monochromatic) X-rays with energy 1,486.6 and 1,253.6 eV, respectively. An XP spectrum is often rich of many signals coming from both main and secondary photoelectron peaks and Auger peaks. Therefore, very often, the use of two different X-ray energies allows the elimination of ambiguities/interferences in the analysis because the kinetic energy of XP signals is *hν* dependent, whereas the kinetic energy of Auger peaks is only related to the electronic structure of the emitting element. For these reasons, many XPS instruments have dual anodes (Al and Mg) that can be excited separately.

3.3.2.2 Monochromatic source

There are a number of reasons for choosing to use an X-ray monochromator on an XPS spectrometer: removal of satellites, elimination of the Bremsstrahlung background, and the reduction in X-ray line width. A monochromator works by filtering a narrower band of X-rays from the resonance peak by X-ray diffraction from a quartz crystal or from bent SiO_2 crystal. In this way, only certain wavelengths can be reinforced into a spot so that the monochromatic X-rays can be directed at the sample. The crystal spacing is such that X-ray wavelengths, which are multiples of the AlKα X-ray resonance, are reinforced following the Bragg condition for X-ray diffraction.

3.3.2.3 Synchrotron radiation source

The synchrotron radiation (SR) is produced when charged particles, for example, electrons, are forced to move in a circular orbit. Continuous Bremsstrahlung X-radiations of different energies are thus emitted; when relativistic velocities are reached, these photons are emitted in a narrow cone in the forward direction, at a tangent to the orbit. Different possibilities are available using SR source because it possesses a number of unique properties: (1) high brightness: the SR is so intense; it is hundreds of thousands of times higher than conventional X-ray anodes; (2) wide energy spectrum: SR is emitted with a wide range of energies, allowing a beam of any energy to be produced; (3) SR is highly polarized. The major advantage of SR is essentially a surface sensitivity better than the sensitivity obtained using AlKα or MgKα. Meanwhile, damage on the sample surface during the measurement may easily occur, probably due to sample heating when SR excitation is used.

3.3.3 Energy analyzers

In XPS, the analysis of the ejected photoelectron energies is the key issue. Therefore, the analyzer is the most important component of an XPS spectrometer. Before discussing about the energy analyzers, few words on the energy resolution are needed. The latter is defined in two ways: the absolute resolution, that is, the full width at half maximum (FWHM) of a chosen peak (ΔE) and the relative resolution, defined as the ratio of ΔE to the KE of the peak position. The absolute resolution is independent of peak position, whereas the relative one should be referenced to a specific KE. When running an XPS experiment, we need the same absolute resolution at all the energies.

There are two types of electron energy analyzer for XPS: the cylindrical mirror analyzer (CMA) and the hemispherical sector analyzer (HSA). The CMA is used when it is not important that the highest resolution is achieved. Meanwhile, spectral resolution plays a key role in XPS, and this led to the development of the HSA as a design of analyzer with sufficiently good resolution. The addition of a transfer lens to the HSA and multichannel detection increase its sensitivity to the point where both high transmission and high resolution are possible, and this type of analyzer may now be used with excellent results.

The CMA consists of two concentric cylinders: the inner cylinder is held at earth potential, whereas the outer is ramped at a negative potential. An electron gun is often mounted coaxially within the analyzer.

The HSA, also known as a concentric hemispherical analyzer (CHA), uses an electric field between two hemispherical surfaces to disperse the electrons according to their kinetic energy. Between the sample and the analyzer, there is usually a lens or a series of lenses. The hemispheres with radii R_1 (inner) and R_2 (outer) are positioned concentrically, and potentials $-V_1$ and $-V_2$ are applied to these spheres, respectively, with V_2 greater than V_1. If electrons arrive at the entrance of the analyzer with $E = eV_0$, they will be focused on the exit of the analyzer only if the following relationship holds:

$$V_0 = \frac{V_1R_1 + V_2R_2}{2R_0}, \tag{3.4}$$

which defines the potential of mean free path analyzer through radius R_0 (medium equipotential surface between R_1 and R_2). Because R_1 and R_2 are fixed, changing V_1 and V_2 will allow, in principle, scanning of electron kinetic energy following mean free path through hemispheres.

The resolution of CHA is given by $R = \Delta E/\text{KE}$. Becase the resolution R is energy dependent, it will not remain uniform across the entire XP spectrum. This can be achieved by retarding the electrons entering energy analyzer to fixed kinetic energy, called the pass energy E_0, so that fixed resolution applies across entire spectrum. For this purpose, either the pass energy can be decreased or R_0 can be increased. The increased resolution for an analyzer is typically ~0.1–1.0 eV.

The CHA is typically operated in two different modes: constant analyzer energy (CAE), also known as fixed analyzer transmission, and constant retard ratio (CRR), also known as fixed retard ratio. In the CAE mode, electrons are accelerated or retarded according to a user-defined pass energy. In the CRR mode, electrons are retarded to some user-defined fraction of their original kinetic energy as they pass through the analyzer. In practice, in the CRR mode, the pass energy is proportional to the kinetic energy, and so the relative resolution is constant throughout the energy range. In the CAE mode of operation, the absolute resolution remains constant throughout the energy range of the scan, ensuring that XPS quantification is more reliable and accentuates the XPS peaks at the low kinetic energy (high BE) end of the spectrum.

3.3.4 Detectors

Once the photoelectrons have been analyzed and transmitted, it is necessary to count the individual electrons arriving at the detector. To achieve this, electron multipliers are used. Although there are many types of electron multipliers, only two types are commonly used: channel electron multipliers (CEMs, or channeltrons) and channel plates.

A CEM is an electron detector able to multiply each electron up to 108 times, thus providing a pulse output suitable for further amplification by conventional electronic circuits. CEM is a bent tube coated with a photoelectric material with a high secondary electron coefficient. The tube is kept at a potential of about 2.5 kV. When an electron strikes the mouth of the tube, a number of secondaries are produced that are accelerated in the channeltron. A single photon is capable of creating an output of 10 million electrons. This process continues down the length of the CEM, creating more and more secondaries along the way. This amplified "cascade" passes through the preamplifier and is thus detected. This pulse is then shaped and ultimately registered by the acquisition electronics. In a similar way, pulses of electrons are produced in the microchannels of a microchannel plate. When this device is used in combination with a position-sensitive anode, electrons can be detected with a lateral resolution of some tens of a micrometer.

A channel plate consists of a disc having an array of small holes. Each of these holes behaves as a small channeltron. The gain of an individual channel is much lower than that of a channeltron; thus, it is common to use a pair of channel plates in tandem. Channel plates are used when it is necessary to detect data in two dimensions; spectrometers have been designed using channel plates to measure signals for parallel acquisition of photoelectron images, XPS line scans, and the parallel acquisition of angle-resolved electrons.

3.3.5 Charge compensation

The electrical properties of the solid materials analyzed in an XPS experiment can vary from the conductive metal to the insulating piece of plastic. Photoemission from an insulating sample causes electrostatic charging to occur in the positive direction. This results in a shift of the peak position in the direction of higher BE, making equivocal the peak attribution. When the photoemission is excited by a non-monochromatic X-ray source, there are usually a sufficient number of low-energy electrons available in the neighborhood of the sample to effectively neutralize the sample surface. When monochromatic X-ray sources are used, these low-energy electrons are not produced in such a large number near the sample, and thus, neutralization does not take place. When analyzing polymers, this effect should be taken into account because they are completely insulating materials. Surface charge compensation is generally accomplished by flooding the surface with low-energy electrons. From the instrumental point of view, there is an additional gun, the flood gun, producing a uniform negative charge of known magnitude at the surface of the sample. This technique minimizes the risk of differential or nonuniform charging. The electron beam used for charge compensation should be of low energy to avoid the risk of damage to the surface of the sample and at the same time supply a high enough flux to adequately compensate for the charging of the sample surface. Nowadays, in modern flood guns, the energy of the electrons is as low as about 1 eV.

3.3.6 Small-area XPS: imaging and mapping

It is often highly desirable to analyze small features of a surface, and to do this instrumental helps are needed. Two ways are possible: (1) limiting the area from which the photoelectrons are collected using the transfer lens and (2) focusing a monochromated beam of X-rays into a small spot on the sample. The first option is referred to as lens-defined analysis: it consists in operating the transfer lens to produce a photoelectron image at some point in the electron optical column where a small aperture is placed; therefore, only electrons emitted from a defined area of the sample can pass through the aperture and reach the analyzer. If the magnification of the lens is M and the diameter of the aperture is d then the diameter of the analyzed area is d/M. Using this technique, commercial instruments can provide high-quality small-area analysis down to about 15 μm, but with the main disadvantage of reducing the detected flux per unit area of the sample. This means longer analysis time, hence a longer exposure of the sample to X-rays, potentially resulting in radiation damage.

The second possibility is the so-called source-defined small-area analysis, when a monochromatic X-ray source is available. The monochromatic X-rays diffracted from a quartz crystal can be focused on the sample using a magnification of unity, which means that the size of the X-ray spot on the sample is approximately equal to the size of the electron spot on the X-ray anode. Analysis areas down to about 10 μm can be achieved in many recent commercial instruments using this method. Because the analysis area is source-defined, the lens can be operated at its maximum transmission, thus dramatically increasing the sensitivity of a spectrometer. This reduces sample damage during analysis and eliminates radiation damage to the surrounding area of the sample(s).

An important practical consequence of small-area XPS is the possibility to register an image or chemical map of the surface. Such an image or map shows the distribution of an element or a chemical state on the surface of the sample. There are two distinct approaches, used by manufacturers, to obtaining XPS maps: serial acquisition, in which each pixel of the image is collected in turn (by X-ray spot, sample stage, or lens scanning all over the defined area to be mapped); parallel acquisition in which data from the whole of the analysis area are collected simultaneously (by additional lenses and a two-dimensional detector).

3.3.7 Ambient-pressure photoelectron spectroscopy

The reasons for strict vacuum conditions for XPS have been itemized in Section 3.3.1. Nevertheless, techniques that provide spectroscopic information of surfaces while exposed to gases or liquids in the atmospheric pressure range are highly desirable. This is the field where ambient-pressure photoelectron spectroscopy (AP-XPS or APPES) developed, producing dramatic changes in the study of liquid and solid surfaces, particularly in areas such as atmospheric, environment, and catalysis sciences [8]. The application of APPES in polymers science is not so widespread and the potentiality of this approach will be only briefly discussed in this chapter.

One characteristic of surfaces in ambient-pressure environments is that they are covered by dense layers of molecules: water, for example, is known to form layers several molecules thick at room temperature in humid environments; metals readily form oxide films several layers thick in oxygen atmospheres; layers of adsorbed molecules can also be produced in UHV, simply cooling the sample to cryogenic temperatures.

Therefore, APPES should rely on instrumental facilities able to circumvent the strong interaction of electrons and ions with the gas. They all involve capturing the particles at distances comparable with their mean free path, which is in the millimeter range for electrons of a few hundred electron volts at millibar-type pressures. This is accomplished by small apertures near the sample and by differential pumping to remove gases leaking through it. All instruments developed to make possible electron spectroscopy under environmental gases are based on this idea, which is accomplished by bringing the sample close to a small aperture at a distance equal or less than λ. Differential pumping is the key to lower the pressure at the other side of the aperture to decrease the collision rate.

As might be expected, all the components discussed above will be present in an ambient-pressure spectrometer, but they are especially designed to operate in a different pressure range [8].

3.4 Chemical information from XPS

The first information that can be derived in characterizing the surface chemistry of the specimen under investigation is the identification of the elements present. To achieve this, it is usual to record a survey, or wide-scan, spectrum over an energy region that will provide peaks for all elements in the periodic table. The individual peaks may be identified thanks to a database generally included in the software. Figure. 3.3 shows a typical wide scan along with the proper element attribution.

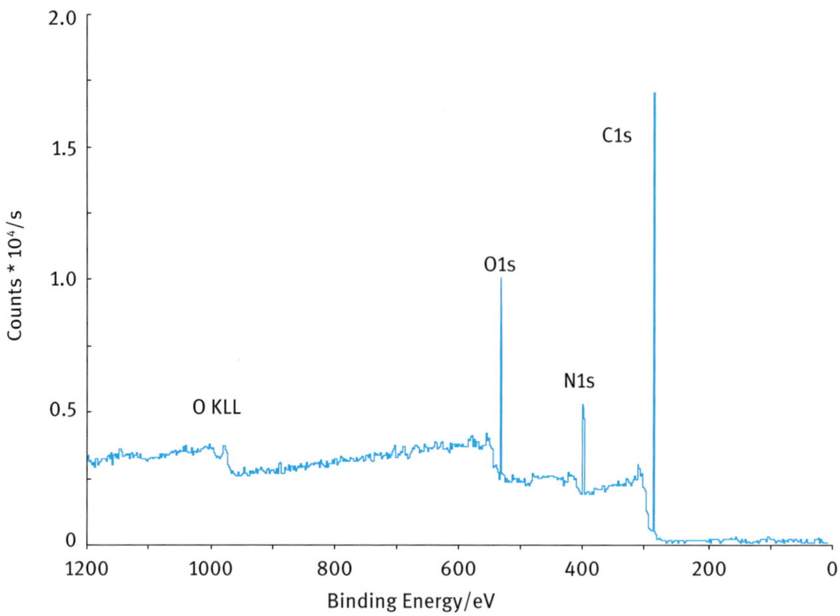

Fig. 3.3: XP wide-scan spectrum with element identification labels.

The labels in the Fig. 3.3 indicate the symbol of the element and the core level that generated the peak (O1s, C1s, and N1s). When the symbol is followed by three letters, then the label is referred to an Auger peak (see Fig. 3.1b for details). Hence, the surface compositional (elemental) analysis can be performed by identifying the peaks with the specific atoms. Quantitative information can be also derived through the measurement of the relative areas of the photoelectron peaks, thus allowing the composition of the sample to be determined. Before calculating the peak areas, however, the background subtraction is needed. Three approaches to defining the spectral background are commonly described in the literature: linear subtraction, Shirley polynomial subtraction, and Tougaard method (see Fig. 3.4); the theory underlying these approaches is beyond the scope of this textbook and can be found elsewhere [9]. However, it is worth noting that for quantitative analysis of polymer spectra the use of the straight-line background or Shirley one is generally recommended.

After background subtraction, the quantification of peak intensity data is performed using peak areas. In particular, the factors affecting the peak intensity are well described in the following equation:

$$I_{i,c} = fN_i\sigma_{i,c}\lambda \cos\alpha FTDA, \tag{3.5}$$

where $I_{i,c}$ is the photoelectron intensity for core level c of element i, f is X-ray flux in photons per unit area per unit time, N_i is the number of atoms of element i per unit

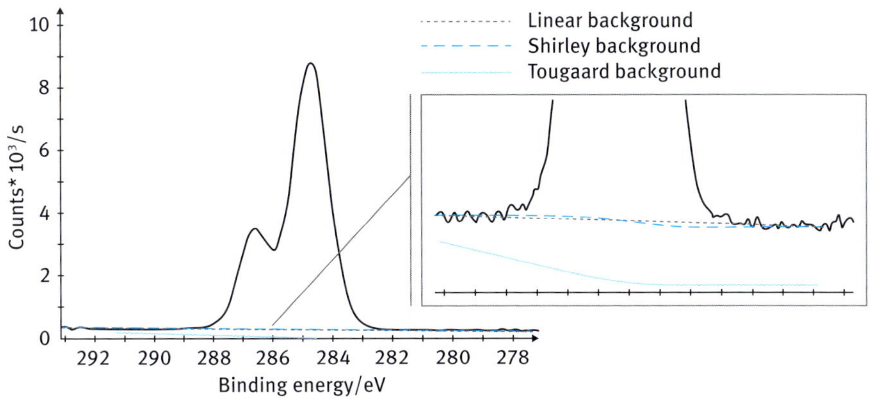

Fig. 3.4: Comparison of experimental background subtraction with the three different methods.

volume, $\sigma_{i,c}$ represents the photoelectric cross section for core level c of element i, λ is the IMFP of the photoelectron in the sample matrix, and α is the angle between the direction of photoelectron and the sample normal. Other parameters are also the analyzer solid angle of acceptance (F), the analyzer transmission function (T), the detector efficiency (D), and the area of sample from which photoelectrons are detected (A).

As can be easily argued, this equation is too complex for usual analyses, and empirical sensitivity factors whose values are specific for different elements have been introduced [10]. Now, peak intensity can be described by the simple form: $I = N \times S$, where S is the sensitivity factor. The result is a semiquantitative information in the form of relative atomic percentages:

$$\text{Atomic \%} = \frac{\frac{I_i}{S_i}}{\Sigma_j \frac{I_j}{S_j}} \times 100\%. \tag{3.6}$$

Because the photoelectrons are strongly attenuated by passage through the sample material itself, the information obtained comes from the sample surface, with a sampling depth of the order of 7–10 nm.

In detail, photoelectrons are created at a depth z below the surface; in straight-line trajectory, only inelastic scattering leads to electron attenuation, which is assumed to follow the Beer-Lambert (BL) law, as does light attenuation in an absorbing medium. Therefore,

$$I_z = I_0 \exp(-z/\lambda \sin\theta), \tag{3.7}$$

where I_z is the intensity from the atoms at depth z, I_0 is the intensity from the surface atoms, and θ is the electron take-off angle to the surface ($\theta = 90° - \alpha$). In this case, λ is the IMFP of the measured electrons, defined as the average distance that an electron with a given energy travels between successive inelastic collisions. In the case of exponential decay, the escape depth d is $\lambda \sin\theta$. Commonly, a surface layer as thick as $3d$ is taken to be the "sampling depth."

Greater sampling depths can be achieved by removing the sample material layer by layer through ion bombardment (down to hundreds of nanometers); variable sampling depth, comprised between near zero (very topmost layers, up to less than 1 nm) and a maximum of $3d$, can be obtained by changing θ. Both modes are defined as depth profiling because they provide the distribution of chemical composition to depths. Ion bombardment is indeed more destructive and is instrumentally performed by an ion gun (positive argon ions accelerated on the sample surface). The application of destructive depth profiling in the polymer analysis has been limited for a long time because of the possible degradation upon ion bombardment along with the loss of chemical information. Recently, this drawback was successfully faced by the development of guns using gas cluster ions instead of monoatomic ions: the bombardment is gentler and the chemical information is preserved.

Surface sensitivity enhancement by variation of θ (take-off angle) is called angle-resolved XPS (ARXPS). Despite the achievable depth regime for this operational mode is limited to $3d < 10$ nm, it is a particularly important technique for polymer surface studies because many important aspects of polymer surface behavior are governed by major compositional or structural variations within this layer thickness. The geometry involved is schematically depicted in Fig. 3.5.

From the instrumental point of view, the take-off angle variation is traditionally achieved tilting the sample, that is, the sample holder. A second possibility has been recently developed: the angular acceptance of the transfer lens is set to provide angular resolution using a two-dimensional detector. In this instrumental arrangement at the output plane of the analyzer, the photoelectron energy is dispersed in one direction (as with a conventional lens analyzer arrangement) and the angular distribution dispersed in the other direction. Such an arrangement provides a parallel angle-resolved collection of spectra, with a number of advantages over the conventional method.

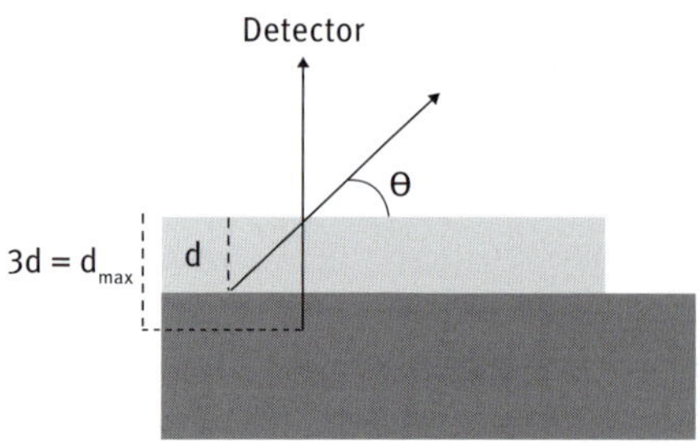

Fig. 3.5: Scheme of geometric parameters involved in angle-resolved XPS experiments.

3.5 Chemical shift and its significance in the analysis of polymers

XPS can be also used to derive information about the atomic oxidation state from the sample surface. As discussed in Section 3.4, core-level binding energies and relative intensities provide atomic identification and relative concentrations. In addition, the exact core-level BE varies slightly depending on the nature of chemical bonding and the presence of neighboring atoms affecting the BE and producing the so-called chemical shift. The BE is determined by the difference between the total energies of the initial-state atom and the final-state ion: chemical bonding will clearly have an influence on both the initial state energy of the atom and the final state energy of the ion created by emission of the photoelectron. These changes can be calculated by quantum chemical methods and are basically due to the redistribution of electrons depending upon the electronegativities of the atoms involved.

Unfortunately, the range of these shifts is small – few electron volts also in the case of C1s – and therefore the energy resolution as well as the peak width are critical in the functional group recognition. Table 3.2 reports a series of core-level BE for C1s in different common functional groups.

Consequently, when the molecule structure involves the same atom in more than one chemical environment, the resulting chemically shifted peaks will probably overlap to produce a complex envelope. Of course, this problem is a crucial issue in the analysis of polymers where many information can be deduced mainly from the C1s core-level spectra due to the different functional groups the carbon is involved in. Extraction of structural information from such spectra requires the component peaks to be resolved and requires knowledge of individual component lineshapes [11]. The approach to do this is the curve fitting technique, whose mathematical basis are well discussed in reference [12].

Before curve fitting, an appropriate background function should be introduced. Afterward, in curve fitting itself a series of components is generated and placed within the measured envelope at appropriate BE positions. The software calculates the sum of the components, and from comparison with the experimental data, it obtains a measure of the "goodness of fit" by the least squares method. This process cannot produce a unique solution, and a great deal of operator inputs are usually required to ensure that the solution is reasonable both from a spectroscopic and chemical point of view. Parameters such as peak lineshape, peak full width at half maximum, peak position, and intensity as well as background position during iteration should be correctly indicated. This way, the fitting process can be "guided" toward a solution that includes the maximum amount of operator understanding of the system.

Two examples of C1s curve fitting are reported in Fig. 3.6: (a) C1s spectrum of a poly lactic acid (PLA) where three components are clearly evident; (b) a more complex case, where the different chemical components constitute an envelope still solvable by accurate curve fitting. In the latter spectrum, it can also be observed the substantial chemical shift of carbon when bound to a very electronegative atom such as fluorine.

Tab. 3.2: C1s typical chemical shifts range for different functional groups

Functional group	BE range (eV)	Functional group	BE range (eV)	Functional group	BE range (eV)
C—C	285.00	$C—NO_2$	285.76	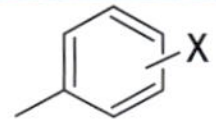	284.44–284.80
C=C, C≡C	284.69–284.76	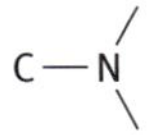	285.56–286.41	C—Si	284.22–284.39
C—OH, C—O—C	286.13–286.75	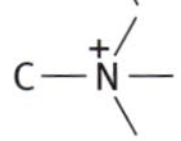	285.99–286.22	C—S	285.21–285.52
C=O, O—C—O	287.81–288.06	C̲—C≡N	286.35–286.46	$C—SO_2$	285.31–285.64
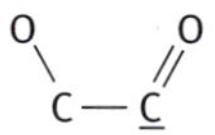	288.64–289.23	—C≡N	286.73	$C—SO_3^-$	285.16
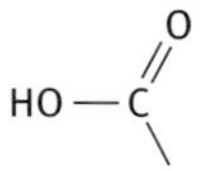	289.18–289.33	$C—ONO_2$	287.62	C—Br	285.74
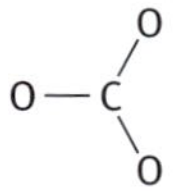	289.30–289.34	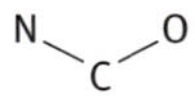	287.78	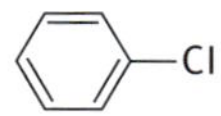	285.99–286.07
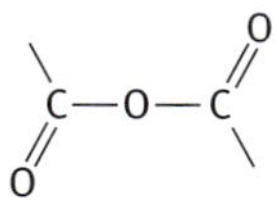	289.36–289.46	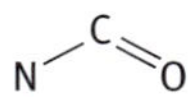	287.97–288.59	C—Cl	287.00–287.03
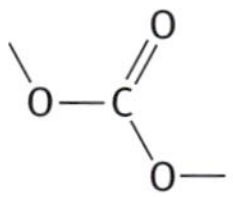	290.35–290.44	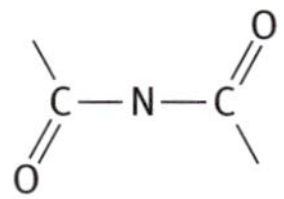	288.49–288.61	$—CCl_2$	288.56
		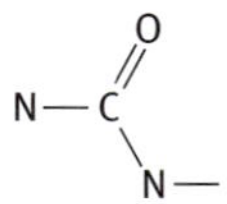	288.84	C—F	287.91
		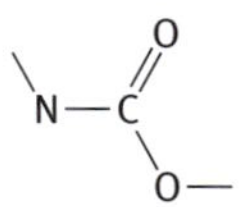	289.60	$—CF_2$	290.90
				$—CF_3$	292.65–292.75

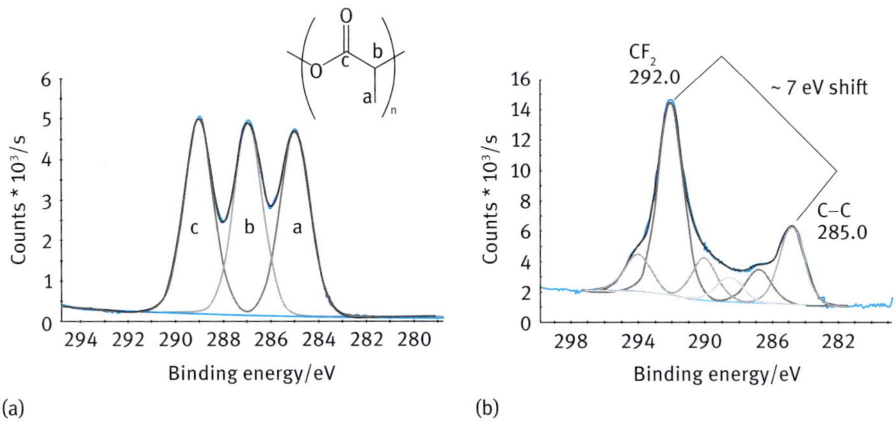

Fig. 3.6: Curve fitting of (a) C1s XP spectrum of PLA polymer and (b) C1s XP spectrum of a fluorinated polymer.

In some cases, the chemical shifts observed in core-level XP spectra are not sufficient to discriminate the surface chemistry of a sample. Especially for XPS analysis of polymers, changes in carbon spectra can be so subtle that other spectral region are necessary to be investigated. Owing to the use of monochromators and high counting rates in recent spectrometers, the study of valence bands have become very useful for identification of materials. Rather than specific molecular orbitals identification, it is used as fingerprints for a sample. In particular, the fingerprints of valence bands may be used in the identification of polymers and/or quantification of polymers mixtures by linear least squares fitting [13] (Fig. 3.7).

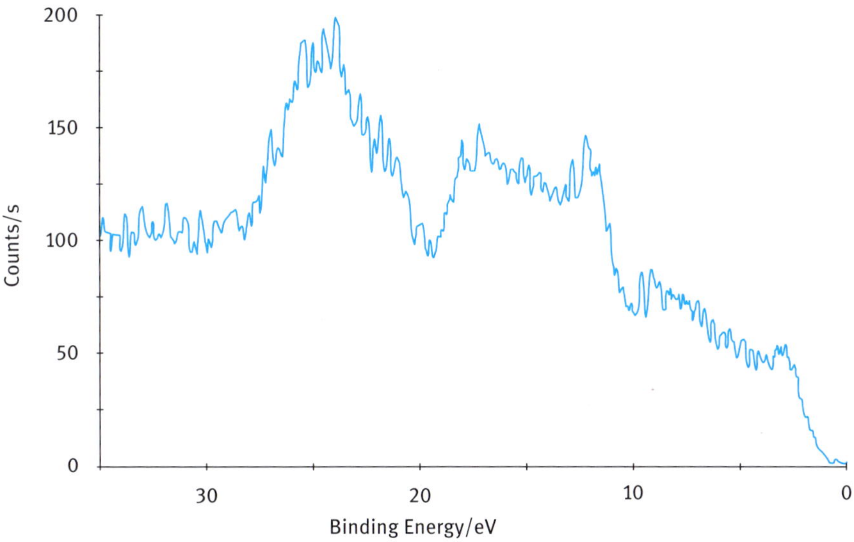

Fig. 3.7: Valence band of a polymer.

Additionally, in the case of aromatic polymer compounds, further features can be very useful: the characteristic shake-up lines. They originate from the photoelectric process described in Section 3.2.1 and involving the energy of the $\pi \rightarrow \pi^*$ transition.

3.6 Chemical derivatization techniques in conjunction with XPS

We have seen in the previous paragraph that polymer surfaces are not always amenable to direct XPS analysis because BE differences between some functional groups may be too small to be distinguishable to spectroscopic probes such as XPS. Curve-fitting procedures are a valuable tool in this case: however, when the BE values of different species overlap, curve-fitting results of core-level photoemission lines may be ambiguous. In such cases, it would be more convenient to have a labeling method to clearly extract the presence of certain reactive functional groups from the spectral envelope. This has led to the development of chemical derivatization techniques: herein, the functional group of interest is labeled with an elemental or molecular tag not present on the original surface [2, 14], thus offering a valid support to peak-fitting analysis.

Chemical derivatization techniques in conjunction with XPS (CD-XPS) can be used both to confirm and exclude the presence of specific functional groups on the surface of polymers as well as to quantify their concentration when they are present and react selectively and stoichiometrically with a specific derivatizing reagent that contains a unique chemical tag.

To provide an accurate estimate of the relative concentration of a particular functional group, the derivatization reaction must be highly selective, particularly when real-world samples, frequently multifunctional (e.g. chemically modified or plasma-modified polymer surfaces) [15], are to be analyzed.

A very simple example is the identification of carbon-carbon double bonds on the surface of a treated polymer when in the presence of fully saturated bonds. Reaction with bromine in carbon tetrachloride is well known from undergraduate chemistry laboratory experiments, and it yields two bromines for every double bond initially present. For surface analysis, however, there are at least two general considerations: (1) solution reactions frequently fail to occur on surfaces; (2) XPS analysis depth is about 10 nm, and any labeling would require penetration and quantitative reaction to at least three times that depth.

The requirements [16] for an effective labeling analysis have been summarized:

- The reaction should preferably be quantitative or should at least always proceed to the same degree.
- Because XPS analyzes samples on the surface to a depth of approximately 10 nm, the concentration distribution of the chemical derivative with depth should be explored. It is preferable if the reaction occurs uniformly throughout the sampling depth.

- The reaction should be rapid and the reaction kinetics should be understood.
- The reaction product should be stable under vacuum and to electron and X-ray sources.
- The reaction product should contain a unique marker atom or group, preferably one that has a high photoemission cross section.
- Cross-reactions of a specific derivatizing reagent with functional groups not under investigation should not occur or should be well understood.
- The chemical reaction should not lead to a rearrangement of the surface structure that is being probed.

Some vapor-phase derivatization reactions meet these criteria. Figure. 3.8 reports some of these reactions and the derivatization method.

For CD-XPS, fluorine atoms are often used as the chemical tags when not present on polymers surfaces; furthermore, the F(1s) core-level transition has a high XP cross section. Some examples of fluorine-based derivatization reactions on model polymers surfaces are reported in the following:

- TFAA derivatization reaction: The vapor-phase trifluoroacetic anhydride (TFAA) derivatization reaction has been widely used to tag hydroxyl groups on surfaces [17, 18]. Carbonyl and carboxyl groups showed low reactivity toward TFAA, evidenced by the low intensity of F1s signals on TFAA-derivatized PVMK and PAA. In particular, the fluorine concentration was equal to 4 wt% of the acid groups in PAA and less than 6 wt% of the ketone groups in PVMK. Meanwhile, TFAA-derivatized PVA showed, within the XPS sampling depth, the complete reactivity of the polymer hydroxyl groups (33 wt%). In Fig. 3.9a, the high-resolution C1s spectra of PVA, before and after the reaction with TFAA, have been reported [19]. The new peaks that appear in the C1s region of an TFAA-derivatized PVA can be attributed to the ester carbon (at 289.5 eV) and to the CF_3 carbon atoms (at 292.6 eV). The experimentally determined CH_x:COC:CO_2:CF_3 peak area ratio of 1.2:1.0:1.0:1.0 for these peaks is close to the 1:1:1:1 ratio estimated for a stoichiometric reaction. TFAA vapor is not selective toward surface hydroxyl groups in polymeric systems containing epoxide groups. In particular, the reactivity of TFAA vapor with epoxides is comparable to its reactivity with hydroxyls. Thus, the vapor-phase reaction with

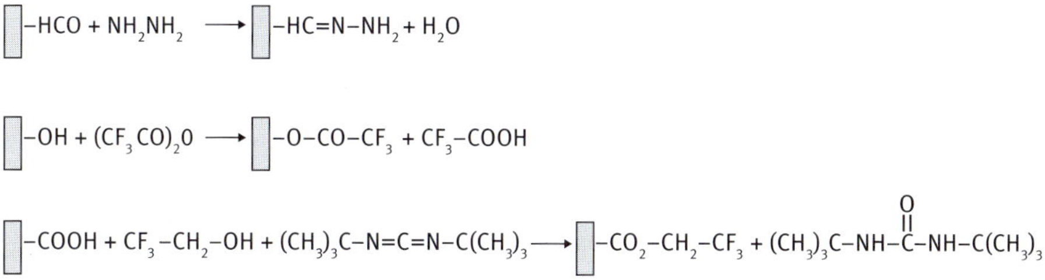

Fig. 3.8: Some vapor-phase derivatization reactions used in CD-XPS.

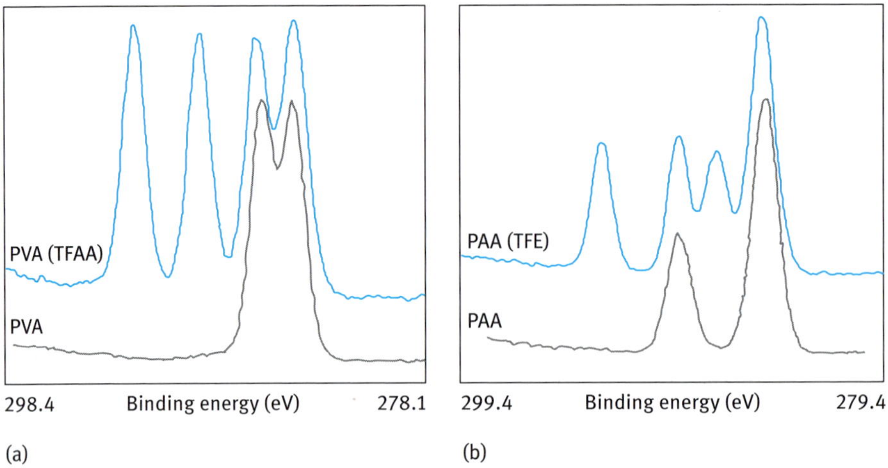

Fig. 3.9: High-resolution C1s spectra of (a) PVA (bottom spectrum) and TFAA-derivatized PVA (top spectrum). (b) High-resolution C1s spectra of (a) PAA (bottom spectrum) and TFE-derivatized PAA (top spectrum). Reprinted with permission Chilkoti A, Ratner BD, Briggs D, *Chem Mater*, 3, 51–61, 1991. Copyright 1991 American Chemical Society [19].

TFAA and subsequent quantitation with XPS cannot be used to assay the relative surface concentration of hydroxyl groups in the presence of epoxide groups [20].

- Trifluoroethanol (TFE) derivatization reaction: The liquid-phase derivatization reaction of carboxyl groups with TFE and dicyclohexylcarbodiimide (DCC) has been reported in literature [21]. Because DCC is a solid at room temperature, this reaction cannot be used for a vapor phase assay. Moreover, the low volatility of DCC and of its reaction product on the surface introduces artifacts in the quantitation of carboxyl groups [22]. Therefore, the use of diterbutylcarbodiimide (Di-tBuC) was preferred to that of DCC due to its higher volatility that allows vapor phase reaction as well as minimizes nonspecific adsorption. Figure 3.9b shows the XPS C1s spectra of PAA before and after TFE derivatization [19]. Upon derivatization, the presence of peaks at 287.2 and 292.6 eV can be ascribed to COC and CF_3 species, respectively. Moreover, the COC:CO_2:CF_3 peak area ratio equal to 1.1:1.1:1.0 suggests the complete reactivity of the PAA acid groups. This finding was confirmed by the fluorine amount detected that was equal to 30 wt%. Investigating a variety of polymers containing different oxygen functional groups, the selectivity of this reaction toward carboxyl groups has been found to be excellent.
- Derivatization methods have been widely used on plasma-treated polymer surfaces [23]. Indeed, the complete knowledge of the surface chemistry of the plasma-treated films is particularly difficult due to their complex chemical nature, which is often multifunctional and displays a range of unsaturation degrees and cross-linking. In this respect, CD-XPS reactions can be exploited to determine surface chemistry changes as a result of different plasma experimental conditions.

Malitesta *et al.* [24] reported an interesting application of CD-XPS to the characterization of complex matrices such as polypyrroles (PPYs) electrosynthesized from aqueous solution. In these polymers, oxygen-carbon and nitrogen-carbon functionalities are contemporaneously present, as well as acidic, basic, charged, and reducible groups. Indeed, CD-XPS allowed to shed light about PPY structure, involving species either not fully resolved or present at low concentration levels, in particular C-OH, COOH, and C=O groups. The derivatizing agents titanium di-isopropoxide-bis(2,4-pentanedionate), TFE, and pentafluorophenylhydrazine were used to label the C-OH, COOH, and C=O groups, respectively. This study represents the first attempt to enlighten unknown structures of electrosynthesized polymers and a significant test for the applicability of CD-XPS to complex samples.

Recently, derivatization reactions have been also developed to quantify functional groups on other carbonaceous surfaces such as carbon nanotubes (CNTs) [25, 26]. CD-XPS provides to gain information about the identity and concentration of oxygen-containing functional groups formed as a result of CNT oxidation. In particular, the quantification of three major oxygen-containing functionalities, hydroxyl, carbonyl, and carboxyl groups can be performed. In three separate reactions, hydroxyl groups are labeled by reaction with TFAA, carbonyl by trifluoroethylhydrazine, and carboxyls *via* a coupling reaction of TFE and a carbodiimide.

An efficient method for surface functional groups recognition involves labeling carboxyl, hydroxyl, and carbonyl species with fluorescent tags [27]. In many ways, fluorescent labeling is complementary to CD-XPS; the fluorescent tags can enable lower concentrations of oxygen functional groups to be detected. However, at higher oxygen functional group concentrations, the bulky fluorescent chromophores are more likely to encounter problems associated with steric hindrance, making quantitative analysis more difficult. Under these circumstances, CD-XPS approach is more effective.

3.7 Polymers surface segregation

As described in Chapter 1, surface dynamics of polymeric materials can be much more complex than those of other rigid materials (i.e. metals, ceramics, etc.), due to the high mobility of macromolecules on the surface. Indeed, polymer blends, copolymers, or additives in polymer systems can show the tendency to rearrange in the nano- or micro-domains, and this feature is particularly evident on surface as a consequence of interfacial interactions between polymer and the external environment.

In particular, for copolymers [28, 29], if there are two or more covalently linked components having different polarity and configuration, certain components could migrate to the surface to satisfy the thermodynamic requirement of minimizing the surface energy of the whole system. In general, components having lower surface energy always aim at enrich the surface: for instance, amphiphilic multiblock copolymers tend to rearrange so that the hydrophilic portion will dominate the interface in a polar hydrophilic medium, whereas the hydrophobic portion tend to dominate in air. Moreover, in going from a dry environment to a wet one, a structural rearrangement of the exposed surface occurs due to the different polarity of the components. In a few cases, if the segmental self-assembly is thermodynamically impervious to restructuring, a locking of the surface can be achieved [30, 31]. The extent to which phase separation occurs in copolymers to form supramolecular structures depends on these inherent critical features (1) compositional dissimilarity between blocks, i.e. difference in solubility parameters or chain interactions; (2) molecular weight of the segments, i.e. block length; (3) differences in crystallinity degree of the segments [32, 33]. The rearrangement of block copolymers can also depend on the way in which the materials are formed. For example, copolymer films obtained by casting using different solvents showed a dependence in segregation according to the solvent used.

Generally, the domain size for block copolymers is smaller than that of polymer blends, because the different blocks in a chain of block copolymers are connected by covalent bonds, whereas the polymer blends are a mixture of at least two polymers or copolymers. Polymer blends can exhibit miscibility or phase separation as well as various levels of mixing in between these extremes (i.e. partial miscibility). Therefore, a larger size scale of phase separated domains can occur in polymer blends with respect to block copolymers. The domain dimensions of block copolymers could be in the nanometric scale. The periodicity of these morphological structures is determined by molecular weight and chemical composition in the block copolymers and typically is in the range of 10 to 100 nm. These nano-heterogeneous structures might generate many of the novel and useful mechanical and surface properties. In this respect, surface segregation of copolymers has been successfully exploited to design novel materials that reconstruct their surface in response to a change in the environment.

XPS technique can allow to experimentally reveal such effects in different copolymer systems: in detailed studies of some di- and tri-block copolymer systems [34–36], supplying depth profiles of block copolymers [33, 37–39], or in random copolymers [40, 41].

A typical example of surface segregation investigation by XPS concerns cast films in chloroform solution of polystyrene (PS)-poly(dimethylsiloxane) (PDMS)-based block copolymers, in AB di-block and ABA and BAB tri-block structures. PDMS surface segregation changes are also observed for some of the block copolymers after annealing the as-cast films. Because of the large differences in surface energy and the large mixing enthalpy of the two different blocks of PS and PDMS, PDMS blocks prefer to migrate to the surface, forming a surface PDMS microdomain and leaving PS blocks in the outermost layer [4]. Surface composition of these amorphous-semicrystalline block copolymers depends also on the way these films are prepared. Mixed solvents can be used to control the competition between crystallization and microphase separation. Blending a small amount of a block copolymer (AB), where the B blocks have a substantially lower surface energy than the A blocks, into a homopolymer (A) has also been demonstrated as an effective way to change the surface composition and lower the surface tension of the polymer blends [42]. The block copolymer added in small amounts acts as a surfactant and segregates to the surface region. The interesting feature is that only the surface properties are changed, whereas the bulk ones remain the same as the homopolymer. Homogeneous mixtures of AB block copolymer and A homopolymer can be obtained in the blend bulk when the concentration of the AB copolymer is low [43, 44]. In this respect, a fundamental approach to monitor how these block copolymers change the surface of the homopolymers is represented by AR-XPS, which supplies the composition profile of the AB/A blends within the technique sampling depth. Andersen *et al.* [45] accurately studied PS-PDMS di-block copolymers using XPS and atomic force microscopy (AFM). Surface segregation of the PDMS blocks was observed for five di-block copolymers with different molecular weights. The films were also annealed above and below the highest glass transition temperature (Tg) of the two polymer blocks, revealing that annealing above the PS Tg induced surface segregation of the dimethylsiloxane chain ends for all the studied copolymers, whereas at room temperature, surface segregation took place only when the amount of dimethylsiloxane in the di-block copolymers was small.

Another class of polymers that can undergo surface rearrangement is represented by polyurethane (PU)-based copolymers because they usually contain hard moieties, consisting of a diisocyanate chain extended with a low-molecular-weight diol or diamine, and soft moieties, which usually are long-chain macrodiols. The hard segments have a high density of urethane groups of high polarity, and for this reason, they are rigid at room temperature, whereas soft segments are flexible at room temperature. The polarity of hard segments produces a strong attraction between them, which causes a high degree of aggregation, and order in this phase, forming crystalline or pseudo-crystalline areas located in a soft and flexible matrix.

They can be synthesized as di-, tri-, or multi-block copolymers, i.e. copolymers where the different macrodiols are polymerized, giving homopolymer subunits and successively linked by junction units (diisocyanate), or random copolymers, i.e. copolymers where the macrodiols were polymerized together, so the sequence of macrodiols residues follows a statistical rule. In both cases, the presence of surface phase separation of the hard and soft segments in PU is very common.

Different authors studied the surface segregation phenomenon of various PU-based copolymers using XPS because this technique, both in conventional and AR modes, can also predict the *in vivo* performance of multiblock copolymer-based biomaterials. In particular, Cometa *et al.* [40] carried out XPS measurements to characterize the surface chemistry of polycaprolactone (PCL)-polyethyleneglycol (PEG) hydro-biodegradable copolymers having different PEG/PCL ratio and different molecular weights, evidencing surface segregation in these common multiphase amphiphilic polymer systems, as also described by different authors [46–49]. Moreover, using three different casting solvents (in terms of polarity and boiling point), XPS analysis evidenced that surface composition was enriched by PCL when the casting solvent was more hydrophobic and/or it evaporated slowly. Therefore, the air-polymer interface was found to be different from the glass-polymer one: when the solution was cast on a Petri dish, the copolymer systems assembled according to the high hydrophilic glass properties (Fig. 3.10).

Finally, the surface segregation of PCL-PEG copolymers, substantiated by a deep XPS characterization, provided a satisfactory explanation of the biological response to the investigated systems: *in vitro* cell adhesion tests have indeed shown no dependence on the PCL/PEG nominal ratio, whereas different molecular weights elicited different cell adhesion. Similarly, Chen and Ruckenstein [50] used AR-XPS to demonstrate that PU surfaces undergo rearrangements when the environment they are exposed to is changed from air to water, by exposing the more polar moieties of the molecular chains. They observed also that organic solvent pretreatment accelerates the rearrangement in water by increasing the mobility of the molecular chains of PU.

Another example of XPS investigation on the surface segregation phenomenon in PU is that of Vaidya and Chaudhury [51]. Segmented polyurethanes (SPU), containing perfluoropolyether (PFPE), polydimethylsiloxane (PDMS), and PEG segments, exhibit different responses to the polarity of the contacting medium, having oleophobic, hydrophobic, and hydrophilic domains. The oleophobic and hydrophobic properties of the SPU are due to the segregation of PFPE segments at the polymer-air interface. In this respect, XPS provides the ability to rewrite the block positioning in the air-polymer interface, as depicted in Fig. 3.11.

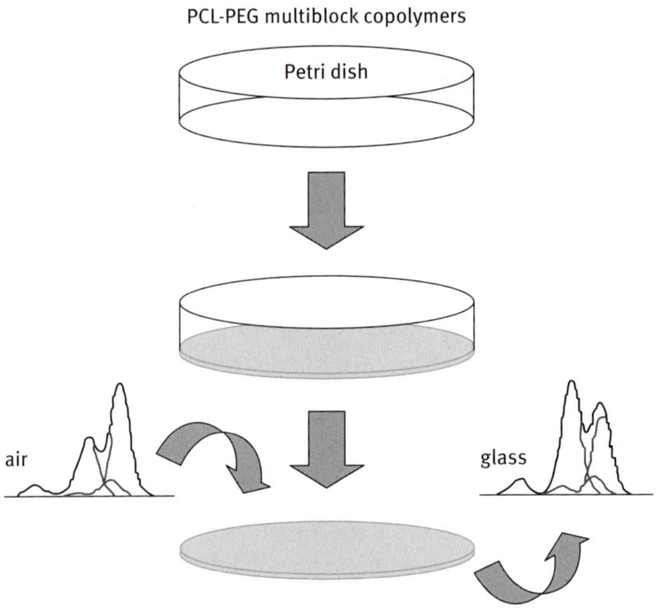

Fig. 3.10: XPS analysis of both air- and glass-exposed sides of PCL-PEG multiblock copolymer films obtained by solvent casting on Petri dishes. The reported spectra refer to C1s regions. Reprinted with permission from Cometa S, Chiellini F, Bartolozzi I, Chiellini E, De Giglio E, Sabbatini L, *Macromol Biosci*, 10, 317–27, 2010. Copyright 2010 Wiley-VCH Verlag GmbH & Co. KGaA, Weinheim [40].

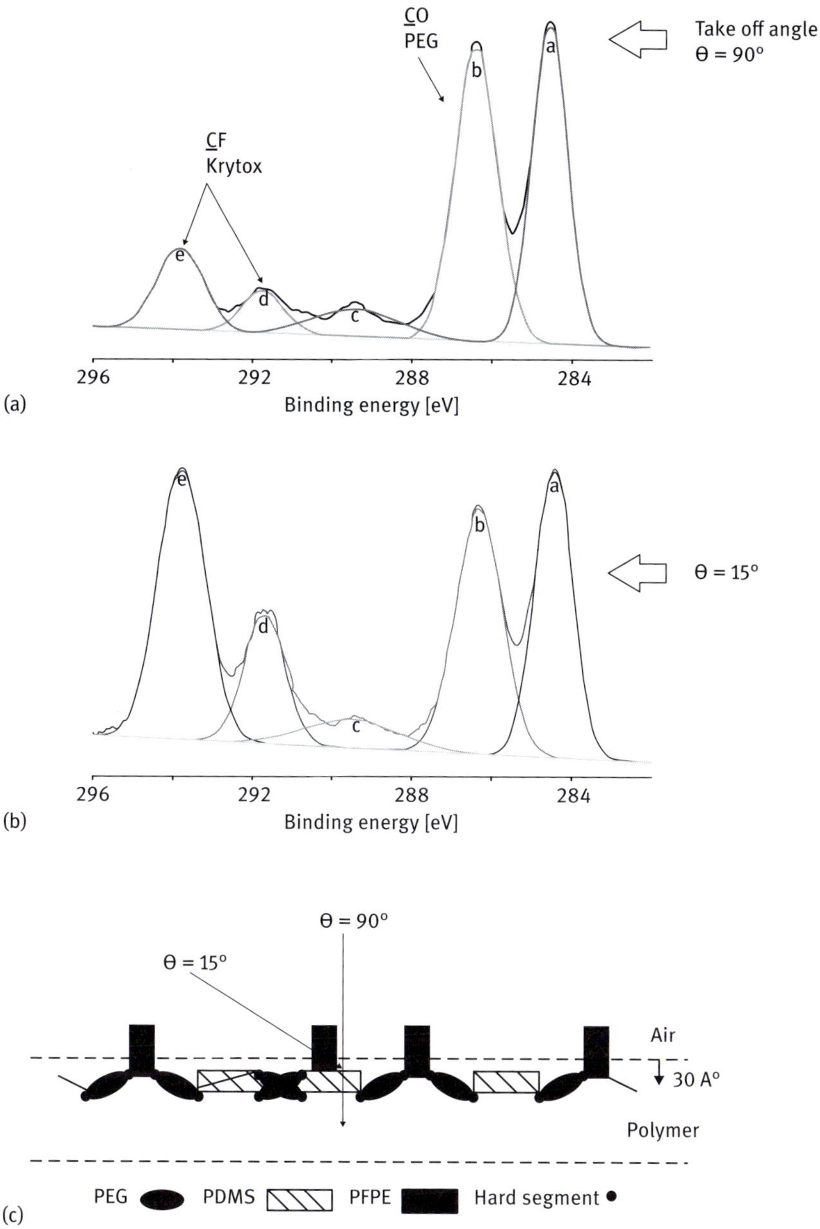

Fig. 3.11: High-resolution C(1s) regions at take-off angles (a) 90° and (B) 15° in the case of PFPE-modified SPU films. (c) Schematics of the structure of PFPE-modified SPU at polymer-air interface. Reprinted from Vaidya A, Chaudhury MK, *J Colloid Interface Sci*, 249, 235–45, 2002. Copyright 2002, with permission from Elsevier [51].

In Fig. 3.11a, the C1s region at a 90° take-off angle in SPU containing 0.86% w/w PFPE film is reported. In this figure, the contributions relevant to the C-H groups, due to the aromatic ring and PDMS segments, and to the C-O bonding present in PEG segments can be recognized, as well as the three peaks corresponding to carbon atoms bonded to fluorine (labeled as CF). To maximize the signal obtained from the surface-enriched segments, detailed spectra were also taken at a 15° take-off angle (Fig. 3.11b), showing that the intensity of CF peaks increases and the relative intensity of the CO peak decreases, compared with those taken at a 90° take-off angle. This indicates that the SPU surface is enriched with PFPE segments, whereas the PEG segments are present close to the surface. These findings suggested that the PFPE segments readily segregate at the polymer-air interface, thus reducing the surface energy of the polymer to very low values. During its segregation, the PFPE groups drag the hydrophilic PEG groups close to the surface (most likely in the subsurface region).

Copolymers containing fluorinated moieties are a typical class where surface segregation occurs. For instance, Pilati *et al.* [52] demonstrated that the addition of fluorinated block copolymers to poly(ethyleneterephthalate) (PET) solutions can be used to prepare PET films with controlled surface morphology, porosity, and chemical composition, and as expected, XPS measurements showed a surface composition much richer in fluorinated segments than expected from bulk composition.

Surface segregation is a common phenomenon in technical polymers when additives are used. These substances are added into the polymer host for different purposes, such as for polymer processing enhancement, compatibilization of different polymers in blends, as well as pigments, antioxidant, plasticizers, and so on [53]. These compounds are sometimes expected to migrate toward the surface, or sometimes stay in bulk according to different requirements. Thermodynamically, the surface enrichment of additives is determined by difference of surface energy and the interaction between the additive and the polymer host. This phenomenon can be easily monitored by XPS.

3.8 Polymers physical treatments/grafting

Polymer treatment by plasma represents one of the most common strategy to improve polymer performances [5]. Plasmas are environmentally friendly and fast, contrary to traditional wet chemistry. Plasmas are very complex media full of highly reactive particles [radicals, ions, electrons, ultraviolet (UV) photons], all of which reach the polymer surface. Because of this complexity, the mechanisms involved in the surface modification of polymer are not yet fully understood. In view of this, CD-XPS provides a successful approach to elucidate the surface structure of plasma-treated polymers [20].

Low- and high-pressure plasma have been used empirically for the treatment of surfaces for many decades, but the plasma treatment of surfaces really started to gain importance with the development of commercially available surface analysis techniques; in particular, XPS and water contact angle (WCA). Plasma treatment of polymers results mainly in two mechanisms: ablation of the polymer and/or grafting of new species on its surface. The two reactions are taking place at the same time, but depending on the experimental conditions, one can prevail on the other. It is worth noting that plasmas modify only the surface, while maintaining unaltered the bulk properties of the polymers. The most common use is the grafting of new functionalities on the sample surfaces. In terms of industrial applications, the most important reason for the use of plasma is undoubtedly the increase of the adhesion properties of a polymer [54–58]. Plasma treatments improve the very poor adhesion properties of a polymer, increasing its surface energy by grafting polar species on its surface. In a similar way, plasma treatments are efficiently used to properly modify other important surface properties such as wettability and biocompatibility. It is worth mentioning that plasma technology can be used for direct polymerization, which allows to deposit polymers on a very large amount of substrates, even another polymer. In this case, a precursor is sent into the plasma, and the reactive species from the plasma will activate the precursor molecules that will react with each other and polymerize. A very large variety of polymers can be created by this technique ranging from diamond-like carbon [59] to PS [60] and polytetrafluoroethylene (PTFE) [61] films. They present several advantages over traditional polymers. For instance, these films exhibit usually a higher degree of cross-linking and therefore have superior mechanical properties that allow to deposit ultrathin films that present good adhesion on various substrates. XPS analysis gives not only information about the atomic composition of the films but also about the degradation of the monomer in the plasma and therefore provides information about the mechanism of the reaction. Weibel *et al.* [62] reports a study on the surface chemical changes produced on PU membranes after acrylic acid (AA) vapor plasma treatment XPS. In all the experimental conditions studied, AA plasma treatment increased the concentrations of carbonyl and ester groups at the PU surface compared with traditional oxygen plasma modification. XPS C1s spectra, reported in Fig. 3.12, showed that a thin film with similar chemical properties to the poly(acrylic acid) (PAA) was formed on top of the PU membrane after 5-W AA plasma treatments.

The UV-ozone (UVO) treatment relies upon the combined effects of UV light and ozone, produced *in situ* from a gas-phase photodissociation of molecular oxygen. UV and ozone photooxidation was first recognized as a potential polymer surface treatment in the early 1980s. This treatment, initially considered as a surface cleaning method [63], has been demonstrated recently to modify the surface chemistry and improve the wetting characteristics of natural and synthetic polymers. UVO treatment causes the surface energy of the polymers to increase through the breaking of the polymer chain by the insertion of oxygen-containing functional groups. It was found that the oxygen levels on UVO-treated polymer surfaces increase quite markedly after relatively short exposure times (<5 min) compared with other UV treatments [64–67]. In particular, such treatment can be useful in a range of technological applications to increase the surface energy of many organic polymers for the substrate to perform its desired task (i.e. printability, wettability, biocompatibility, etc.). A principal advantage

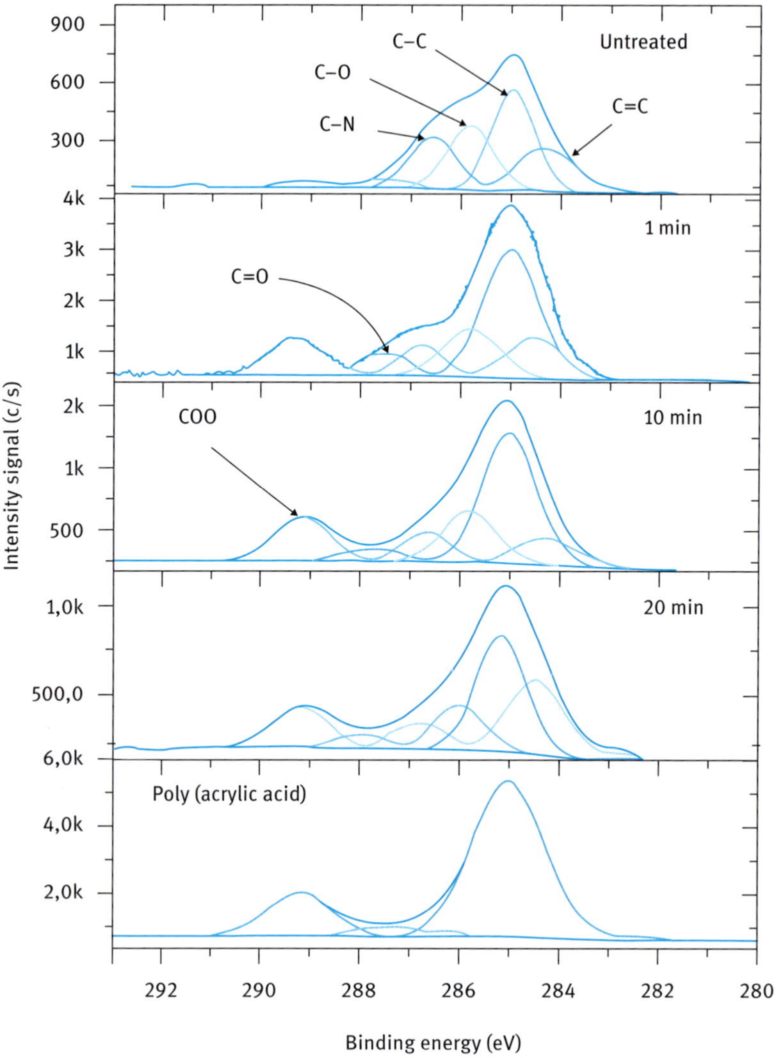

Fig. 3.12: High-resolution XPS spectra of the C1s peak of an untreated PU membrane, AA plasma-treated PUs at several plasma treatment times, and synthesized PAA. Plasma power: 5 W. The C=O and COO groups are shown in different spectra for better presentation. Reprinted from Weibel DE, Vilani C, Habert AC, Achete CA, *J Membr Sci*, 293, 124–32, 2007. Copyright 2007, with permission from Elsevier [62].

of the UVO method is that it can be applied under ambient conditions as a continuous treatment method, or in small batches with a very high degree of control. The individual and combined effects of UV light and ozone have been studied for a variety of surfaces, including polyethylene, polypropylene, PET, poly(ether ether ketone), and PS. Teare *et al.* [68] investigated the correlation between surface oxygen chemistry and topographical changes of PET and PS surfaces by XPS, contact angle measurement, and AFM. The study led to a greater understanding of the changes in structure and chemistry that occur on the modified polymer surfaces. It was observed that the PET and PS surfaces undergo considerable reorganization with exposure to the UVO source. The surface roughness, polar surface energy, and grain size increased with exposure time. With PET, the longer the exposure to UVO, the more oxidized the surface becomes. The XPS results indicated that the point of oxidation was predominantly the ester section of the polymer chain. In both polymers, carboxylic acid and ester functionalities were formed upon treatment. The oxidation was seen to proceed in two stages: insertion of oxygen to form ether and hydroxyl groups; followed by the dominant stage, formation of unsaturated carbon to oxygen bonds. The UVO-PS surfaces are more stable to rearrangement with time than the UVO-PET surfaces due to the nonpolar nature of the underlying bulk polymer. Quoc Toan Le *et al.* [69] have characterized the modification of polyarylene films after exposure to UVO treatment. The obtained results showed that UVO induced different types of oxygen functional groups at the film surface. XPS data indicated that mainly C-O, C=O, COOH, and/or COOR are formed after treatment. The formation of such functional groups were responsible for the rapid change in hydrophobicity-hydrophilicity of the surface as evidenced by a decrease in the WCA.

The ion beam co-sputtering (IBS) technique can be used successfully to deposit different polymeric and inorganic materials in an extremely controlled manner by selecting appropriate targets and adjusting sputtering deposition conditions. In this context, the incorporation of metal nanoclusters into polymeric matrices is a research topic of considerable interest because the metallic species can dramatically influence the chemical and physical properties of the resultant composite material. The high research and industrial interest in metal nanoparticle (NP)/fluoropolymer composites (metal-CFx) is due to their wide range of applications, for example, in microelectronics, nonlinear optics, optical recording, and in the development of biocompatible coatings. Gold-fluoropolymer (CFx(Au))-nanostructured composites have been successfully fabricated using technologies such as plasma deposition or ion co-sputtering [70, 71]. The latter is particularly interesting, being a cost-effective and environmental friendly technology, allowing also a higher degree of control over important deposition parameters such as the sputtering energy and the metal loading.

Cioffi *et al.* [72] reported a systematic spectroscopic and morphological characterization of gold-fluoropolymer (CFx(Au)) nanocomposites, deposited by ion beam co-sputtering. These composites are technologically relevant because they exhibit vapor-sensing properties based on the swelling phenomenon. XPS analysis assessed that gold co-deposition induces a progressive defluorination of the polymeric chains, leading to a significant increase in polymer chain branching and in the concentration of unsaturated fluorinated carbons. Meanwhile, angle-resolved XPS indicated a nonhomogeneous in-depth distribution of the gold particles, with the outer surface being less rich in NPs.

Cu-CFx composite coatings deposited by the IBS technique are demonstrated to inhibit the growth of many target microorganisms and are therefore proposed as bioactive membranes [73]. XPS provided useful information on the surface composition of the Cu-CFx films, which is shown to change markedly as a function of the bulk copper loading. The ion-beam properties (energy and current), and hence the growth rate ratio, were chosen to obtain copper-fluorocarbon composites with a copper volume fraction in the interval $0 \leq \Phi \leq 0.25$. Fluorine percentage, and the abundance of highly fluorinated carbon groups in the C1s spectra, decreases when Φ increases. In these composites, copper surface concentration is always lower than the value expected on the basis of the bulk metal content. This is attributed to the penetration of metallic clusters into the soft outer layer of the fluoropolymeric matrix.

Regardless the co-sputtered inorganic material, the higher is the metal loading, the more the nanocomposite surface chemistry is affected. Typical C1s spectra relative to gold/fluoropolymer composite are reported in Fig. 3.13 as a function of the metal loading; six photoelectron components and a shake-up weak shoulder have been used to fit the signal. The attribution of the different species has been reported in the following. Starting from the left side of the figure, a weak shake-up signal (BE, ~296.0 ± 0.3 eV) can be seen, due to fluorinated-unsaturated groups, whose intensity slightly increases with Φ. The peak falling at 293.8 ± 0.2 eV (293.3 ± 0.1 eV for $\Phi = 0.25$) is attributed to CF_3 chain terminations; CF_2 groups are responsible for the third peak (BE = 291.8 ± 0.3 eV for $\Phi \leq 0.15$; BE = 291.4 ± 0.2 eV for $\Phi = 0.25$); the relative intensity of this chemical environment decreases as Φ increases. The spectral region falling at lower BE values comprises three peaks whose attribution is not unequivocal because an overlap of signals from fluorinated and oxygenated species occurs. A general inspection of the C1s spectra indicates that at higher Φ values ($\Phi = 0.15$ and $\Phi = 0.30$), the structure itself of the fluoropolymer matrix dramatically changes: there is a very high degree of branching and the reduced abundance of strongly electron-attractive fluorine substituents leads to an appreciable decrease of the chemical shift of some photoelectron peaks. At low metal content ($\Phi \leq 0.10$), the typical C1s spectrum of a fluoropolymer with a low degree of branching and low abundance of oxygenated species can be recorded. Similar findings were observed also on Cu-CFx and Pd-CFx [73, 74] composites: in general, the higher the metal NP concentration, the greater the extent of fluoropolymer branching.

In literature, annealing treatments of polymers have been suggested to produce more uniform and less porous films [75, 76]. This phenomenon likely originates from a viscous flow in the polymer chains leading to smoother and more compact polymer films. This finding also clarifies why the polymer thickness decreases after annealing. Usually, annealing is performed at a temperature greater than the polymer glass transition temperature (Tg). De Giglio *et al.* [77] reported a study in which an annealing treatment was applied to polymethyl methacrylate (PMMA) to improve the performance of the polymer coating as barrier films against corrosion of titanium-based orthopedic implants. PMMA coatings were annealed at 200°C in air for 10 min, and XPS was exploited to ascertain possible changes in polymer surface chemistry

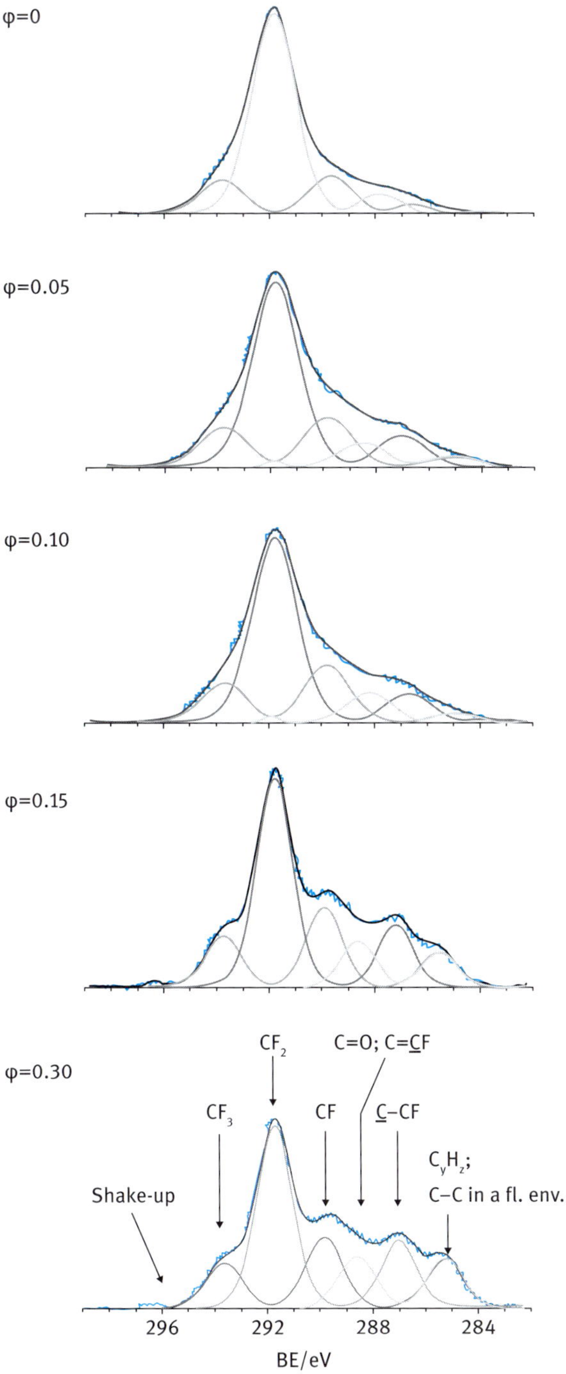

Fig. 3.13: C1s XPS region of Au-CFx nanocomposites, as a function of the different gold loading Φ. Reprinted with permission from Cioffi N, Losito I, Torsi L, *et al.*, *Chem Mater*, 14, 804–11, 2002. Copyright 2002 American Chemical Society [74].

after the annealing treatment. A comparison between C1s spectra of annealed and unannealed PMMA films showed that no alteration of surface chemical composition was introduced by the annealing treatment. Moreover, SEM investigation showed that the morphology of the annealed films was improved with respect to uniformity and porosity. The final results of this investigation indicated that the annealing process produces coatings with considerable anticorrosion performances.

The covalent grafting of target molecules on polymer surfaces is a widely used strategy to address the polymer response toward specific applications. In this context, XPS technique represents a useful tool to characterize step by step the polymer grafting reactions involved. For example, the grafting of bioactive molecules on a polymer surface is one of the most used strategies in the biomaterials research field to improve the biological response of the investigated system. In this context, the covalent grafting of peptides containing the Arg-Gly-Asp (RGD) minimal recognition sequence has been attempted onto different polymer substrates to promote cell adhesion, thanks to the formation of cell membrane integrin-RGD bonds. In these investigations, XPS proved to be an efficient tool to assess the presence and availability of the peptide on the polymer surface [78–80]. Another example of polymer grafting is the photo-induced graft polymerization of 2-methacryloyloxyethyl phosphorylcholine on a PDMS substrate using UV irradiation and benzophenone as photoinitiator [81]. The aim of this study was the development of a substrate that exhibits a strong resistance to protein and subsequently hinders the following unfavorable biological responses initiated by nonspecific protein adsorption. Increasing UV irradiation time and monomer concentration, an increased density of the grafted polymer was observed. Resistance to protein adsorption, water wettability, and lubrication characteristics were tunable with the graft density.

Interesting results have been reported in a study in which an annealing treatment and a grafting reaction have been both applied to improve the biological performances of a PAA coating electrosynthesized on titanium substrates [82]. The annealing treatment provided optimal anti-corrosion performances to the PAA coating while preserving the surface polymer composition, as demonstrated by XPS analyses performed before and after the annealing treatment. Meanwhile, the presence of carboxylic moieties on the polymer film surface allows the grafting of biomolecules promoting positive interface reactions with the surrounding biological system. As an example, a simple model compound, the amino acid residue 4-fluorophenylalanine, was grafted onto the annealed PAA: the fluorine target atom enabled a simple monitoring of the reaction effectiveness as well as of the reaction yield by estimating the peak area of the F1s XPS signal.

3.9 Polymers aging

As it happens to all materials, interaction with the environment induces, with time, chemical modifications on polymer surfaces, which are referred to as the aging phenomenon.

It must be emphasized, however, that aging is particularly prominent and often remarkably accelerated in those polymers or polymer coatings obtained by plasma treatments. One commonly observed phenomenon, for example, is the deterioration of the beneficial surface properties (such as hydrophilicity) upon storage in air. The problem is complex because of the variety of processes that may be involved, e.g. the diffusion and reaction of free radicals [83] created by exposure to plasma, the adsorption of polar [84] and nonpolar contaminants [85], and the reorientation of polar groups [86, 87] away from the surface. As a matter of fact, although the use of plasmas for the modification of polymer surfaces currently represents a mature technology, research is still needed to forecast the evolution of the treated surfaces with time.

Vandencasteele *et al.* [5] demonstrated, by contact angle measurements, that the storage in water of a PTFE surface, treated by a CO_2-plasma, causes a significant improvement in wettability. Meanwhile, when the samples are stored in a hydrophobic environment such as air, a hydrophobic recovery occurs, which is not happening when the samples are stored in a hydrophilic environment such as water [88].

Touzin *et al.* [89] carried out a study, during a period of 4 weeks, on the stability of a plasma-polymerized coating deposited on 316L stainless steel cardiovascular stents. The deposition of an ultrathin (from 10 to 100 nm) plasma-polymerized fluorocarbon coating (Teflon-like) has been proposed to improve the long-term performance and safety of these devices (i.e. corrosion prevention). The stability of the plasma-polymerized coating has been investigated using deionized (DI) water as a simple aging medium. After different periods of immersion in DI water, the characterization of the morphology and chemistry of the coating was performed. In particular, a constant decrease in the fluorine content and an increase in the oxygen and carbon contents were observed. The XPS F1s/C1s area ratio decreases gradually, and this trend was in agreement with the decreasing of contact angle values. In addition, an increase in the O/C ratio is observed during aging. The loss in fluorine is then compensated by an increase in the concentration of oxygen. A rapid incorporation of oxygen between the second and the third week of aging, associated with an oxidation of the coating surface, was detected. In Fig. 3.14, the high-resolution XPS spectra of C(1s) regions relevant to plasma polymer coatings before and after their immersion in DI water at different periods are reported. Before immersion, negligible oxygen amount was detected on the coating surface; therefore, the C1s spectrum of the coating was curve fitted with five spectral components assigned to C-H or -C-C- (285 eV), -C-CF (286.5 eV), -C-CF (288.8 eV), $-CF_2$ (291.4 eV), and $-CF_3$ (293.8 eV) groups. After immersion, the C1s spectrum was still fitted using five components, but the increase of the signal intensity in the low binding energies (and the concomitant increase of oxygen peak intensity) suggests that oxygen components such as C-O (286.4–287 eV) and C=O (288–288.4 eV), overlapping the -C-CF and -C-CF contributions, should be considered. Before immersion, the coating mainly consists of CFx groups with a significant percentage of CF_3 end groups, leading to a hydrophobic coating; defluorination and oxygen incorporation upon aging causes a loss of hydrophobicity.

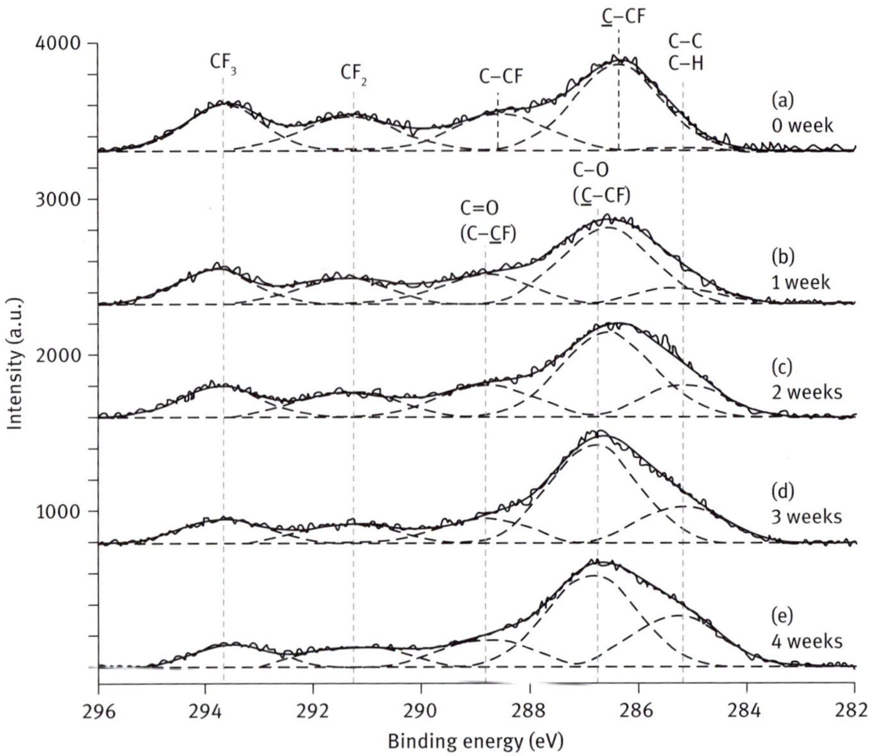

Fig. 3.14: High-resolution C1s X-ray photoelectron spectra of plasma polymer coatings: (a) before, (b) after 1 week, (c) after 2 weeks, (d) after 3 weeks, and (e) after 4 weeks of immersion in DI water. Reprinted from Touzin M, Chevallier P, Lewis F, *et al.*, *Surf Coat Technol*, 202, 4884–91, 2008. Copyright 2008, with permission from Elsevier [89].

Tanikella *et al.* [90] also observed a decrease of CFx and an increase of the C-C/C-H peak intensity contribution in similar system, and this finding was attributed to the formation on the surface of the deposited coatings of a layer of low-molecular-weight oxidized material, which can leave the surface of the coating upon aging, thus leading to the decrease in the CF_3 and CF_2 concentration.

An aging phenomenon was also observed by Tran *et al.* [91], when a plasma-polymerized fluoropolymer coating was exposed under humid and UV conditions for up to 105 h to test its stability.

Paynter [92] reports the results of an XPS study of the surface modifications carried out by the exposure of PS and PET to plasmas containing three working gases: helium, 10% oxygen/90% helium, and 33% nitrogen/67% hydrogen. Oxygen and nitrogen plasmas are known to produce a more hydrophilic surface, and the helium plasma was used to evaluate the possibility of the incorporation of oxygen and nitrogen from the ambient. PET is a widely used polymer that contains oxygen, whereas PS contains no oxygen but, like PET, contains aromatic carbon rings. In a previous work [93], the surface chemical changes resulting from exposure of the PS and PET samples to helium, nitrogen, and oxygen plasmas were evaluated and the formation of amide groups on polymers exposed to both oxygen and nitrogen plasmas were evidenced.

The apparent surface composition, determined by XPS, of PS and PET samples exposed to plasmas containing helium, nitrogen, and oxygen has been observed to evolve following storage in air for a period of 360 days. The observed changes were found to be in excess with respect to those on a control sample not exposed to plasma. Peak-fitting of the C1s signal allowed the characterization of the less-stable surface chemical groups in each case. Results led to conclude that evolution of the surface chemical composition depended upon the plasma gas used, and even the order of application of oxygen and nitrogen plasmas was found to have a discernible influence.

References

[1] Briggs D. New developments in polymer surface analysis. Polymer 1984;25:1379–91.

[2] Batich CD. Chemical derivatization and surface analysis. Appl Surf Sci 1988;32:57–73.

[3] Merrett K, Cornelius RM, McClung WG, Unsworth LD, Sheardown H. Surface analysis methods for characterizing polymeric biomaterials. J Biomater Sci-Polym E 2002;13:593–621.

[4] Ha C-S, Gardella JA Jr. X-ray photoelectron spectroscopy studies on the surface segregation in poly(dimethylsiloxane) containing block copolymers. J Macromol Sci, Part C—Polym Rev 2005;45:1–18.

[5] Vandencasteele N, Reniers F. Plasma-modified polymer surfaces: characterization using XPS. J Electron Spectrosc 2010;178–9:394–408.

[6] Briggs D, Seah MP. Practical surface analysis, auger and x-ray photoelectron spectroscopy. Chichester (UK): Wiley; 1990.

[7] Siegbahn K, Nordling C, Fahlman A, et al. ESCA: atomic molecular and solid state structure studied by means of electron spectroscopy. Uppsala (Sweden): Almquist & Wiksells; 1967.

[8] Salmeron M, Schlögl R. Ambient pressure photoelectron spectroscopy: a new tool for surface science and nanotechnology. Surf Sci Rep 2008;63:169–99.

[9] Tougaard S, Jansson C. Comparison of validity and consistency of methods for quantitative XPS peak analysis. Surf Interface Anal 1993;20:1013–46.

[10] Wagner CD, Davis LE, Zeller MV, Taylor JA, Raymond RH, Gale LH. Empirical sensitivity factors for quantitative analysis by electron spectroscopy for chemical analysis. Surf Interface Anal 1981;3:211–25.

[11] Briggs D. Surface analysis of polymers by XPS and static SIMS. Cambridge (UK): Cambridge University Press; 1998.

[12] Sherwood PMA. Data analysis in X-ray photoelectron spectroscopy. In: Briggs D, Seah MP eds. 2nd edn. Chichester (UK): Wiley; 1990.

[13] Moulder JF, Stickle WF, Sobol PE, Bomben KD. Handbook of x-ray photoelectron spectroscopy. Eden Prairie (MN): Perkin Elmer; 1992.

[14] Windawi H, Ho F. Applied electronic spectroscopy for chemical analysis: New York (NY): Wiley-Interscience; 1982.

[15] Baker AD, Brundle CR. Electron spectroscopy: theory, techniques and applications. Vol IV. London (UK): Academic Press; 1981.

[16] Ratner BD, Castner DG. Advances in X-ray photoelectron spectroscopy instrumentation and methodology: instrument evaluation and new techniques with special reference to biomedical studies. Colloid Surf B 1994;2:333–46.
[17] Gerenser LJ, Elman JF, Mason MG, Pochan JM. E.s.c.a. studies of corona-discharge-treated polyethylene surfaces by use of gas-phase derivatization. Polymer 1985;26:1162–66.
[18] Pochan JM, Gerenser LJ, Elman JF. An e.s.c.a. study of the gas-phase derivatization of poly(ethylene terephthalate) treated by dry-air and dry-nitrogen corona discharge. Polymer 1986;27:1058–62.
[19] Chilkoti A, Ratner BD, Briggs D. Plasma-deposited polymeric films prepared from carbonyl-containing volatile precursors: XPS chemical derivatization and static SIMS surface characterization. Chem Mater 1991;3:51–61.
[20] Chilkoti A, Ratner BD. An X-ray photoelectron spectroscopic investigation of the selectivity of hydroxyl derivatization reactions. Surf Interface Anal 1991;17:567–74.
[21] Everhart DE, Reilley CN. Chemical derivatization in electron spectroscopy for chemical analysis of surface functional groups introduced on low-density polyethylene film. Anal Chem 1981;53:665–76.
[22] Nakayama Y, Takahagi T, Soeda F, Hatada K, Nagaoka S, Ishitani A. XPS analysis of NH_3 plasma-treated polystyrene films utilizing gas phase chemical modification. J Polym Sci Pol Chem 1988;26:559–72.
[23] Povstugar VI, Mikhailova SS, Shakov AA. Chemical derivatization techniques in the determination of functional groups by X-ray photoelectron spectroscopy. J Anal Chem 2000;55:405–16.
[24] Malitesta C, Losito I, Sabbatini L, Zambonin PG. Applicability of chemical derivatization X-ray photoelectron spectroscopy (CD-XPS) to the characterization of complex matrices: case of electrosynthesized polypyrroles. J Electron Spectrosc 1998;97:199–208.
[25] Wepanick KA, Billy AS, Bitter JL, Fairbrother DH. Chemical and structural characterization of carbon nanotube surfaces. Anal Bioanal Chem 2010;396:1003–14.
[26] Zschoerper NP, Katzenmaier V, Vohrer U, Haupt M, Oehr C, Hirth T. Analytical investigation of the composition of plasma induced functional groups on carbon nanotube sheets. Carbon 2009;47:2174–85.
[27] Dementev N, Feng X, Borguet E. Fluorescence labeling and quantification of oxygen-containing functionalities on the surface of single-walled carbon nanotubes. Langmuir 2009;25:7573–77.
[28] Hsu SH, Tang CM, Lin CC. Biocompatibility of poly(ε-caprolactone)/poly(ethylene glycol) diblock copolymers with nanophase separation. Biomaterials 2004;25:5593–601.
[29] Rasmont A, Leclere Ph, Doneux C, et al. Microphase separation at the surface of block copolymers, as studied with atomic force microscopy. Colloid Surf B 2000;19:381–95.
[30] Andruzzi L, D'Apollo F, Galli G, Gallot B. Synthesis and structure characterization of liquid crystalline polyacrylates with unconventional fluoroalkylphenyl mesogens. Macromolecules 2001;34:7707–14.
[31] Bertolucci M, Galli G, Chiellini E, Wynne KJ. Wetting behavior of films of new fluorinated styrene-siloxane block copolymers. Macromolecules 2004;37:3666–72.
[32] Losito I, Sabbatini L, Gardella JA Jr. Electron, ion, and mass spectroscopy. In: Brady RF Jr, ed. Comprehensive desk reference of polymer characterization and analysis. New York (NY): Oxford University Press; 2003:375–407.
[33] Schmidt JJ, Gardella JA Jr, Salvati L Jr. Surface studies of polymer blends. 2. An ESCA and IR study of poly(methyl methacrylate)/poly(vinyl chloride) homopolymer blends. Macromolecules 1989;22:4489–95.
[34] Clark DT, Dilks A, Peeling J, Thomas HR. Application of ESCA to studies of structure and bonding in polymers. Faraday Discuss 1975;60:183–95.

[35] Thomas HR, O'Malley JJ. Surface studies on multicomponent polymer systems by x-ray photoelectron spectroscopy. polystyrene/poly(ethylene oxide) diblock copolymers. Macromolecules 1979;12:323–29.
[36] O'Malley JJ, Thomas HR, Lee GM. Surface studies on multicomponent polymer systems by x-ray photoelectron spectroscopy. polystyrene/poly(ethylene oxide) triblock copolymers. Macromolecules 1979;12:996–1001.
[37] Chen X, Gardella JA Jr, Kumler PL. Fourier-transform infrared and electron spectroscopy for chemical analysis studies of block copolymers of styrene and dimethylsiloxane. Macromolecules 1992;25:6621–30.
[38] Chen X, Gardella JA Jr, Kumler PL. Surface ordering of block copolymers. Macromolecules 1992;25:6631–37.
[39] Vargo TG, Hook DJ, Gardella JA Jr, Eberhardt MA, Meyer AE, Baier RE. A surface spectroscopic and wettability study of a segmented block copolymer poly(etherurethane). Appl Spectrosc 1991;45:448–56.
[40] Cometa S, Chiellini F, Bartolozzi I, Chiellini E, De Giglio E, Sabbatini L. Surface segregation assessment in poly(e-caprolactone)-poly(ethylene-glycol) multiblock copolymer films. Macromol Biosci 2010;10:317–27.
[41] Weng L-T, Ng K-M, Cheung ZL, Lei Y, Chan C-M. Quantitative analysis of styrene-pentafluorostyrene random copolymers by ToF-SIMS and XPS. Surf Interface Anal 2006;38:32–43.
[42] Dwight DW, McGrath JE, Riffle JS, Smith SD, York GA. ADXPS/STEM studies of surface and bulk microphase behavior in block copolymers and graft copolymers and their blends. J Electron Spectrosc 1990;52:457–73.
[43] Kinning DJ, Thomas EL, Fetters LJ. Morphological studies of micelle formation in block copolymer-homopolymer blends. J Chem Phys 1989;90:5806–25.
[44] Leibler L, Orland H, Wheeler JC. Theory of critical micelle concentration for solutions of block copolymers. J Chem Phys 1983;79:3550–57.
[45] Andersen TH, Tougaard S, Larsen NB, Almdal K, Johannsen I. Surface morphology of PS-PDMS diblock copolymer films. J Electron Spectrosc 2001;121:93–110.
[46] Yoon SC, Ratner BD. Surface and bulk structure of segmented poly(ether urethanes) with perfluoro chain extenders. 2. FTIR, DSC, and x-ray photoelectron spectroscopic studies. Macromolecules 1988;21:2392–400.
[47] Helfand MA, Mazzanti JB, Fone M, Reamey RH. Effect of acetate distribution on surface segregation in poly(vinyl alcohol-co-vinyl acetate) copolymer films. Langmuir 1996;12:1296–302.
[48] Bottino FA, Di Pasquale G, Pollicino A. XPS study on surface segregation in poly(ethylene-iso/terephthalate)-perfluoropolyether block copolymers. Macromolecules 1998;31:7814–19.
[49] Andruzzi L, Hexemer A, Li X, et al. Control of surface properties using fluorinated polymer brushes produced by surface-initiated controlled radical polymerization. Langmuir 2004;20:10498–506.
[50] Chen J-H, Ruckenstein E. Solvent stimulated surface rearrangement of polyurethanes. J Colloid Interface Sci 1990;135:496–507.
[51] Vaidya A, Chaudhury MK. Synthesis and surface properties of environmentally responsive segmented polyurethanes. J Colloid Interface Sci 2002;249:235–45.
[52] Pilati F, Montecchi M, Fabbri P, et al. Design of surface properties of PET films: effect of fluorinated block copolymers. J Colloid Interface Sci 2007;315:210–22.
[53] Chen S, Tanaka H. Surface analysis of paper containing polymer additives by X-ray photoelectron spectroscopy I: application to paper containing dry strength additives. J Wood Sci 1998;44:303–9.
[54] Plasma surface modification of polymer for improved adhesion: a critical review. In: Strobel M, Lyons CS, Mittal KL, eds. Plasma surface modification of polymers: relevance to adhesion, Utrecht (The Netherlands): VSP; 1994.

[55] Hegemann D, Brunner H, Oehr C. Plasma treatment of polymers for surface and adhesion improvement. Nucl Instrum Methods B 2003;208:281–6.
[56] O'Kell S, Henshaw T, Farrow G, Aindow M, Jones C. Effects of low-power plasma treatment on polyethylene surfaces. Surf Interface Anal 1995;23:319–27.
[57] Vandencasteele N, Fairbrother H, Reniers F. Selected effect of the ions and the neutrals in the plasma treatment of PTFE surfaces: an OES-AFM-contact angle and XPS study. Plasma Process Polym 2005;2:493–500.
[58] Vandencasteele N, Merche D, Reniers F. XPS and contact angle study of N2 and O2 plasma-modified PTFE, PVDF and PVF surfaces. Surf Interface Anal 2006;38:526–30.
[59] Ozeki K, Hirakuri KK. The effect of nitrogen and oxygen plasma on the wear properties and adhesion strength of the diamond-like carbon film coated on PTFE. Appl Surf Sci 2008;254:1614–21.
[60] Merche D, Poleunis C, Bertrand P, Sferrazza M, Reniers F. Synthesis of polystyrene thin films by means of an atmospheric-pressure plasma torch and a dielectric barrier discharge. IEEE T Plasma Sci 2009;37:951–60.
[61] Reniers F, Vandencasteele N, Burry O. Procédé pour déposer une couche fluorée à partir d'un monomère précurseur. EP 2 098 305 A1 (2009) Bruxelles (BE).
[62] Weibel DE, Vilani C, Habert AC, Achete CA. Surface modification of polyurethane membranes using acrylic acid vapour plasma and its effects on the pervaporation processes. J Membr Sci 2007;293:124–32.
[63] Vig JR. UV/ozone cleaning of surfaces. J Vac Sci Technol A 1985;3:1027–34.
[64] Mathieson I, Bradley RH. Improved adhesion to polymers by UV/ozone surface oxidation. Int J Adhes Adhes 1996;16:29–31.
[65] Mathieson I, Bradley RH. Surface oxidation of poly ether ether ketone films using ultraviolet/ozone. J Mater Chem 1994;4:1157–57.
[66] Hill JM, Karbashewski E, Lin A, Strobel M, Walzak MJ. Effects of aging and washing on UV and ozone-treated poly(ethylene terephthalate) and polypropylene. J Adhes Sci Technol 1995;9:1575–91.
[67] MacManus LF, Walzak MJ, McIntyre NS. Study of ultraviolet light and ozone surface modification of polypropylene. J Polym Sci A 1999;37:2489–501.
[68] Teare DOH, Ton-That C, Bradley RH. Surface characterization and ageing of ultraviolet-ozone-treated polymers using atomic force microscopy and x-ray photoelectron spectroscopy. Surf Interface Anal 200;29:276–83.
[69] Quoc Toan Le, Whelan CM, Struyf H, et al. Polymer dielectric treatment using ultraviolet-ozone: a photoelectron spectroscopy and ellipsometric porosimetry study. J Appl Phys 2004;96:3807–10.
[70] D'Addabbo A, Valentini A, Convertino A. Swelling of CFx and CFx(Au) films. J Appl Phys 2000;87:2039–42.
[71] Bassi A, Valentini A, Convertino A. Gold doping effects on swelling of teflon-like thin films. Appl Phys A 2000;71:109–11.
[72] Cioffi N, Losito I, Torsi L, et al. Analysis of the surface chemical composition and morphological structure of vapor-sensing gold-fluoropolymer nanocomposites. Chem Mater 2002;14:804–11.
[73] Cioffi N, Ditaranto N, Torsi L, et al. Analytical characterization of bioactive fluoropolymer ultra-thin coatings modified by copper nanoparticles. Anal Bioanal Chem 2005;381:607–16.
[74] Cioffi N, Torsi L, Farella I, et al. Deposition and analytical characterization of fluoropolymers modified by palladium nanoparticles. Thin Solid Films 2004;449:25–33.
[75] Cram SL, Spinks GM, Wallace GG, Brown HR. Electropolymerised acrylic coatings for polymer-metal adhesion enhancement. J Adhes Sci Technol 2003;17:1403–23.

[76] Malgas GF, Motaung DE, Arendse CJ. Temperature-dependence on the optical properties and the phase separation of polymer-fullerene thin films. J Mater Sci 2012;47:4282–89.
[77] De Giglio E, Cometa S, Sabbatini L, Zambonin PG, Spoto G. Electrosynthesis and analytical characterization of PMMA coatings on titanium substrates as barriers against ion release. Anal Bioanal Chem 2005;381:626–33.
[78] Biltresse S, Attolini M, Marchand-Brynaert J. Cell adhesive PET membranes by surface grafting of RGD peptidomimetics. Biomaterials 2005;26:4576–87.
[79] Pourcelle V, Devouge S, Garinot M, Preat V, Marchand-Brynaert J. PCL-PEG-Based Nanoparticles grafted with GRGDS peptide:preparation and surface analysis by XPS. Biomacromolecules 2007;8:3977–83.
[80] De Giglio E, Sabbatini L, Colucci S, Zambonin G. Synthesis, analytical characterization and osteoblast adhesion properties on RGD-grafted polypyrrole coatings on titanium substrates. J Biomater Sci Polym Ed 2000;11:1073–83.
[81] Goda T, Matsuno R, Konno T, Takai M, Ishihara K. Photografting of 2-methacryloyloxyethyl phosphorylcholine from polydimethylsiloxane: tunable protein repellency and lubrication property. Colloid Surf B 2008:64–72.
[82] De Giglio E, Cometa S, Cioffi N, Torsi L, Sabbatini L. Analytical investigations of poly(acrylic acid) coatings electrodeposited on titanium-based implants: a versatile approach to biocompatibility enhancement. Anal Bioanal Chem 2007;389:2055–63.
[83] Chen J-R. Study on free radicals of cotton and wool fibers treated with low-temperature plasma. J Appl Polym Sci 1996;62:1325–9.
[84] Gil'man AB, Drachev AI, Kuznetsov AA, et al. Effect of charge states on wettability of polyimide films modified ina low-frequency glow-discharge plasma. High Energy Chem 1996;30:335–8.
[85] Paynter RW. Angle-resolved x-ray photoelectron spectroscopy studies of the evolution of plasma-treated surfaces with time. Surf Interface Anal 1999;27:103–13.
[86] Morra M, Occhiello E, Garbassi F. Surface characterization of plasma-treated PTFE. Surf Interface Anal 1990;16:412–17.
[87] Yasuda T, Okuno T, Yoshida K, Yasuda H. A study of surface dynamics of polymers. II. Investigation by plasma surface implantation of fluorine-containing moieties. J Polym Sci B 1988;26:1781–94.
[88] König U, Nitschke M, Pilz M, Simon F, Arnhold C, Werner C. Stability and ageing of plasma treated poly(tetrafluoroethylene) surfaces. Colloid Surf B 2002;25:313–24.
[89] Touzin M, Chevallier P, Lewis F, et al. Study on the stability of plasma-polymerized fluorocarbon ultra-thin coatings on stainless steel in water. Surf Coat Technol 2008;202:4884–91.
[90] Tanikella RV, Agraharam S, Allen SAB, Hess DW, Kohl PA. Moisture absorption studies of fluorocarbon films deposited from pentafluoroethane and octafluorocyclobutane plasmas. J Electron Mater 2002;31:1096–103.
[91] Tran ND, Dutta NK, Choudhury NR. Weatherability and wear resistance characteristics of plasma fluoropolymer coatings deposited on an elastomer substrate. Polym Degrad Stab 2006;91:1052–63.
[92] Paynter RW. XPS studies of the ageing of plasma-treated polymer surfaces. Surf Interface Anal 2000;29:56–64.
[93] Paynter RW. XPS studies of the modification of polystyrene and polyethyleneterephthalate surfaces by oxygen and nitrogen plasmas. Surf Interface Anal 1998;26:674–81.

Filippo Mangolini and Antonella Rossi

4 Attenuated total reflection-Fourier transform infrared spectroscopy: A powerful tool for investigating polymer surfaces and interfaces

4.1 Principles of Fourier transform infrared spectroscopy

Infrared (IR) spectroscopy is widely used in research and industry as a simple, reliable, and cost-effective technique for the structural elucidation and chemical analysis of both organic and inorganic compounds. One of the great advantages of IR spectroscopy is the possibility of analyzing samples in different states, i.e. liquids, solids, and gases.

IR spectrometers have been available since the 1940s. Although numerous technical developments have been made over the years, the most significant advance in IR spectroscopy results from the introduction of Fourier transform IR (FTIR) spectrometers, which have drastically improved the quality of the spectra, minimizing, at the same time, the spectral acquisition time [1].

The basic principle of IR spectroscopy is the interaction of electromagnetic (EM) radiation in the IR region of the spectrum (wavelength range, 0.78–1,000 μm; wavenumber range, 10–12,820 cm^{-1}) with a molecular system. Such an interaction induces transitions between vibrational levels [1–6].

The basic model for describing molecular vibrations is the harmonic oscillator, which assumes the atoms in a molecule to be spheres of mass m_i and the chemical bonds connecting them as massless springs of spring constant k_i [1–6].

Taking into consideration a diatomic molecule (masses of the two atoms m_1 and m_2), the restoring force of the spring F is proportional to the displacement x of the atoms from their equilibrium position (Hooke's law):

$$F = -kx, \tag{4.1}$$

where k is force constant of the spring (in N/m), which is a measure of bond strength.

The vibrational frequency ($\tilde{\nu}$ wavenumber, in cm^{-1}) of the harmonic oscillator in terms of classical mechanics is then given by:

$$\tilde{\nu} = \frac{1}{2\pi c}\sqrt{\frac{k}{\tilde{m}}}, \tag{4.2}$$

where c is the speed of light and $\tilde{m}$ is the reduced mass ($\tilde{m} = m_1 m_2/(m_1 + m_2)$).

In the case of a harmonic oscillator, only changes of ±1 in the vibrational quantum number are allowed ($\Delta n = \pm 1$). However, the actual variation of the potential energy as a function of the displacement of the atoms from their equilibrium position is described in a more proper way using an anharmonic function. An effect of the anharmonicity is to relax the selection rule (i.e. it allows transitions with $|\Delta n|>1$). Additional peaks, such as overtone and combination bands, can thus appear in the mid-IR spectrum along with absorption bands due to fundamental transitions ($\Delta n = \pm 1$).

The motion of atoms during the vibration is usually described in terms of normal modes of vibration (q_i), in which all atoms in the molecule vibrate with the same frequency and all of them pass through their equilibrium positions simultaneously. To be IR-active, a change in the molecular dipole (μ_j) associated with the normal mode of vibration (IR selection rule) must occur (i.e. $\partial \mu / \partial q_i \neq 0$).

With respect to the direction of the vibrational movement, we may distinguish between stretching and deformation vibrations. The latter can be subdivided into bending modes, twisting and scissoring modes, wagging modes, and rocking modes. Further subdivision refers to the symmetry of the vibration (i.e. symmetric or antisymmetric, in-plane or out-of-plane) (Fig. 4.1).

4.2 Theory of attenuated total reflection FTIR spectroscopy

Since the pioneering work of Harrick [7, 8] and Fahrenfort [9], who independently described the theoretical principles and the experimental applications of attenuated total reflection (ATR) IR spectroscopy, an ever-increasing number of publications, exploiting this technique for a wide variety of applications, has appeared in the literature [1, 8, 10–14]. Several monographs have already been published on the application of ATR/FTIR spectroscopy in different research fields [8, 11, 12, 14, 15], such as polymer science [16], electrochemistry [17], and the science of biological interfaces [18]. This analytical method was also demonstrated to have potential for fundamental *in situ* investigations in other fields of research, where the phenomena occurring at the solid-liquid interface play an important role, such as catalysis [19–22], environmental chemistry [23], and tribology [24–32].

To explain the theory of ATR/FTIR spectroscopy, the physical principles of the propagation of IR radiation through a planar interface between two isotropic media will be first outlined.

4.2.1 Propagation of IR radiation through a planar interface between two isotropic media

When an EM wave propagating in an isotropic medium (medium 1) with refractive index n_1 strikes an interface with a second isotropic medium (medium 2) with refractive index n_2, the wave splits into two components: the reflected and the transmitted (refracted) components (Fig. 4.2) [14, 33].

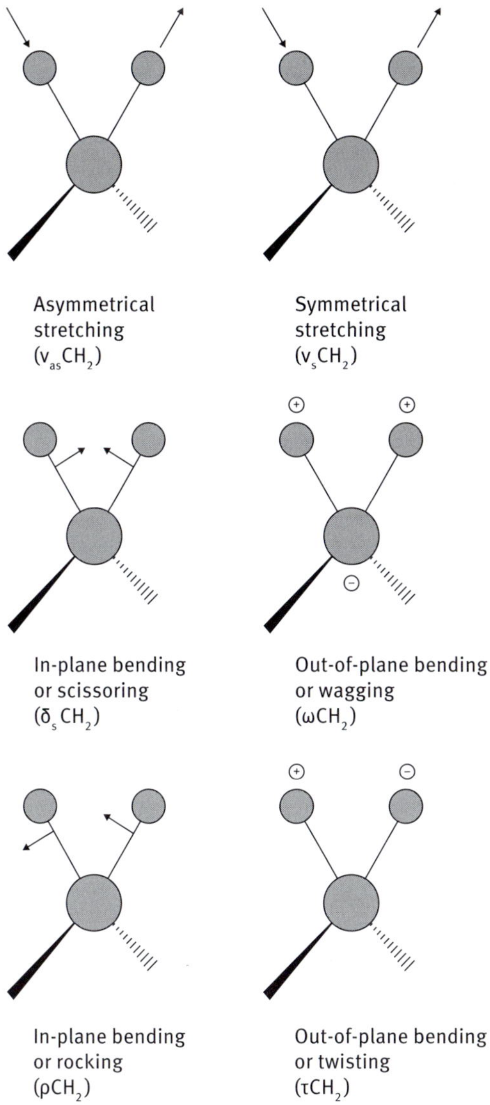

Fig. 4.1: Vibrational modes for a nonlinear group, CH_2 (+ indicates motion from the plane of the page toward the reader; – indicates motion from the plane of the page away the reader) (wavenumber of the characteristic vibrations of the CH_2 group from [1]).

The refraction angle (φ_2) is related to the incidence angle (φ_1) by Snell's law [33]:

$$n_1 \sin \varphi_1 = n_2 \sin \varphi_2, \tag{4.3}$$

where n_1 and n_2 are the refractive indices of the input and output media, respectively.

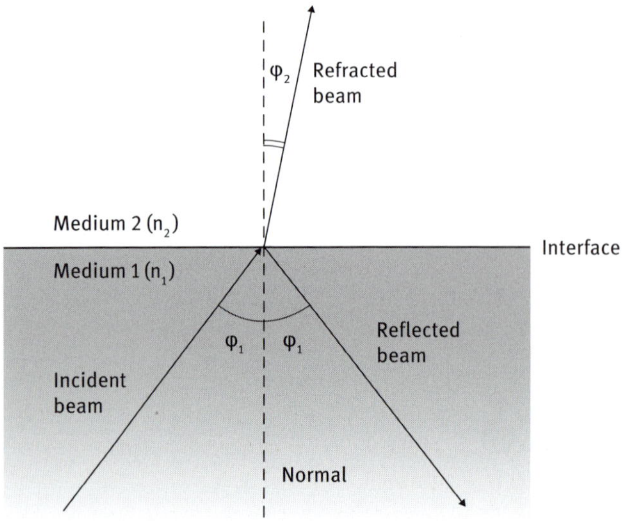

Fig. 4.2: External reflection and refraction of light at interface of two phases ($n_2 > n_1$).

The intensities of the reflected (r) and transmitted (t) beams (for both the s-polarized, in which the electric field is perpendicular to the plane of incidence, and p-polarized component, in which the electric field is parallel to the plane of incidence) can be calculated starting from Maxwell's equations considering the continuity of the electric and magnetic field across the boundary, i.e. the interface. The coefficients, that interconnect the amplitudes of the waves, are the so-called Fresnel amplitude coefficients:

$$r_{12}^{p} = \frac{\tan(\varphi_1 - \varphi_2)}{\tan(\varphi_1 + \varphi_2)}; \qquad t_{12}^{p} = \frac{2n_1/\cos\varphi_2}{n_2/\cos\varphi_2 + n_1/\cos\varphi_1} \tag{4.4}$$

$$r_{12}^{s} = -\frac{\sin(\varphi_1 - \varphi_2)}{\sin(\varphi_1 + \varphi_2)}; \qquad t_{12}^{s} = \frac{2\sin\varphi_2\cos\varphi_1}{\sin(\varphi_1 + \varphi_2)}, \tag{4.5}$$

in which the superscripts s and p refer to the polarization of the light.

If the angle of incidence exceeds the critical angle (φ_c), total internal reflection occurs.

$$\varphi_c = \arcsin\left(\frac{n_2}{n_1}\right) \tag{4.6}$$

Upon inspecting the optical field near the interface, a tail of the EM field is found to penetrate into the rarer medium as an inhomogeneous wave.

$$E \propto \exp\left[i\left(\omega t - k_x x + k_z z\right)\right] \tag{4.7}$$

The analysis of the wavevector, which can be used for describing such a wave, shows that the component along the vertical direction (k_z) is imaginary, implying that the electric field vector decays exponentially in the optically rarer medium, as depicted in Fig. 4.3. Because of that, this wave is usually referred to as *evanescent*, from the Latin *evanescere*, meaning "to vanish."

Starting from the Fresnel coefficients (Eq. 4.4 and Eq. 4.5), it can be demonstrated that the radiation energy is totally reflected regardless of the polarization state. However, the Fresnel amplitude transmission coefficients ($t^{s,p}$) have nonvanishing values, and the wavevectors of the components parallel to the interface are real, indicating the presence of an energy flow parallel to the boundary between the two phases. As a consequence of this energy flow, the reflected wave is laterally displaced (Goos-Hänchen shift).

However, when the contacting media are absorbing (i.e. they have a nonzero extinction coefficient, $k \neq 0$), the Fresnel coefficients have to be adapted by replacing the real refractive indices with the complex ones ($\hat{n} = n + ik$):

$$r_{12}^{p} = \frac{\hat{\varepsilon}_2 \xi_1 - \hat{\varepsilon}_1 \xi_2}{\hat{\varepsilon}_2 \xi_1 + \hat{\varepsilon}_1 \xi_2}; \qquad t_{12}^{p} = \frac{2\hat{n}_1 \hat{n}_2 \xi_1}{\hat{n}_2^2 \xi_1 + \hat{n}_1^2 \xi_2} \tag{4.8}$$

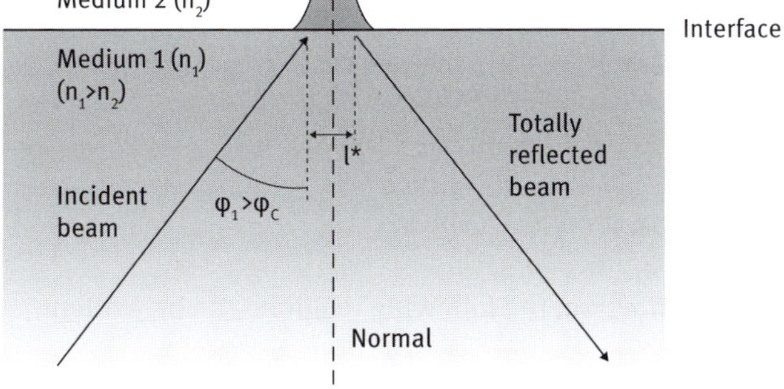

Fig. 4.3: Schematic representation of the physical principle of total internal reflection of light at the interface between two media ($n_1 > n_2$). Here, l^* denotes the Goos-Hänchen shift.

$$r^s_{12} = \frac{\xi_1 - \xi_2}{\xi_1 + \xi_2}; \qquad t^s_{12} = \frac{2\xi_1}{\xi_1 + \xi_2}, \tag{4.9}$$

where $\hat{\varepsilon}_1$ and $\hat{\varepsilon}_2$ are the permittivity values of the input and output medium, respectively, and ξ_i is the generalized complex index of refraction:

$$\xi_i = \hat{n}_i \cos\varphi_i = \sqrt{\hat{n}_i^2 - \hat{n}_1^2 \sin^2\varphi_1}, \tag{4.10}$$

where φ_i is the complex angle of refraction.

Because the intensity of the evanescent wave, and consequently of the reflected radiation, is attenuated at the wavelengths of the absorption of the optically rarer medium, the name *attenuated total reflection* is usually used for indicating such a physical principle. By exploiting this phenomenon, the layer next to the interface can then be probed.

4.2.2 Propagation of IR radiation through stratified media

The propagation of IR radiation through stratified media is of interest for many engineering applications [14]. In the limiting case of an inhomogeneous (in depth) layer, the propagation of an EM wave could be investigated by slicing the layer into several infinitely thin layers, which are assumed to be homogeneous [14].

Abelès and Hansen [14, 34, 35] developed a matrix method that allows the calculation of the overall reflectance of any combination of absorbing and non-absorbing isotropic layers at any angle of incidence as well as of the mean square electric field (MSEF) at any position within a certain layer of an N-isotropic-phase system (Fig. 4.4).

The basic concept of this matrix method involves the construction of a characteristic matrix M_i for each layer ($z_{j-1} < z < z_j$) and for the first and last layers ($z > z_{N-1}$), being assumed to be semi-infinite and non-absorbing. From the physical point of view, this characteristic matrix M_i for each layer relates the tangential amplitudes of the electric and magnetic field vectors at the input and output film boundaries (for both s- and p-polarized light).

$$M^s_j = \begin{pmatrix} \cos\beta_j & -\frac{i}{p_j}\sin\beta_j \\ -ip_j \sin\beta_j & \cos\beta_j \end{pmatrix}; \qquad M^p_j = \begin{pmatrix} \cos\beta_j & -\frac{i}{q_j}\sin\beta_j \\ -iq_j \sin\beta_j & \cos\beta_j \end{pmatrix} \tag{4.11}$$

By defining the generalized complex index of refraction (ξ_i) as:

$$\xi_i = \hat{n}_i \cos\varphi_i = \sqrt{\hat{n}_i^2 - \hat{n}_1^2 \sin^2\varphi_1}, \tag{4.12}$$

where φ_i is the complex angle of refraction, the following relations can be written:

$$p_j = \xi_j; \qquad q_j = \frac{\xi_j}{\hat{n}_j^2}; \qquad \beta_j = 2\pi\left(\frac{d_j}{\lambda}\right)\xi_j, \tag{4.13}$$

where d_j is the thickness if the j-th layer and λ is the wavelength.

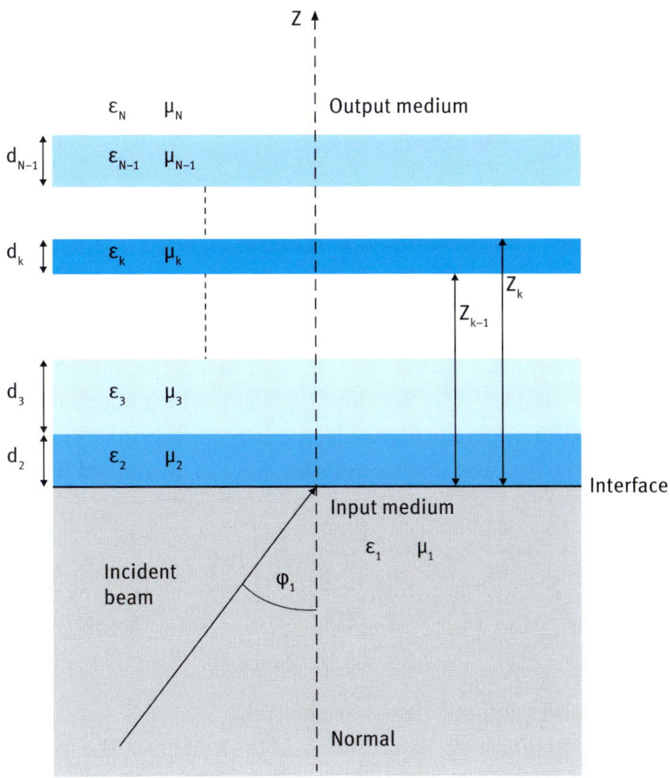

Fig. 4.4: Scheme of a layered system with N phases and $N-2$ interfaces. ε and μ are the permittivity and permeability of the material in the n-th layer, respectively.

The characteristic matrix of the whole multilayer structure is then calculated as the product of the elementary matrices:

$$\hat{M} = M_2 M_3 M_4 \ldots.. M_{N-1} = \begin{pmatrix} m_{11} & m_{12} \\ m_{21} & m_{22} \end{pmatrix} \tag{4.14}$$

Using the elements of the characteristic matrix of the overall system, the Fresnel amplitude reflection and transmission coefficients of the N-isotropic-phase system can be obtained:

$$r^s = \frac{(m_{11} + m_{12}p_N)p_1 - (m_{21} + m_{22}p_N)}{(m_{11} + m_{12}p_N)p_1 + (m_{21} + m_{22}p_N)}; \quad r^p = \frac{(m_{11} + m_{12}q_N)q_1 - (m_{21} + m_{22}q_N)}{(m_{11} + m_{12}q_N)q_1 + (m_{21} + m_{22}q_N)} \tag{4.15}$$

$$t^s = \frac{2p_1}{(m_{11} + m_{12}p_N)p_1 + (m_{21} + m_{22}p_N)}; \quad t^p = \frac{2q_1}{(m_{11} + m_{12}q_N)q_1 + (m_{21} + m_{22}q_N)} \tag{4.16}$$

The overall reflectance and transmittance of the stratified medium can be finally derived as:

$$R^{s,p} = |r^{s,p}|^2 \tag{4.17}$$

$$T^s = \frac{\mu_1 \operatorname{Re}(\hat{n}_N \cos\varphi_N)}{\mu_N n_1 \cos\varphi_1}|t^s|^2; \qquad T^p = \frac{\mu_N \operatorname{Re}(\hat{n}_N \cos\varphi_N/\hat{n}_N^2)}{\mu_N n_1 \cos\varphi_1/n_1^2}|t^p|^2. \tag{4.18}$$

The matrix formalism developed by Hansen could also be used for calculating the MSEF at any position within a certain layer of a N-isotropic-phase system [14, 35].

To evaluate the MSEF at a position z within the k-th layer of a N-isotropic-phase system, the column vectors $Q_1^{s,p}$ and $Q_{N-1}^{s,p}$ (Eq. 4.19 and Eq. 4.20), which characterize the amplitudes of the tangential components of the electric (s-polarization) and magnetic (p-polarization) fields at the first and last boundaries, have to be defined first.

$$Q_1^s = \begin{pmatrix} E_{y1}^0 \\ H_{x1}^0 \end{pmatrix} = \begin{pmatrix} 1 + r^s \\ p_1(1 - r^s) \end{pmatrix}; \qquad Q_{N-1}^s = \begin{pmatrix} E_{y(N-1)}^0 \\ H_{x(N-1)}^0 \end{pmatrix} = \begin{pmatrix} t^s \\ p_N t^s \end{pmatrix} \tag{4.19}$$

$$Q_1^p = \begin{pmatrix} H_{y1}^0 \\ E_{x1}^0 \end{pmatrix} = \begin{pmatrix} 1 + r^p \\ q_1(1 - r^p) \end{pmatrix} n_1; \qquad Q_{N-1}^p = \begin{pmatrix} H_{y(N-1)}^0 \\ E_{x(N-1)}^0 \end{pmatrix} = \begin{pmatrix} t^p \\ q_N t^p \end{pmatrix} n_1, \tag{4.20}$$

where E and H are the electric and magnetic fields, respectively.

For calculating the MSEF magnitude at the position z of the k-th layer, the matrix $N_j^{s,p}$, which is the reciprocal of the characteristic matrixes of the same layers $M_j^{s,p}$, should be generated. For the k-th phase, the matrix $N_k^{s,p}$ is constructed as:

$$N_k^s(z) = \begin{bmatrix} \cos[2\pi\nu\xi_k(z - z_{k-1})] & \frac{i}{p_k}\sin[2\pi\nu\xi_k(z - z_{k-1})] \\ ip_k \sin[2\pi\nu\xi_k(z - z_{k-1})] & \cos[2\pi\nu\xi_k(z - z_{k-1})] \end{bmatrix}; \tag{4.21}$$

$$N_k^p(z) = \begin{bmatrix} \cos[2\pi\nu\xi_k(z - z_{k-1})] & \frac{i}{q_k}\sin[2\pi\nu\xi_k(z - z_{k-1})] \\ iq_k \sin[2\pi\nu\xi_k(z - z_{k-1})] & \cos[2\pi\nu\xi_k(z - z_{k-1})] \end{bmatrix}. \tag{4.20}$$

For the final phase ($k = N$), the resultant vectors $Q_N^{s,p}(z)$ are defined as (at $z \geq z_j = \sum_{h=2}^{N-1} d_h$):

$$Q_N^{s,p}(z) = N_N^{s,p}(z) Q_{N-1}^{s,p} \tag{4.23}$$

By defining the resultant vector $Q_k^{s,p}(z)$ as

$$Q_k^{s,p}(z) = \begin{pmatrix} U_k^{s,p}(z) \\ V_k^{s,p}(z) \end{pmatrix} \tag{4.24}$$

and introducing the quantity $W_k(z) = n_1 \sin\varphi_1 U_k^p(z)/\hat{\varepsilon}_k$ ($\hat{\varepsilon}_k$ complex dielectric constant), the MSEF magnitude can be calculated as:

$$\langle E_k^{s2}(z)\rangle = \langle E_{yk}^2(z)\rangle = |U_k^s(z)|^2 \tag{4.25}$$

$$\langle E_k^{p2}(z)\rangle = \langle E_{xk}^2(z)\rangle + \langle E_{zk}^2(z)\rangle = |V_k^p(z)|^2 + |W_k^p(z)|^2 \tag{4.26}$$

The formalism developed by Abelès and Hansen [36] has also been generalized for anisotropic layers, making it possible to carry out simulations of uniaxial ($\hat{n}_x = \hat{n}_y \neq \hat{n}_z$) and biaxial ($\hat{n}_x \neq \hat{n}_y \neq \hat{n}_z$) layers.

The matrix method described (Section 4.2.2) was demonstrated to be a powerful tool for clarifying unknown optical behaviors: by numerically calculating the dependence of the reflectance on various parameters, such as angle of incidence, layer thickness, electronic configuration, and optical constants, physico-chemical changes occurring at the interface between the internal reflection element (IRE) and the outer medium could be explained [21, 22, 30, 35].

4.2.3 Penetration depth and effective thickness

As pointed out in Section 4.2.1, by solving Maxwell's equations while considering the continuity of the electric and magnetic field across the interface, an EM wave is found to penetrate into the rarer medium and its electric field amplitude falls off exponentially with the distance from the boundary [8, 11, 12, 14]:

$$E = E_0 \exp(-\gamma z), \tag{4.27}$$

where E_0 is the electric field amplitude at the interface ($z = 0$) and γ is the exponential constant (also called electric field amplitude decay coefficient), which is defined as:

$$\gamma = \frac{2\pi n_1 \sqrt{\sin^2\varphi - n_{21}^2}}{\lambda}, \tag{4.28}$$

where φ is the angle of incidence.

The quantity d_p, usually referred to as penetration depth, was introduced by Harrick [8] to assess the penetration of the evanescent wave in the rarer phase. This parameter is defined as the distance along the normal to the boundary between the two media required for the electric field amplitude to fall to e^{-1} of its value at the interface (E_0) and is given by:

$$d_p = \frac{1}{\gamma} = \frac{\lambda_1}{2\pi n_1 \sqrt{\sin^2\varphi - n_{21}^2}}, \tag{4.29}$$

where λ_1 is the wavelength in the denser medium and n_{21} is the ratio of the refractive index of the rarer medium divided by that of the denser phase ($n_{21} = n_2/n_1$). Note that some authors have defined the penetration depth by replacing the factor 2 in the denominator of Eq. 4.29 by a factor 4 [10, 37].

The penetration depth divided by the wavelength λ_1 is plotted as a function of the angle of incidence in Fig. 4.5 for a number of interfaces (i.e. different n_{21} values) [8, 38]. The depth of penetration increases as the angle of incidence decreases and becomes infinitely large as φ approaches the critical angle φ_c. At a fixed angle, d_p is larger for better index matching, i.e. as $n_{21} \to 1$. As a consequence of the wavelength dependence of the penetration depth, the bands at longer wavelengths tend to be relatively more intense than those at shorter wavelengths. Moreover, a broadening of the absorption bands on their high-wavelength side together with a shift to longer wavelengths is usually observed in comparison with transmission measurements.

According to Harrick, even if the reflection loss due to the interaction of the evanescent wave with the rarer medium could be exactly calculated starting from either Maxwell's or Fresnel's equation, such calculations do not provide any physical insight into the absorption mechanism or into the interaction of the penetrating EM field with the absorbing phase [8]. Because of this, the concept of the effective thickness (d_e) was introduced to express the strength of the coupling of the exponentially decaying wave with the rarer phase and to compare internal reflection spectra to the one acquired in transmission mode.

In conventional transmission IR spectroscopy, the Beer-Lambert (BL) law is used to relate the transmittance (T) to the sample thickness (d) [1]:

$$T = \frac{I}{I_0} = \exp(-\alpha d), \tag{4.30}$$

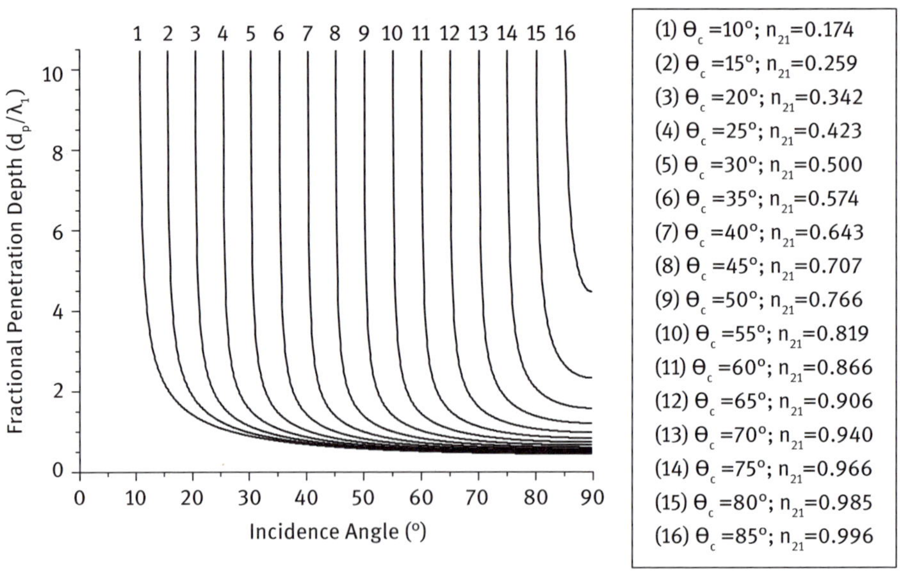

Fig. 4.5: Fractional penetration depth of EM field in an optically rarer medium in the case of total internal reflection *vs* angle of incidence for a number of interfaces [8, 38].

where I and I_0 are the transmitted and incident intensity, respectively, and α is the absorption coefficient. Under the assumption of low absorption ($\alpha d < 0.1$), the BL law can be expressed as:

$$T = \frac{I}{I_0} \cong 1 - \alpha d \tag{4.31}$$

In a similar way, the reflectivity in internal reflection spectroscopy can be written as:

$$R = 1 - \alpha d_e, \tag{4.32}$$

where d_e is the effective thickness. For multiple reflections (N number of reflections), the reflected power is given by:

$$R^N = (1 - \alpha d_e)^N. \tag{4.33}$$

In the case of low absorptions ($\alpha d_e << 1$),

$$R^N \cong 1 - N\alpha d_e, \tag{4.34}$$

i.e. the reflection loss is increased by the number of reflection N.

The comparison of the expressions obtained in case of the low-absorption approximation for transmission (Eq. 4.31) and reflection (Eq. 4.34), where it should be noted that the same absorption coefficient α appears, clearly suggests that the effective thickness represents the actual thickness of a film that would be required in a transmission measurement to measure an absorption value equal to that obtained in a reflection acquisition using a semi-infinite bulk sample [8].

The expressions derived by Harrick, which are different in the case of a semi-infinite bulk material and in the case of a thin film, not only give an insight into the nature of the interaction of the EM wave in internal reflection measurements but are helpful for choosing the angle of incidence and the refractive index of the IRE.

In the case of bulk materials, the sample thickness is much larger than the penetration depth, implying that the electric field exponentially falls to a very low value within the rarer medium. The low-absorption approximation for the effective thickness can then be calculated from the electric fields for zero absorption:

$$d_e = \frac{n_{21}}{\cos\varphi}\int_0^{\infty} E^2 dz = \frac{n_{21}E_0^2 d_p}{2\cos\varphi}, \tag{4.35}$$

where the electric field E in the rarer medium is expressed as in Eq. 4.27 and a unit electric field amplitude is assumed for the incident wave in the denser phase.

Because of the dependency of the electric field amplitude in the rarer medium on the polarization of the EM wave, the effective thickness for parallel and perpendicular polarization turn out to be different. Equation 4.35 also shows that the strength of the coupling of the evanescent wave to the absorbing rarer medium is determined by the penetration depth, which is proportional to the wavelength. Therefore, the effective thickness depends on the wavelength of the radiation as well, which explains the appearance of more intense absorption bands at longer wavelengths. Two peaks having the same intensity in transmission spectra will have unequal intensities in internal reflection spectra. Such a wavelength dependence of the effective thickness also results in greater absorption on the high-wavelength side of a certain band, which then appears distorted.

In the case of thin films, the penetration depth is much higher than the layer thickness (d), i.e. $d_p >> d$. As a consequence of this, the electric field of the radiation can be assumed to be constant over the film thickness, and the corresponding effective thickness can be expressed as:

$$d_e = \frac{n_{21} E_0^2 d}{\cos\varphi}. \tag{4.36}$$

By comparing Eqs. 4.35 and 4.36, it is evident that the same factors controlling the effective thickness for bulk materials control the effective thickness for thin films, except the penetration depth. Because d_e, in the case of thin films, is not dependent on the penetration depth and, therefore, is not proportional to the wavelength, the internal reflection spectra of thin layers exhibit neither stronger absorption bands at longer wavelengths nor peaks distorted on the longer-wavelength side. Internal reflection spectra of thin films resemble those obtained *via* transmission mode more closely than do internal reflection spectra of bulk materials.

The equations proposed by Harrick for the penetration depth (Eq. 4.29) and for the effective thickness (Eq. 4.35 and Eq. 4.36) were defined in the case of total internal reflection, i.e. under the assumption of non-absorbing media ($k = 0$). However, it has been demonstrated that these simplified expressions can be satisfactorily applied to weakly absorbing materials, such as organic molecules [8, 11]. Besides Harrick, Hansen [39] also derived an expression for the effective thickness on the basis of a Taylor expansion of Fresnel's equations in the absorption coefficient. The resulting equations were found to be a good approximation, which, however, was again valid only for weakly absorbing materials.

In the case of ATR, i.e. in the presence of an absorbing phase, a new parameter (d'_p), which describes the exponential decay of the EM wave in the rarer medium, could be defined upon inserting the complex index of refraction ($\hat{n} = n + ik$) into the characteristic equation of the evanescent wave (Eq. 4.29) [40–42]:

$$d'_p = \frac{\lambda}{2\pi n_1} \frac{1}{\omega_2^+}, \tag{4.37}$$

where

$$\omega_2^{\pm} = \frac{1}{\sqrt{2}} \sqrt{\sqrt{\nu_2^2 + \mu_2^2} \pm \nu_2}$$

$$\nu_2 = \sin^2\varphi - n_{21}^2(1 - \kappa_2^2),$$

$$\mu = 2n_{21}^2\kappa_2 \tag{4.38}$$

and κ is the absorption index ($\kappa = k/n$).

The values of d_p and d_p' were found to be roughly the same for angles of incidence well above the critical angle (φ_c). In contrast, although the depth of penetration defined as in Eq. 4.29 increases infinitely at the critical angle, d_p' stays finite.

Even if the foregoing equations were defined to establish if a system could be described as consisting of two or three layers, the formalism developed by Müller and colleagues [40–42] was based on the assumption that the second and third layers have lower refractive index than the IRE. Therefore, internal reflection occurs at the interface between the first/denser (IRE) phase and the second one.

In a similar way to the penetration depth, Müller and colleagues [40–42] also suggested different expressions for the effective thickness in the case of absorbing samples.

4.2.3.1 Penetration depth in ATR/FTIR spectroscopic analysis of polymers

In the following, the penetration depth of the evanescent wave in a typical ATR/FTIR experiment will be discussed. A model system, consisting of a germanium ATR crystal (refractive index $n_{Ge} = 4.0$) and a bulk specimen with refractive index ($n_{specimen}$) equal to 1.46 (a typical value for the refractive index of polymers [43]), is considered in the calculations (Fig. 4.6a). The penetration depth (d_p), calculated using the formula proposed by Harrick (Eq. 4.29) [8], is displayed as function of the wavenumber in Fig. 4.6b. For the model system considered in the calculations, d_p increases from 164 nm at 4,000 cm^{-1} to 1,095 nm at 600 cm^{-1}. This strong wavelength dependence of the penetration depth with the incidence angle results in clear differences between ATR/FTIR and FTIR transmission spectra. Section 2.4 will discuss in details the differences between the outcomes of transmission FTIR and ATR experiments. As a consequence of the wavelength dependence of the penetration depth of the evanescence wave, the surface sensitivity in ATR/FTIR spectroscopy is not constant over the typical mid-IR region, a factor to be considered when interpreting ATR/FTIR results.

As pointed out in Section 4.2.2, the matrix formalism developed by Hansen [14, 35] could also be used for evaluating the MSEF at any position within a certain layer of a N-isotropic-phase system (Eq. 4.11 through Eq. 4.26).

In the case of a two-layered system (Ge ATR crystal/specimen), the MSEF magnitude can be numerically calculated at different positions within the rarer medium (i.e. the specimen under investigation) and along the normal to the boundary between the two phases (z-axis). By fitting the MSEF values using an exponential function ($E = E_0\exp(-kz)$), where the fitting parameters are the decay constant (k) and the electric field amplitude at the interface (E_0), the penetration depth can

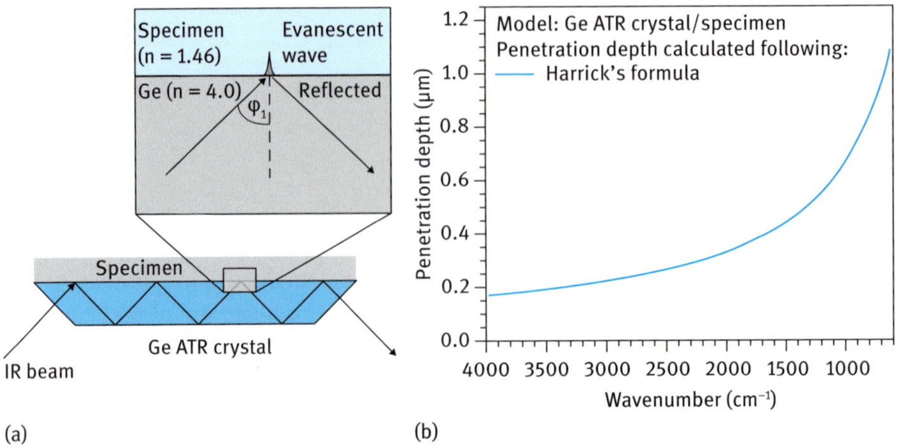

Fig. 4.6: (a) Schematic of the two-layered system (Ge ATR crystal/specimen). (b) Penetration depth (μm) as a function of the wavenumber in the case of a two-layered system (Ge ATR crystal/specimen). In the calculations a specimen with refractive index equal to 1.46, a typical value for the refractive index of polymers [43], was considered. The d_p values were computed using the formula introduced by Harrick [8] (Eq. 4.29).

be determined ($d_p = k^{-1}$). Because in the case of *p*-polarized light the calculation of the MSEF magnitude requires the knowledge of the complex dielectric constant, the MSEF magnitude was only evaluated for *s*-polarized radiation. The penetration depth values, determined using the aforementioned approach for the case of a two-layered system consisting of a Ge ATR crystal and a specimen with a bulk refractive index equal to 1.46, are in good agreement with those calculated by the formula proposed by Harrick (Eq. 4.29) (Fig. 4.7).

4.2.4 Transmission FTIR *vs* ATR/FTIR spectroscopy

Since the development of the first IR spectrometers, transmission-sampling techniques have been widely used for acquiring IR spectra, thanks to their simplicity and to the introduction of few artifacts into the spectra [1]. Consequently, most of the published tables reporting characteristic group frequencies and most of the commercially available software libraries are derived from transmission FTIR experiments.

In the last decades, ATR/FTIR has become the most common sampling technique in FTIR spectroscopy, as demonstrated by the profusion of ATR accessories on the market. The IR spectra obtained by the ATR/FTIR technique turned out to be not identical to those obtained by transmission. Studying the physical principles of ATR/FTIR spectroscopy helped in understanding the resulting relative shifts in band intensity and absolute shifts in frequency, which characterize the ATR/FTIR spectra. The interpretation of these spectra on the basis of published frequency

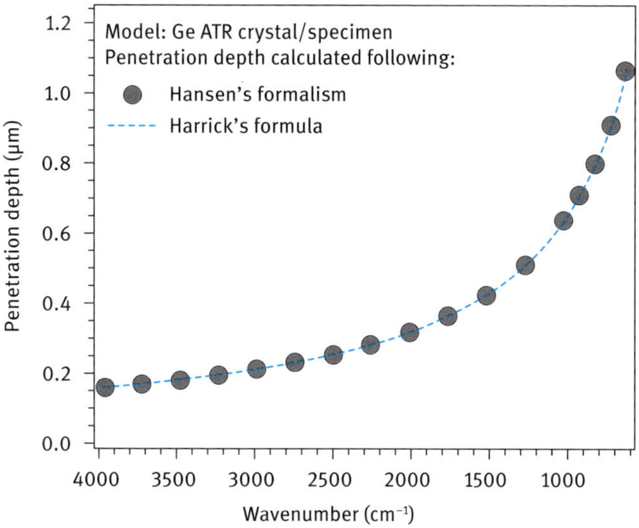

Fig. 4.7: Penetration depth (μm), calculated using the formalism developed by Hansen (Eq. 4.11 through 4.26) [14, 35], as a function of the wavenumber in the case of a two-layered system (Ge ATR crystal/specimen with refractive index equal to 1.46). The d_p values calculated using the formula introduced by Harrick (Eq. 4.29) [8] are included for comparison.

tables or software libraries might then be misleading. To overcome such a problem, most spectroscopic software packages provide algorithm for correcting the intensity and frequency shifts [11, 44]. In the following, the differences between ATR/FTIR and transmission FTIR spectra are outlined and discussed in detail.

As pointed out in Section 4.2.1, Harrick introduced the concept of the effective thickness (d_e) to express the strength of the coupling of the exponentially decaying wave within the rarer phase in ATR/FTIR spectroscopy and to compare internal reflection spectra with those acquired in transmission mode [8]. In the case of bulk materials, the effective thickness is proportional to the penetration depth (Eq. 4.35). As a consequence of the wavelength dependence of the penetration depth, the effective thickness increases with the wavelength, resulting in the appearance of more intense absorption bands at longer wavelengths (lower wavenumbers). Two peaks having the same intensity in transmission spectra will have unequal intensities in internal reflection spectra. Such a wavelength dependence of the effective thickness also induces a broadening of the absorption bands on their high-wavelength side together with a shift to longer wavelengths [8]. Conversely, in the case of thin films, because the penetration depth does not affect the effective thickness (Eq. 4.36), no relative distortion or shift of the absorption bands is observed. The resulting ATR/FTIR spectra of thin films closely resemble those acquired by transmission [8].

Dispersion effects (see Section 2.3.1) should also be considered when comparing ATR/FTIR to transmission FTIR spectra. In ATR/FTIR spectroscopy, the variation of the refractive index of a sample near am absorption band results in the detection of absorption bands whose intensity appears higher than it should normally be, in distortions of the shape of the peaks and in small displacements of the band maximum. Such effects become more pronounced for molecular vibrations showing strong absorption bands (i.e. high values of the extinction coefficient) [13] and for incident angles close to the critical value. To minimize the distortion of the peaks appearing in ATR/FTIR spectra, incident angles well above the critical angle should be used [45] (see Section 4.3.1 for more details).

4.3 Experimental methods in ATR/FTIR spectroscopy

4.3.1 Internal reflection elements

Internal reflection elements (IREs), or ATR crystals, are optical elements transparent in the IR region (or in limited ranges of the IR region of the EM spectrum) used in ATR/FTIR spectroscopy for establishing the conditions necessary to obtain ATR/FTIR spectra of the materials under investigation [8, 11, 12].

The IR radiation, which propagates in the IRE through total internal reflection, interacts with the sample, which is in contact with the reflecting surface.

The acquisition of an ATR/FTIR spectrum and the information provided by its acquisition strongly depends on the characteristics of the IRE, i.e. angle of incidence, number of reflections, aperture, number of passes, surface preparation, and the material out of which the IRE is constructed.

The number of reflections is calculated from simple geometrical considerations. The so-called skip distance, equal to $t \tan \varphi$ (where t is the thickness of the IRE and φ is the angle of incidence), defines the distance the beam advances for each reflection. The total number of reflections for a single pass IRE is given by:

$$N = \frac{l}{t} \cot \varphi, \tag{4.39}$$

where l is the IRE length, measured, in the case of a single-pass crystal, from the center of the entrance aperture to the center of the exit aperture.

A number of IREs have been developed and reported in the literature. The geometry of the reflection element is crucial because it determines the ease with which ATR/FTIR spectra can be obtained. A thorough discussion of the different types of IREs with their uses, advantages, and disadvantages can be found in [8].

Because in the case of thin films, particularly monomolecular layers, single-reflection IREs do not provide spectra of sufficient contrast, multiple-reflection elements are commonly used. Among the different geometries, single-pass (i.e. where the incident IR radiation enters the plate, propagates *via* multiple internal reflections, and leaves through the opposing aperture) trapezoids and parallelograms with an angle of incidence of 45° are the most widely used IREs.

Besides geometrical issues, the choice of the IRE for a specific analytical study should be carried out, considering the material the crystal is made of and the surface preparation. As for the former, transparency, hardness, brittleness, conductivity, and chemical inertness are all factors that play a fundamental role in the accessibility of an ATR crystal, which is suitable for a specific investigation. The material types available include glass-like materials, pressed ones, polycrystalline, and high purity single crystals. The latter are mainly required for surface studies because grain boundaries in polycrystalline materials may result in radiation losses. Hardness and brittleness play an important part in the ease of preparation of the IRE as well as in the resistance to surface damage. Conductivity and chemical inertness are two additional factors that have to be taken into consideration in the determination of the species that can be studied.

When choosing the optical material for an IRE, the optical properties of the analyte should be also carefully considered. In fact, it has to be emphasized that the refractive index of a medium strongly varies in the vicinity of an absorption band [8, 13, 33, 45]. This variation is known as dispersion. Due to such a change of the real part of the refractive index of the phase under investigation in correspondence of an absorption band, it may happen that some peaks do not satisfy the total reflection condition (Eq. 4.6). This would results in bands whose intensity appears higher than it should normally be, in distortions of the shape of the peaks, and in small displacements of the band maximum. Because of the foreseeable effects of dispersion on ATR/FTIR spectra, ATR crystals with a refractive index considerably higher than that of the sample should be chosen [8, 13, 45].

Because a change in the real part of the refractive index of the sample under investigation induces a variation of the critical angle (Eq. 4.6), dispersion effects in ATR/FTIR spectra are dependent on the incident angle of the IR radiation. To minimize such effects, angles of incidence well above the critical angle should be used [45]. The rule of thumb is $\varphi \geq \varphi_c + 15°$.

There are several reasons that can contribute to the poor performance of an ATR crystal such as poor optical material (defects in the quality of the material), inadequate surface polishing (which causes scattering losses), poor tolerance on lengths and angles, and non-parallelism of the reflecting surfaces. A procedure for checking the quality of an IRE is suggested in [38]. In the last few years, the input-output apertures of ATR crystals are coated by an anti-reflection material to enhance their optical efficiency [14].

4.3.2 Internal reflection attachments

As in the case of other reflection-absorption IR spectroscopic techniques, the acquisition of ATR/FTIR spectra requires special accessories that are mounted in the sample compartment of a commercial spectrometer and that hold the IRE. The optical design of an internal reflection attachment (IRA) is primarily determined by the configuration of the IRE. As a consequence of this, the following types of IRA are usually distinguished: (1) variable-angle single internal reflection; (2) fixed-angle multiple internal reflection; (3) variable-angle multiple internal reflection [38]. All the accessories contain two systems of mirrors: the first one focuses and directs the IR beam into the IRE and the second one collects the reflected radiation as it leaves the IRE and directs it toward the IR detector.

The principal requirements for ATR/FTIR accessories are the high accuracy in setting the angle of incidence and the low level of radiation losses. To simplify routine analysis of bulk samples, ATR accessories with optimized optical configurations have been developed over the last few years. Although these attachments turn out to be the IR analytical tool of choice for investigating bulk samples due to the predetermined angles of incidence, they are less effective for studying thin films.

The major problem of ATR spectroscopy is to achieve an optimal optical contact between the IRE and the sample. Among the various approaches that have been applied to this problem [14], the following ones are the most commonly used:

- The simplest and most common approach involves an IRE made from the material of the film substrate. This is possible for dielectrics (e.g. sapphire, calcite) and weakly doped semiconductors (e.g. germanium, silicon).
- When the IRE of a particular material cannot be made, it has become common practice to coat the IRE with a film of the desired material (by means, for example, radiofrequency/direct current magnetron reactive sputtering).
- In the case of liquids, liquid cells are commercially available, allowing the IRE/solution interface to be continuously monitored.
- For polymer surfaces, the optical contact can be improved by pressing the sample against the IRE.

4.3.3 Metal underlayer ATR/FTIR spectroscopy

Despite the plethora of analytical techniques that can be used for the study of surface chemistry [46], relatively little attention has been paid, over the years, to the development of spectroscopic methods that allow the investigation of the chemical processes occurring at solid-liquid interfaces. Vibrational spectroscopy, providing information about the metal-adsorbate bonds, the local orientation of the molecules, and intermolecular interaction within the adsorbate layer, has been a powerful tool in the study of solid-liquid interfaces [14].

Recently, ATR/FTIR spectroscopy proved to be a suitable analytical method for probing solid-gas and solid-liquid interfaces *in situ* [15, 20, 21, 47]. Because common technologically relevant materials are generally not suitable for IREs, thin metal films deposited by physical vapor deposition or sputtering onto the IRE, have become an

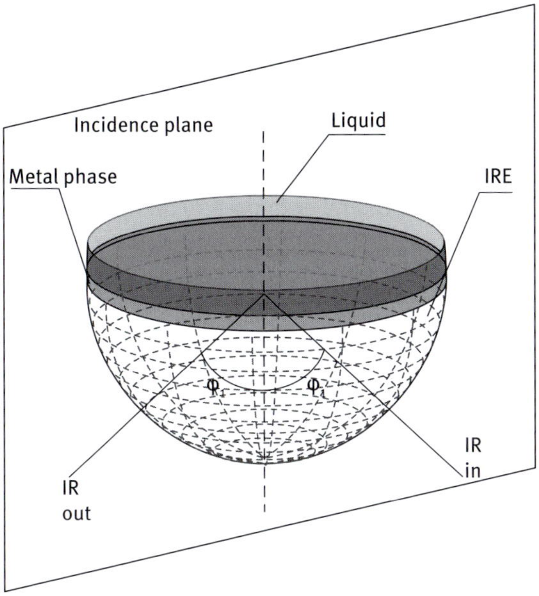

Fig. 4.8: Schematic representation of the MUATR configuration.

important type of model system. The combination of an IRE and such a thin metal film is usually referred to as metal underlayer (Kretschmann) ATR (MUATR) configuration [14] (Fig. 4.8).

Reproducibility of film preparation and mechanical stability of the resulting metal layer are two fundamental issues for practical applications. Cleaning the IRE before depositing any metal film plays a decisive role in determining reproducibility. Despite this, metal layers deposited on IREs may be still mechanically unstable and peel off [48, 49]. To overcome this problem, a metal oxide support material, such as Al_2O_3 (which has a low absorption coefficient in the mid-IR region), is commonly evaporated first. Such a deposition also results in covering or displacing the contaminants, which might be present on the IRE surface. The presence of a 50- to 100-nm-thick Al_2O_3 layer, deposited by electron beam physical vapor deposition, on a Ge IRE was found not to affect the reflectivity of the ATR system or enhance the stability of the overlying platinum film under catalytic reaction conditions [21].

The metal film deposited onto the IRE should not be too thick, so that the electric field is not completely attenuated by it and the EM wave can still reach the metal-liquid or metal-gas interface [21]. In the case of a Ge/Pt/vacuum system, it has been shown that for increasing platinum layer thicknesses, the reflectance first decreases, reaches a minimum, and then starts increasing [21]. In the latter regime, the film acts as a mirror and is therefore useless for spectroscopy. It has to be emphasized that, in the case of multiple reflection elements, the total reflectivity is the reflectivity of a single reflection to the power of the number of reflections N ($R_{TOT} = R^N$). Meanwhile, the intensity of an absorption band linearly increases with the number of reflections ($A_{TOT} = AN$). Hence, the experimental setup should be optimized to attain a suitable signal-to-noise ratio.

The deposition of thin films onto IREs can also result in the formation of inhomogeneous layers composed of islands smaller than the wavelength of the radiation concerned. The morphology of these layers strongly depends on the substrate (IRE), the metal and the deposition conditions (temperature, background gases, etc.) [21, 50, 51]. These metal films consisting of islands have been widely used in IR spectroscopy because the intensity of the characteristic absorption bands of molecules in contact with the metal can be considerably enhanced compared with the signals in the absence of the metal film. This effect is commonly referred to as surface-enhanced IR absorption (SEIRA) [50, 51]. Since the pioneering work of Hartstein *et al.* [52] in the early 1980s, the SEIRA phenomenon has been exploited in chemical and biochemical IR sensors in the study of electrode-electrolyte interfaces, Langmuir-Blodgett (LB) films, and self-assembled monolayers [14, 50, 51]. Over the years, the same effect has been reported with materials different from those used by Hartstein *et al.* [52] (Au and Ag), such as Cu, In, Li, Sn, Pb, Fe, Pt, Pd, and Pt-Fe alloys [51]. The enhancement factor was, however, found to strongly depend on the morphology of the islands, their size and shape, the spacing between the islands, and the material of which they are made. In general, less noble metals show smaller enhancements than noble metals in air. Enhancement factors of up to 1,000 have been reported for Au films. Research on the mechanism of the enhancement led to a general consensus that two different mechanisms, i.e. EM and chemical, contribute to the total enhancement. A detailed discussion about these mechanisms is beyond the aim of this chapter and can be found in [50, 51].

It has to be emphasized that in the case of thin films exhibiting island structure, the ATR system can still be modeled as being layered, but the optical constants of the bulk material and mass thickness have to be replaced by the effective optical constants and optical thickness of the island film. According to the Maxwell-Garnett [53], the Bruggeman [54], and other effective medium theories [55], the effective optical constants depend on several parameters, such as packing density of the particles, optical constants of the surroundings and polarizability of the mutually interacting metal islands. As a consequence of this, they may significantly differ from the bulk values. However, in the case of thick films, the optical constants approach the bulk values: for silver, this is achieved at a thickness of around 10 nm [56], whereas for iron, it was found to be at around 3 nm on MgO(001) surfaces [57].

The presence of a metal layer on the IRE has been found to drastically affect the spectrum of the medium under investigation and of the adsorbate [21, 28]. As already pointed out in Section 4.2.4, ATR/FTIR spectra show distortions of the shape of the peaks and small displacements of the band maximum due to the variation of the refractive index of a sample near an absorption band. Although, in the case of IRE/adsorbate/solvent systems, distortions of the peaks are minimized using incident angles well above the critical angle [45], in the case of IRE/metal/adsorbate/solvent systems, strongly distorted bands appear in the ATR/FTIR spectra even far from the critical angle [21].

Piras and colleagues [24, 28] reported that in the case of a Ge/Fe film (10 nm)/zinc dialkylditiophosphate (ZnDTP) system, the ATR/FTIR spectra show a shift of the characteristic P-O-(C) and C-O absorption band of ZnDTP to lower wavenumbers and a distortion of the bands on the lower-wavenumber side for an incident angle well above the critical angle. Besides optical arguments, the author also suggested that the adsorption of ZnDTP on the iron surface might contribute to the observed shift of absorption bands [28].

The band shape of IR absorption bands measured by ATR/FTIR for the case of thin films has been experimentally and theoretically investigated by Bürgi and Baiker [19, 21]. The theoretical results showed that strongly distorted absorption bands appeared at incident angles well above the critical angle in the case of both bulk media and thin adsorbate films: a negative-going peak at high wavenumbers and an increased background at low wavenumbers were always observed. At the same time, the band maximum shifted to lower energy. The theoretical modeling also revealed that the band shape in MUATR spectroscopy is mainly influenced by the optical properties of the metal film: the distortion of the bands increases with increasing thickness, increasing refractive index and decreasing absorption index of the metal layer. Such theoretical results were found to be in good agreement with the measured spectra in the case of thick (10 nm) metallic Pt layers.

4.3.3.1 Effect of metal underlayer on penetration depth in ATR/FTIR spectroscopic analysis of polymers

In the following, the penetration depth of the evanescent wave in a typical MUATR/FTIR experiment will be discussed. A model system, consisting of a germanium ATR crystal (refractive index $n_{Ge} = 4.0$) coated with a thin iron layer (10 nm) and a bulk specimen with refractive index ($n_{specimen}$) equal to 1.46 (a typical value for the refractive index of polymers [43]), is considered in the calculations.

Because the iron refractive index (n_{Fe}) is higher than that of germanium ($n_{Ge} = 4.0$) over the whole spectral region (Fig. 4.9a), when the IR radiation strikes the Ge/Fe interface, the EM wave splits into the reflected and transmitted (refracted) components. The angle of refraction (φ_2) is related to the incidence angle (φ_1) by Snell's law (Eq. 4.3), whereas the intensities of the reflected and refracted beams are interconnected by the so-called Fresnel amplitude coefficients (Eq. 4.4 and Eq. 4.5). It has to be emphasized that because the iron refractive index varies with the wavelength (Fig.4.9 a), the angle of refraction (φ_2) is not constant over the mid-IR region.

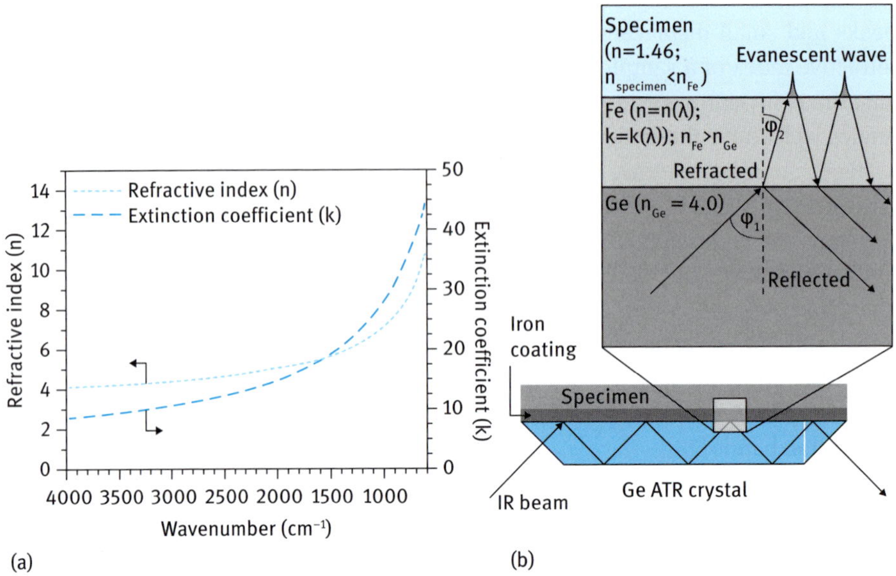

Fig. 4.9: (a) Metallic iron optical constants (refractive index, *n*, and extinction coefficient, *k*) [58] used for modeling the ATR/FTIR system used in the present work. (b) Schematic of the three-layered system (Ge ATR crystal/Fe/specimen with refractive index equal to 1.46, a typical value for the refractive index of polymers [43]).

At the iron/specimen interface, the IR radiation is totally reflected only if the angle of incidence (equal to the angle of refraction, φ_2, of the IR beam as it passes from Ge to Fe) exceeds the critical angle (φ_c) (Eq. 4.6). Due to the frequency dependence of the iron refractive index (Fig. 4.9a), the critical angle also varies with the wavelength.

The penetration depth in the case of the three-layered system (Ge ATR crystal/Fe/specimen) can be determined using the formula proposed by Harrick (Eq. 4.29) [8] and taking into account the frequency dependence of the iron refractive index and of the incidence angle of the IR beam at the Fe/specimen interface (equal to the angle of refraction of the IR beam as it passes from Ge to Fe):

$$d_p = \frac{\lambda}{2\pi n_1\sqrt{\sin^2\varphi_2 - n_{21}^2}},$$

where

$$n_1 = n_{Fe}(\lambda);\ n_{21} = n_2/n_1 = n_{specimen}/n_{Fe}(\lambda);\ \varphi_2 = \arcsin(\sin\varphi_1 {}^\star n_{Ge}/n_{Fe}(\lambda));\ \varphi_1 = 45°. \quad (4.40)$$

Because of the increase of the iron refractive index with the wavelength, the angle of refraction (φ_2) significantly decreases as the wavelength increases, but it always exceeds the critical angle (φ_c) (Fig. 4.10). Therefore, the IR beam undergoes total reflection at the iron/specimen interface.

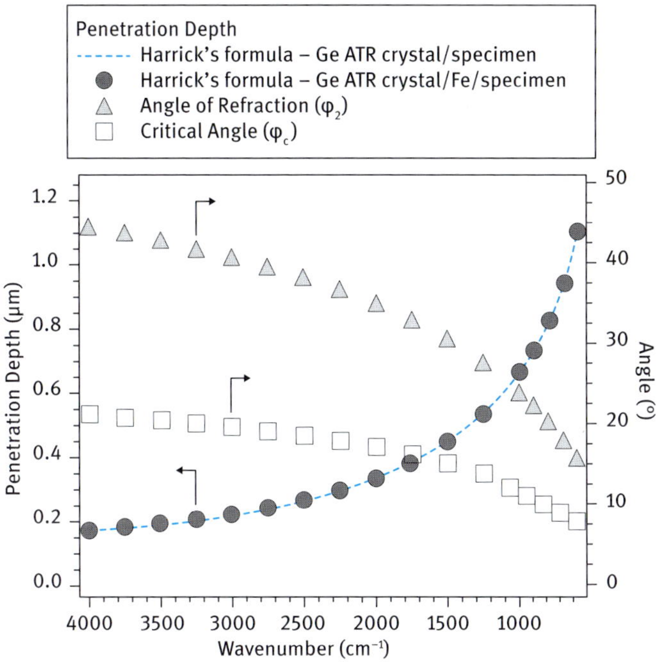

Fig. 4.10: Penetration depth (μm), calculated using the formula introduced by Harrick [8], as a function of the wavenumber in the case of the three-layer system Ge ATR crystal/Fe/specimen with refractive index equal to 1.46, a typical value for the refractive index of polymers [43]. As a consequence of the variation of the iron reflective index with the wavelength (Fig. 4.9), the angle of incidence of the IR beam at the Fe-specimen interface (equal to the angle of refraction of the IR beam as it passes from Ge to Fe) and the critical angle for total reflection at the Fe-specimen interface also change with the wavenumber.

The penetration depth values, determined using the formula proposed by Harrick (Eq. 4.29) [8], in the case of the three-layered system Ge ATR crystal/Fe/specimen are comparable with those calculated in the case of the corresponding ATR/FTIR system without any iron coating (i.e. two-layered system, Ge ATR crystal/specimen).

As already pointed out in Section 4.2.3.1, the matrix formalism developed by Hansen could also be used for evaluating the MSEF at any position within a certain layer of a *N*-isotropic-phase system (Eq. 4.11 through Eq. 4.26) [14, 35].

In the case of a three-layered system (Ge ATR crystal/metal underlayer/specimen), the MSEF magnitude can be numerically calculated at different positions within the rarer medium (i.e. specimen) and along the normal to the boundary between the two phases (*z*-axis). By fitting the MSEF values using an exponential function ($E = E_0\exp(-kz)$), where the fitting parameters are the decay constant (k) and the electric field amplitude at the interface (E_0), the penetration depth can be determined ($d_p = k^{-1}$). Here, because in the case of *p*-polarized light the calculation of the MSEF magnitude requires the knowledge of the complex dielectric constant, the MSEF magnitude was only evaluated for *s*-polarized radiation.

The penetration depth values in the case of the three-layered system Ge ATR crystal/Fe/specimen (with refractive index of the specimen equal to 1.46) are comparable with those calculated for the same ATR/FTIR system without any iron coating (i.e. for the two-layered system Ge ATR crystal/specimen with refractive index equal to 1.46) (Fig. 11a).

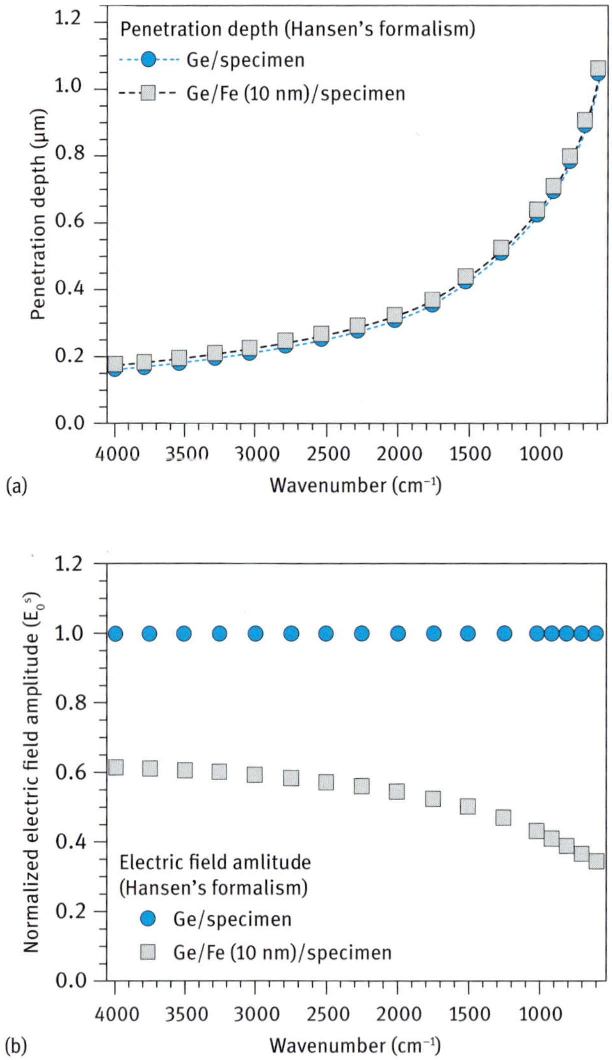

Fig. 4.11: (a) Penetration depth (µm), calculated using the formalism developed by Hansen [14, 35], as a function of the wavenumber in the case of a three- and two-layered system (Ge ATR crystal/ Fe (10 nm)/specimen and Ge ATR crystal/specimen). (b) Normalized electric field amplitude at the Fe/specimen (two-layered system) and iron/specimen (three-layered system) interface (E_0^s) as a function of the wavenumber. In the calculations, a refractive index equal to 1.46, a typical value for the refractive index of polymers [43], was considered for the rarer medium (specimen).

The numerical calculation of the MSEF does not only lead to the determination of the penetration depth of the evanescent wave in the rarer medium, but it also allows the evaluation of the electric field amplitude at the interface between the denser and the rarer medium (E_0). Compared with the case of a two-layered system (Ge ATR crystal/specimen), a significant reduction of the amplitude of the electric field at the interface between the denser and the rarer medium is observed when an iron coating is present on the IRE (Fig. 11b).

4.4 Potential and limitations of ATR/FTIR spectroscopy

The capability of yielding surface chemical information, information about the kinetics of adsorption and reaction at solid-liquid interfaces, and the possibility of performing ATR/FTIR analysis using metal-coated IREs has made ATR/FTIR spectroscopy the analytical technique of choice in different research fields, such as polymer science [16], electrochemistry [17, 59], tribology [24–26, 28–32], the science of biomembranes and biofilms [18], catalysis [15, 20, 21], and environmental chemistry [23].

In the case of solid-liquid interfaces (e.g. during polymerization studies), because ATR/FTIR spectroscopy simultaneously provides information about the chemistry of both dissolved species, such as reactants and products in the bulk phase, and species adsorbed on the IRE/metal-coated IRE, such as reaction intermediates, it is unspecific in the sense that a signal cannot *a priori* be assigned to dissolved or adsorbed compounds. Moreover, because ATR/FTIR probes the solid-liquid interface, where surface reactions occur, the concentrations of the species produced at the interface and dissolved in the bulk phase might differ from those measured by other bulk analytical methods, such as gas chromatography.

Recently, ATR/FTIR spectroscopy has been successfully applied to the study of supported metal catalysts. Besides providing real-time monitoring capabilities, it also yields information about changes of the state of metal films and supported metal particles. Based on reflectivity calculations, Bürgi *et al.* [22] showed that small changes of the optical constants of the metal (Pd), deriving from an increase or decrease of the concentration of free electrons, result in changes in reflectivity, in agreement with the measured ATR/FTIR spectra. This change in the concentration of free electrons could be due to the application of an electrochemical potential, formation of an adsorbed layer with a concomitant change of the work function, hydrogen adsorption, or oxidation of the metal-supported catalyst.

Based on this overview, it can be easily deduced that the principal challenge of ATR/FTIR spectroscopy is associated with the complexity of the results. Because all the chemical species present at the metal-liquid interface, either adsorbed on the surface or dissolved in the medium, contribute to the measured ATR/FTIR spectrum, the identification of a particular species and of its role in the surface chemistry become a major challenge.

Besides the difficulties in interpreting the outcomes of ATR/FTIR analysis, practical challenges also have to be faced [28]. The practical limitations of this technique are mainly due to the IRE and to the IR detector. As for the former, the most widely used IREs, i.e. Ge, Si, ZnSe, have a limited useful spectral range, which does not allow the acquisition of ATR/FTIR spectra over the whole mid-IR region. Moreover, the investigation of surface reactions can also require the acquisition of spectra at high temperatures, precluding then the use of semiconductor materials, such as Ge, whose useful transmission range becomes narrower with increasing temperature as a consequence of the generation of free electrons. Meanwhile, ATR/FTIR spectroscopy usually requires the use of detectors having high sensitivity, such as the mercury-cadmium-telluride detectors, which, however, have wavenumber cutoffs between 800 and 600 cm^{-1}.

4.5 Applications of ATR/FTIR spectroscopy

4.5.1 ATR/FTIR spectroscopy in polymer science

IR spectroscopy is one of the most powerful analytical techniques for the elucidation of the structure and composition of polymers and polymer surfaces [1, 16, 60–67]. Among the various experimental methods in IR spectroscopy, ATR/FTIR spectroscopy has been the analytical technique of choice for the characterization of a wide variety of elastomeric materials as well as for the *in situ* investigation of polymerization processes.

The power of ATR/FTIR spectroscopy in probing the surface structure and composition of polymers was already highlighted in the 1980s by Dothee *et al.* [68], who investigated commercial polyethylene films and suggested that the crystallinity of polyethylene was higher in the near-surface region than in the bulk of the material. The influence of uniaxial and biaxial elongations on the surface orientation of polypropylene was studied by Mirabella [69, 70] in a series of pioneering works. Using two distinct features in ATR/FTIR spectra, information about the surface orientation of the helix axis of polypropylene could be obtained. The inspiring work of Mirabella stimulated further studies [71, 72] concerning the stretching of polymers and the influence of the applied mechanical strain on the molecular orientation. In particular, Walls carefully examined the surface composition and orientation of uniaxially drawn polyethylene terephthalate (PET) films using polarized ATR/FTIR spectroscopy [72].

The versatility of ATR/FTIR spectroscopy also opened the path for the investigation of surface modifications and reactions of polymers in liquids as well as at elevated temperatures. The reaction of polyvinylidene fluoride in aqueous sodium hydroxide and in the presence of a phase transfer catalyst (namely, tetrabutylammonium hydrogen sulfate) was shown to lead to surface dehydrofluorination [73, 74], which significantly affected the electrical conductivity of the material [75, 76]. Meanwhile, ATR/FTIR spectroscopy could also be used to shed light on the influence of thermal history on the surface modification of polyvinyl chloride (PVC): significant differences in the ATR/FTIR spectra of PVC were observed between hot-pressed PVC specimens and solvent-casted PVC samples [16].

A particular field of research for which ATR/FTIR spectroscopy has been highly attractive deals with multicomponent systems and hybrid materials, in which interfacial interactions, stability, and mutual compatibility are crucial. For the investigation of these systems, an inherent drawback of ATR/FTIR measurements is the lack of an internal calibration method, which makes the construction of calibration curves necessary for performing quantitative analysis [16]. In one particular example, Brandrup *et al.* [77] investigated the migration of one component (namely polyamide) in polymeric systems (i.e. epoxy) to the film-air or film-substrate interface due the lower surface tension of polyamide compared with its epoxy counterpart.

In light of their importance for real technological applications, the interactions of polymer surfaces with solid substrates have been widely reported in the literature. ATR/FTIR spectroscopy was effectively used by Nguyen *et al.* [78] for monitoring the diffusion of water through pigmented and alkyd coatings, which explained the observed interfacial degradation. Meanwhile, Urban *et al.* [79] examined the influence of gas-plasma modifications on silicone elastomers. Among the plasma treatments, ammonia-plasma turned out to be the only process inducing an increase of the storage modulus, whereas argon or carbon dioxide treatments resulted in a decrease in the same property.

A powerful aspect of ATR/FTIR spectroscopy relies on the ability of conducting orientation studies of (macro)molecules on solid surfaces. The use of polarized light can, in fact, provide valuable insights into the structure of adsorbates and molecules in the near-surface region as well as how it is affected by mechanical strain, temperature, and substrate surface chemistry. For example, the orientation of molecules containing hydrophobic and hydrophilic component was found to strongly depend on the wettability of the surface the molecule interact with [16]. The dependence of the orientation of isotactic polypropylene on the applied uniaxial stress was investigated for the first time by Flournoy *et al.* [80]: the dichroic ratios in the analysis of the polymer by transmission FTIR and ATR/FTIR spectroscopy with parallel-polarized light was equal, thus suggesting that the orientation of the polymer molecules is the same near the surface region and in the bulk of the material.

ATR/FTIR spectroscopy has also been the analytical technique of choice for investigating polymerization processes *in situ*. Using this approach, information about the kinetics of the polymerization reaction(s), crystallization processes, chain packing, and orientation could be obtained. A pioneering study in this respect was performed by Ishida and Scott [81], who investigated the polymerization reactions of nylon 6. Information about reaction mixtures can also obtained by ATR/FTIR over short time intervals: the rate of consumption of free isocyanate groups and the competitive formation of urethane, isocyanurate, and urea linkages could be deduced from the spectra [66]. Curing processes could also be followed *in situ* by ATR/FTIR spectroscopy: Snyder and Fuerniss [82] reported the *in situ* investigation of the cross-linking of polyester-photoresists, which were attached to an ATR crystal.

4.5.2 *In situ* ATR/FTIR spectroscopy of tribochemical phenomena

Lubricants for mechanical systems (e.g. internal combustion engines, gear boxes, gas turbines) are usually based on mineral or synthetic oils. To improve or impart new and useful properties, additives for lubricating oils have been used since the 1920s. Among them, chemical compounds that are able to adsorb and/or react with the sliding surfaces to produce easily removable, low-shear-strength tribochemical reaction layers that prevent direct contact between the sliding surfaces, are often used to reduce wear in the boundary-lubrication regime, where sliding speeds are too low and loads are too high for a full-fluid lubricating film to be maintained [83–88].

Zinc dialkyldithiophosphates (ZnDTPs), widely used in many different kinds of industrial lubricants, represent one of the most successful classes of antiwear additives ever synthesized [83, 85, 88–90]. The layers formed on steel surfaces by ZnDTPs under purely thermal conditions and tribological conditions in the boundary-lubrication regime were found to primarily consist of patchy glassy phosphate with a layered structure [83, 89, 90]: the first layer above the metallic steel substrate is thought to be made of iron oxide/sulfides and is covered by two poly(thio)phosphate layers. The innermost one consists of short-chain iron/zinc poly(thio)phosphates and the outermost one of long-chain zinc poly(thio)phosphates. Iron-zinc oxides/sulfides are also present in the layers as inclusions. A soft, weakly bound alkyl phosphate film was found to be present on top of the layers.

Despite the remarkable effectiveness of ZnDTPs under a wide range of conditions [83, 90], the large amounts of phosphorus, sulfur, and zinc in this class of additives, which are known to lead to blocking filters and catalyst degradation in exhaust after-treatment systems for gasoline and diesel automobile engines [91], have brought the environmental consequences of using ZnDTPs into focus. Consequently, the permissible levels of sulfated ash, phosphorus, and sulfur in oil formulations have been progressively limited in modern engine-lubricant specifications [85]. Among the different chemistries proposed as a replacement or supplement to ZnDTPs, ashless equivalents to ZnDTPs, such as thiophosphates and phosphates without any metal, have been recently investigated quite extensively [92].

Most studies dealing with antiwear additives have focused on the characterization of films formed under purely thermal conditions and tribological conditions in the boundary-lubrication regime to understand their mechanism of action. The analytical studies were mostly performed *ex situ*, i.e. outside the tribometer and after the friction experiment [93]. Although this approach gives the possibility of combining various complementary modern surface analytical techniques [83], the specimens tend to be characterized in a very different environment from that found within the tribosystem. The inevitable air exposure during the transfer of the sample from the tribometer to the analytical instrument, even when this is carried out as rapidly as possible, as well as the solvent washing needed before performing the analysis, might render the surface unrepresentative of the original surface material.

The recognition of these limitations, together with the need to study the kinetics of thermal film and tribofilm formation and how they are influenced by different experimental parameters (e.g. temperature, load, sliding speed), has encouraged tribologists in the last decade to develop *in situ* techniques that allow the investigation of these layers as they form, in their "natural state" [94, 95]. Among the various *in situ* approaches used for tribological interface studies [93, 95], molecular spectroscopies (i.e. IR and Raman spectroscopy) were found to provide a valuable insight into the chemistry taking place under steady-state conditions [24–32, 96–100]. Such techniques often require one of the rubbing surfaces to be transparent to the wavelength of radiation concerned.

ATR/FTIR spectroscopy was demonstrated to be a powerful analytical tool for the *in situ* chemical characterization of boundary lubricant layers formed by antiwear and extreme-pressure additives between iron and steel surfaces [24–32, 96–100]. By coupling an ATR/FTIR spectrometer with a tribotester, Piras and colleagues [24–26, 28] investigated the oil (either pure ZnDTP or a solution of ZnDTP in base oil)-metal interface from "underneath," through a thin metallic film (MUATR/FTIR configuration), which represents one surface of the tribopair. By periodically acquiring ATR/FTIR spectra, changes in the lubricant chemistry and the growth of reaction films could be investigated as a function of different parameters, such as temperature. In the case of pure ZnDTP, no thermal or tribochemical reactions with iron were detected at temperatures up to 353 K after 38 or 90 h, respectively [24, 26, 28]. At 423 K, a rearrangement of the molecule was observed both under purely thermal conditions and following tribological tests: in the former case, the formation of P-O-P bonds, characteristic of polyphosphates, was found, whereas in the latter, a phosphate film was formed on iron. The comparison of these results with those obtained using an uncoated germanium ATR crystal allowed the authors to highlight the role of iron in the surface chemistry of ZnDTP: iron seemed not to be involved in the thermal reaction but affects the formation of tribofilms. The tribochemical reaction, in fact, was found to occur only on iron surfaces. The same authors also investigated the effect of blending ZnDTP with a synthetic oil, i.e. poly-α-olefin (PAO) [25, 26]: the rate of thermal decomposition of ZnDTP blend turned out to be lower than that of the pure additive.

The surface chemistry of tributyl thiophosphate (TBT) on air-oxidized iron was investigated by Rossi *et al.* [27] combining ATR/FTIR tribometry with XPS and temperature-programmed reaction spectroscopy (TPRS). Both thermal films and tribofilms were found to consist of iron polyphosphate and sulfate. Under tribological conditions in the boundary-lubrication regime, however, the amount of sulfate and polyphosphate in the film was higher. The comparison of these results with those obtained using tributyl phosphate (TBP) allowed the authors to clarify the role of sulfur in this class of compounds: it facilitates the formation of long-chain polyphosphates and lowers the temperature of the chemical and tribochemical reactions by around 50 K. A reaction mechanism between TBT and air-oxidized iron surface was proposed on the basis of these findings: after the breakage of the P=S bond to give tributyl phosphite, butoxy species can be produced through the scission of the P-O bond, in accordance with [83, 101–105].

The authors of this chapter used ATR/FTIR spectroscopy for the *in situ* investigation of the reactivity of triphenyl phosphorothionate (TPPT), an ashless antiwear additive, on air-oxidized iron under both purely thermal and tribological conditions in the boundary-lubrication regime [29–32]. The ATR/FTIR spectra acquired during thermal tests of a 0.044 mol/dm^3 solution of TPPT in PAO on iron-coated (10 nm) germanium ATR crystal (Fig. 4.12) supported a rearrangement of the molecule upon heating at 423 K leading to the formation of pyrophosphates, organophosphates, and sulfates [30]. At the same time, the oxidation of the base oil during thermal testing induces the production of oxygenated compounds, which can adsorb onto the iron surface and react with it to yield carbonates and carboxylates. The corrosion of the iron coating by the aggressive species formed during the thermo-oxidative aging of PAO provides the cations necessary for the formation of orthophosphate compounds. *Ex situ* angle-resolved XPS analyses corroborated the ATR/FTIR results and provided further information about the in-depth distribution of the surface reaction products.

Reflectivity calculations were performed to determine the optical response of the overall ATR/FTIR system to the formation, at the metal-oil interface, of a reaction layer, which is chemically and optically different from all the other layers [30]. The matrix formalism developed by Hansen (see Section 4.2.2) [14, 35] was used for evaluating the overall reflectance of a model simulating the growth of a reaction layer consisting of a mixture of iron phosphate and sulfate (Fig. 4.13). The qualitative comparison of the reflectivity calculations with the experimental results suggests that these baseline changes in the ATR/FTIR spectra acquired during thermal testing arise from the formation of a thin film at the metal-oil interface, which is chemically and optically different from the other layers.

Although the coupling of an ATR/FTIR spectrometer with a tribotester allowed Piras and colleagues [24–26, 28] to investigate the oil-metal interface from "underneath," through a thin metallic film (MUATR/FTIR configuration), which represents one surface of the tribopair, the absence of any force sensor did not make it possible to measure the friction forces during tribological experiments and to correlate them with the chemical changes occurring at the buried sliding interface. The authors of the present chapter developed a new tribometer (Fig. 4.14) incorporating a load cell, thus allowing spectroscopic and tribological measurements to be carried out concurrently [29].

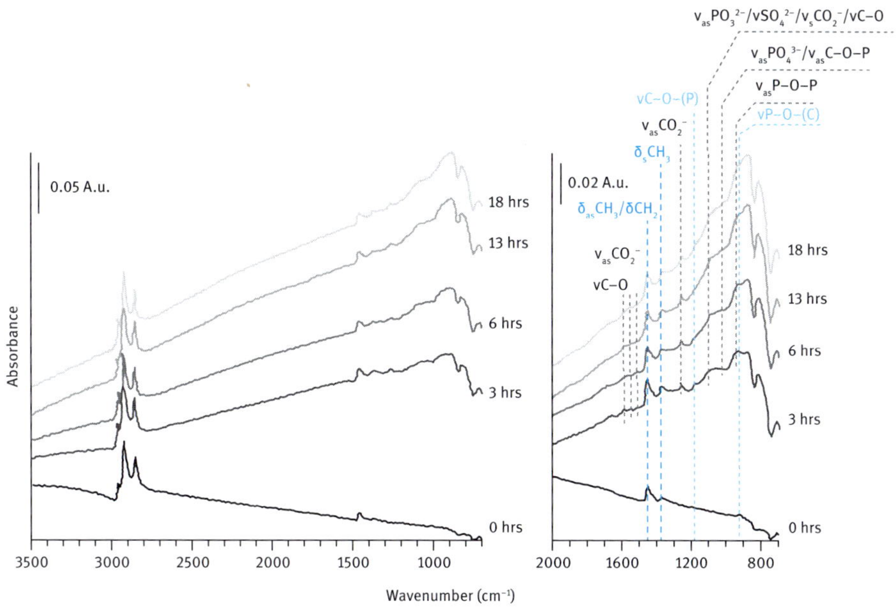

Fig. 4.12: ATR/FTIR spectra collected during a thermal test at 423 K in the presence of a 0.044 mol/dm^3 solution of TPPT in PAO on a Ge ATR crystal coated with iron (10 nm) [30]: (left) whole spectral region and (right) fingerprint region. The spectrum of the unheated solution (0 h) showed the characteristic peaks of TPPT [red vertical lines; peaks at 1,185 cm^{-1} νC-O-(P), 931 cm^{-1} νP-O-(C)] and PAO (blue vertical lines; peaks at 2,953 cm^{-1} $\nu_{as}CH_3$, 2,917 cm^{-1} $\nu_{as}CH_2$, 2,870 cm^{-1} ν_sCH_3, 2,849 cm^{-1} ν_sCH_2, 1,454 cm^{-1} $\delta_{as}CH_3/\delta CH_2$, 1,375 cm^{-1} δ_sCH_3) [2, 106–108]. Upon heating at 423 K, new bands were detected in the fingerprint region: in the 1,600- to 1,500-cm^{-1} region, where the $v_{as}CO_2^-$ vibration of carboxylate salts as well as the $\nu_{as}CO_2^-$ and νC-O vibrations of monodentate and bidentate carbonate complexes, respectively, occur [5, 108–112], three peaks appeared at 1,588, 1,551, and 1,512 cm^{-1}. An absorption band was detected at 1,263 cm^{-1} and assigned to the $\nu_{as}CO_2^-$ (bidentate) vibration in carbonate complexes [5, 108–112]. A broad peak appeared at 1,087 cm^{-1} and exhibited a shoulder on the lower-wavenumber side, which increased in intensity with the heating time and became a well-defined peak at 1,022 cm^{-1} after 13 h of heating at 423 K. These two bands at 1,087 and 1,022 cm^{-1} can be assigned to the stretching vibration of PO_3^{2-}/SO_4^{2-} ($\nu_{as}PO_3^{2-}/\nu SO_4^{2-}$) and of PO_4^{3-}/C-O-P ($\nu_{as}PO_4^{3-}/\nu_{as}$C-O-P) groups, respectively [108, 113, 114]. A further contribution to the peak at 1,087 cm^{-1} might come from the characteristic $\nu_sCO_2^-$ (bidentate)/ νC-O (monodentate) vibrations of carbonate complexes [5, 109–112]. The appearance of an absorption band at 939 cm^{-1}, assigned to the ν_{as}P-O-P, suggests the formation of pyrophosphate groups at the iron-oil solution interface [108, 113]. The ATR/FTIR spectra were also characterized by the presence of two strong peaks at 881 and 822 cm^{-1}, which correspond to the bending vibration of the carbonate anion (δC-O) [5, 109–112]. The intensity of the peaks detected in the fingerprint region increased upon heating at 423 K. At the same time, a progressive change of the baseline of the ATR/FTIR spectra was observed.

The extended ATR/FTIR tribometer was used for the *in situ* investigation of the reactivity of TPPT on air-oxidized iron under tribological conditions in the boundary-lubricant regime [31, 32]. The tribological results showed that TPPT is not a friction-modifier additive. The ATR/FTIR spectra, acquired *in situ*, provided an important

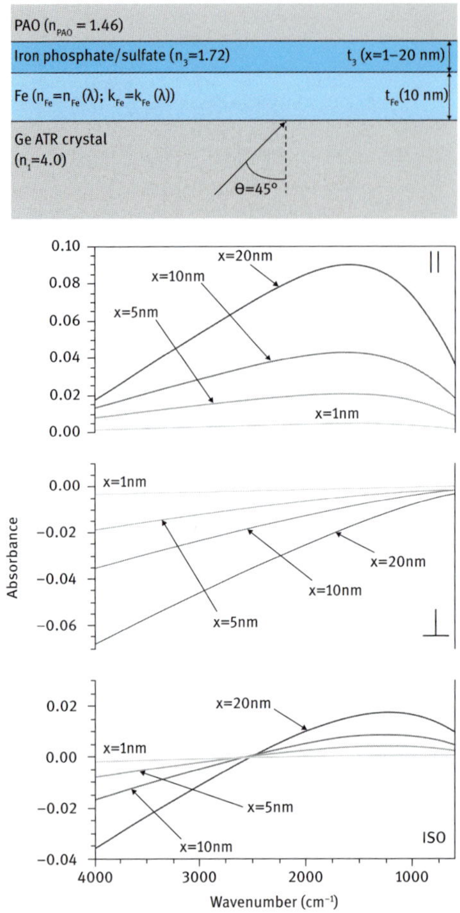

Fig. 4.13: Calculated ATR/FTIR absorbance for parallel, perpendicular, and isotropically polarized light in the case of the growth of a reaction layer consisting of iron phosphate and iron sulfate on the iron-coated Ge ATR crystal [30]. In the model, the first (germanium ATR crystal) and last (PAO) layers were assumed to be semi-infinite, transparent (i.e. $k = 0$), and with refractive index of 4.0 and 1.46 [115, 116], respectively. Because of the strong variation of the optical properties of metals in the mid-IR region, the dependence of the refractive index and extinction coefficient of iron on the wavelength had to be considered (Fig.4.9a) [58]. As for the reaction layers, the average value of the refractive index of the two compounds [115] constituting the phase was calculated. Calculations were performed for parallel and perpendicular polarized light, and the results are given in absorbance ($A = -\log(R/R_0)$, where R and R_0 are the reflectivity of the sample and the reference, respectively. The reflectivity of the reference was obtained considering the presence of no reaction layer on top of the iron-coated Ge ATR crystal). Starting from the reflectivity of parallel ($R_{\parallel}$) and perpendicular ($R_{\perp}$) polarized light, the reflectivity of isotropically polarized light was estimated as $R = mR_{\parallel} + (1 - m) R_{\perp}$, where m, which defines the polarization state of the spectrometer and should be, in principle, frequency dependent, was assumed to be equal to 0.5 over the whole spectral range. The reflectivity calculations predict a significant variation of the ATR signals for parallel- and perpendicular-polarized light as well as isotropically polarized light with the thickness of the reaction layer.

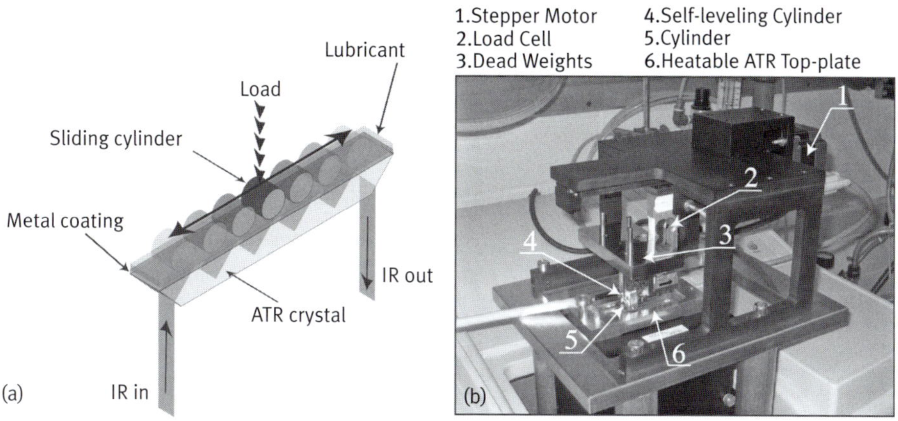

Fig. 4.14: (a) Principle of ATR/FTIR tribometry: a fixed cylinder slides in a reciprocating motion across the metal-coated ATR crystal surface. By periodically acquiring ATR/FTIR spectra during tribological tests, the chemical changes taking place at the lubricant-metal interface due to thermal or tribochemical reactions can be investigated as a function of critical tribological parameters such as applied load, sliding speed, temperature, and sliding speed and correlated with friction measurements. (b) Picture of the *in situ* ATR/FTIR tribometer mounted on top of the ATR/FTIR [29].

molecular level insight supporting the reaction of the molecule upon tribotesting at 423 K on air-oxidized iron surface with the formation of pyrophosphates, organophosphates, and sulfates (Fig. 4.15). At the same time, as a consequence of the thermal reaction, the oxidation of the base oil (PAO) during tribological testing resulted in the production of oxygenated compounds, which adsorbed onto the iron surface and reacted with it to form carbonates and carboxylates. The corrosion and wear of the iron coating induced, respectively, by the aggressive species formed during the thermo-oxidative aging of PAO and by the sliding process, provided the cations necessary for the production of orthophosphate groups. The decrease in absorbance of the characteristic bands of the reaction layer appearing in the ATR/FTIR spectra upon sliding also suggested that the rate of tribofilm formation was lower than that of tribofilm removal or that, because the iron coating was worn away upon sliding, the film could not form anymore. The formation of short-chain iron polyphosphates, organophosphates, and iron sulfates, together with carbonates and carboxylates, was corroborated by *ex situ* XPS analyses.

Based on the ATR/FTIR results, it was suggested that the formation of a tribofilm containing a mixture of organic and inorganic phosphates and sulfates provides a coating able to mitigate wear processes. The results of the *in situ* investigations indicated that the composition of the boundary layers formed by TPPT differs from that of the tribofilms generated by ZnDTP, which contain zinc and sulfide ions, and demonstrated the usefulness of *in situ* ATR/FTIR tribological testing, allowing the measurement of the friction force during the tribological tests and its correlation with FTIR results.

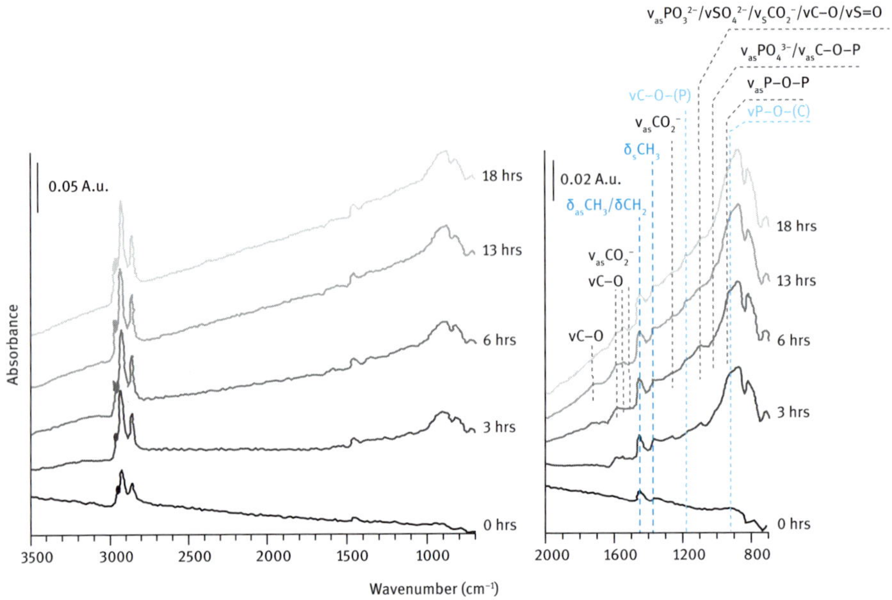

Fig. 4.15: ATR/FTIR spectra collected during a tribological test (normal load, 6.9 N; sliding speed, 0.5 mm/s; stroke, 45 mm; countersurface, 18Cr10Ni steel) at 423 K in the presence of a 0.044 mol/dm^3 solution of TPPT in PAO on a Ge ATR crystal coated with iron (10 nm) [31, 32]: (left) whole spectral region and (right) fingerprint region. The spectrum of the unheated solution (0 h) showed the characteristic peaks of TPPT [red vertical lines; peaks at 1,185 cm^{-1} νC-O-(P), 931 cm^{-1} νP-O-(C)] and PAO (blue vertical lines; peaks at 2,953 cm^{-1} $\nu_{as}CH_3$, 2,917 cm^{-1} $\nu_{as}CH_2$, 2,870 cm^{-1} $\nu_s CH_3$, 2,849 cm^{-1} $\nu_s CH_2$, 1,454 cm^{-1} $\delta_{as}CH_3/\delta CH_2$, 1,375 cm^{-1} $\delta_s CH_3$) [2,106–108]. Upon sliding at 423 K for 3 h, new bands were detected in the fingerprint region. In the 1,600- to 1,500-cm^{-1} region, where the $\nu_{as}CO_2^-$ vibration of carboxylate salts as well as the $\nu_{as}CO_2^-$ and νC-O vibrations of monodentate and bidentate carbonate complexes are found [5, 108–112], respectively, three peaks appeared at 1,588, 1,550, and 1,511 cm^{-1}. An absorption band at 1,265 cm^{-1} was also detected and assigned to the $\nu_{as}CO_2^-$ (bidentate) vibration in carbonate complexes [5, 108–112]. The spectrum acquired after 3 h of sliding at 423 K in the presence of a solution of TPPT in PAO also showed the appearance of a broad band at 1,098 cm^{-1}, which could be assigned to the stretching vibration of PO_3^{2-}/SO_4^{2-} ($\nu_{as}PO_3^{2-}/\nu SO_4^{2-}$) groups [108, 113, 114]. A contribution to this peak at 1,098 cm^{-1} might come from the $\nu_s CO_2^-$ (bidentate)/ νC-O (monodentate) vibrations of carbonate complexes [5, 109–112] and from the characteristic stretching vibration of the S=O group (νS=O) in organic sulfoxides [2, 108, 117]. The detection of an absorption band at 938 cm^{-1}, assigned to the ν_{as}P-O-P [108, 113], suggests the formation of pyrophosphate groups at the iron-oil interface. The peak at 938 cm^{-1} was found to be asymmetric and to have a shoulder on the high-wavenumber side (at *ca* 1,020 cm^{-1}), which could be assigned to the stretching vibration of PO_4^{3-}/C-O-P ($\nu_{as}PO_4^{3-}/\nu_{as}$C-O-P) groups [108, 113, 114]. The ATR/FTIR spectra also showed the presence of two strong peaks at 880 and 820 cm^{-1}, which correspond to the bending vibration of the carbonate anion (δC-O) [5, 109–112]. The intensity of the absorption bands appearing in the fingerprint region of the ATR/FTIR spectra after 3 h of sliding was found to decrease upon rubbing at 423 K. After 13 h, a new weak peak, which became more intense after 18 h, was detected at 1,722 cm^{-1} and assigned to the C=O stretching vibration (νC=O) [2, 107, 108]. At the end of the tribological test, the characteristic νC-O-(P) vibration of the TPPT molecule at 1,187 cm^{-1} could still be clearly detected.

Acknowledgments

F.M. acknowledges support from the Marie Curie International Outgoing Fellowship for Career Development within the Seventh European Community Framework Programme under contract no. PIOF-GA-2012-328776. Professor Nicholas D. Spencer of ETH Zurich is gratefully acknowledged for having supported the work presented here.

References

[1] Stuart BH. Infrared spectroscopy: fundamentals and applications. 1st edn. New York (NY): Wiley; 2004.
[2] Colthup NB, Daly LH, Wiberley SE. Introduction to infrared and Raman spectroscopy. 3rd edn. London (UK): Academic Press; 1990.
[3] Griffiths PR, de Haseth JA. Fourier transform infrared spectroscopy. 1st edn. New York (NY): Wiley; 1986.
[4] Kellner R, Mermet J-M, Otto M, Valcárcel M, Widmer HM, eds. Analytical chemistry: a modern approach to analytical science. 2nd edn. New York (NY): Wiley; 2004.
[5] Nakamoto K. Infrared and raman spectra of inorganic and coordination compounds, part A: theory and applications in inorganic chemistry. 5th edn. New York (NY): Wiley; 1997.
[6] Smith BC. Fundamentals of Fourier transform infrared spectroscopy. Boca Raton (FL): CRC Press; 1996.
[7] Harrick NJ. Study of physics and chemistry of surfaces from frustrated total internal reflections. Phys Rev Lett 1960;4:224.
[8] Harrick NJ. Internal reflection spectroscopy. New York (NY): Interscience Publishers; 1967.
[9] Fahrenfort J. Attenuated total reflection: a new principle for the production of useful infra-red reflection spectra of organic compounds. Spectrochim Acta 1961;17:698–709.
[10] Hansen WN, Delahay P, Tobias CW, eds. Advances in electrochemistry and electrochemical engineering. New York (NY): Wiley; 1973.
[11] Mirabella FM, ed. Internal reflection spectroscopy: theory and applications. New York (NY): Marcel Dekker; 1993.
[12] Mirabella FM, ed. Modern techniques in applied molecular spectroscopy. New York (NY): Wiley; 1998.
[13] Suëtaka W, Yates JTJ. Surface infrared and Raman spectroscopy: methods and applications. New York (NY): Plenum Press; 1995.
[14] Tolstoy VP, Chernyshova I, Skryshevsky VA. Handbook of infrared spectroscopy of ultrathin films. 1st edn. New York (NY): Wiley; 2003.
[15] Fringeli UP, Baurecht D, Siam M, et al. ATR spectroscopy of thin films. In: Nalwa HS, edn. Handbook of thin film materials. San Diego (CA): Academic Press; 2002:191–229.
[16] Urban MW. Attenuated total reflectance spectroscopy of polymers: theory and practice. Washington (DC): American Chemical Society; 1996.
[17] Hansen WN. Internal reflection spectroscopy in electrochemistry. In: Muller RH, ed. Optical techniques in electrochemistry. New York (NY): Wiley; 1973:1–60.
[18] Fringeli UP, Günthard H, Baurecht D. Biophysical infrared modulation spectroscopy. In: Gremlich HU, Yan B, eds. Infrared and Raman spectroscopy of biological materials. New York (NY): Marcel Dekker; 2000:143–92.
[19] Bürgi T. ATR-IR spetroscopy at the metal-liquid interface: influence of film properties on anomalous band-shape. Phys Chem Chem Phys 2001;3:2124–30.

[20] Bürgi T. In situ spectroscopy of catalytic solid-liquid interfaces and chiral surfaces. Chimia 2003;57:623–7.

[21] Bürgi T, Baiker A. Attenuated total reflection infrared spectroscopy of solid catalysts functioning in the presence of liquid-phase reactants. In: Bruce CG, Helmut K, eds. Advances in catalysis. Amsterdam (Netherlands): Academic Press; 2006:227–83.

[22] Bürgi T, Wirz R, Baiker A. In situ attenuated total reflection infrared spectroscopy: a sensitive tool for the investigation of reduction-oxidation processes on heterogeneous Pd metal catalysts. J Phys Chem B 2003;107:6774–81.

[23] Eggleston CM, Hug S, Stumm W, Sulzberger B, Dos Santos Afonso M. Surface complexation of sulfate by hematite surfaces: FTIR and STM observations. Geochim Cosmochim Acta 1998;62:585–93.

[24] Piras FM, Rossi A, Spencer ND. Growth of tribological films: in situ characterization based on attenuated total reflection infrared spectroscopy. Langmuir 2002;18:6606–13.

[25] Piras FM, Rossi A, Spencer ND. In situ attenuated total reflection (ATR) spectroscopic analysis of tribological phenomena. In: Dowson D, ed. Boundary and mixed lubrication: science and applications. Proceedings of the 28th Leeds-Lyon Symposium, 2002. Lyon (France): Elsevier; 2002:199–206.

[26] Piras FM, Rossi A, Spencer ND. Combined in situ (ATR FT-IR) and ex situ (XPS) study of the ZnDTP-iron surface interaction. Tribol Lett 2003;15:181–91.

[27] Rossi A, Piras FM, Kim D, Gellman AJ, Spencer ND. Surface reactivity of tributyl thiophosphate: effects of temperature and mechanical stress. Tribol Lett 2006;23:197–208.

[28] Piras FM. In situ attenuated total reflection tribometry. Zurich (Switzerland): ETH Zurich; 2002.

[29] Mangolini F, Rossi A, Spencer N. In situ attenuated total reflection (ATR/FT-IR) tribometry: a powerful tool for investigating tribochemistry at the lubricant/substrate interface. Tribol Lett 2012;45:207–18.

[30] Mangolini F, Rossi A, Spencer ND. Chemical reactivity of triphenyl phosphorothionate (TPPT) with iron: an ATR/FT-IR and XPS investigation. J Phys Chem C 2011;115:1339–54.

[31] Mangolini F, Rossi A, Spencer ND. Tribochemistry of triphenyl phosphorothionate (TPPT) by in situ attenuated total reflection (ATR/FT-IR) tribometry. J Phys Chem C 2012;116:5614–27.

[32] Mangolini F. Reactivity of environmentally compatible lubricant additives: an in situ and ex situ investigation. Zurich (Switzerland): ETH Zurich; 2011.

[33] Hecht E. Optics. San Francisco (CA): Pearson Education; 2002.

[34] Abèles F. Investigation on propagation of sinusoidal electromagnetic wave in stratified media application to thin film. Ann Phys (Paris) 1950;5:596–640.

[35] Hansen WN. Electric fields produced by the propagation of plane coherent electromagnetic radiation in a stratified medium. J Opt Soc Am 1968;58:381–90.

[36] Yamamoto K, Ishida H. Interpretation of reflection and transmission spectra for thin films: reflection. Appl Spectrosc 1994;48:775–87.

[37] Hirschfeld T. Relationships between the Goos-Hänchen shift and the effective thickness in attenuated total reflection spectroscopy. Appl Spectrosc 1977;31:243–4.

[38] International A. Standard practices for internal reflection spectroscopy. In: Molecular spectroscopy. West Conshohocken (PA): ASTM International; 2001.

[39] Hansen WN. Expanded formulas for attenuated total reflection and the derivation of absorption rules for single and multiple ATR spectrometer cells. Spectrochim Acta 1965;21:815–33.

[40] Müller G, Abraham K, Schaldach M. Quantitative ATR spectroscopy: some basic considerations. Appl Opt 1981;20:1182–90.

[41] Müller GJ, Abraham-Fuchs K. Matrix dependence in single and multilayer IRS spectra. Optik 1991;88:83–95.

[42] Müller GJ, Abraham-Fuchs K. Matrix dependence in single and multilayer internal reflection spectra. In: Mirabella FM, ed. Internal reflection spectroscopy theory and applications. New York (NY): Marcel Dekker; 1993;120–2.

[43] Callister WDJ. Materials science and engineering an introduction. 7th edn. New York (NY): Wiley; 2007.

[44] Nunn S, Nishikida K. Advanced ATR correction algorithm. Waltham (MA): Thermo Fisher Scientific; 2008.

[45] Belali R, Vigoureux J-M, Morvan J. Dispersion effects on infrared spectra in attenuated total reflection. J Opt Soc Am B 1995;12:2377–81.

[46] Bubert H, Jenett H, eds. Surface and thin film analysis: principles, instrumentation, applications. 1st edn. New York (NY): Wiley; 2002.

[47] Ryczkowski J. IR spectroscopy in catalysis. Catal Today 2001;68:263–381.

[48] Ishida KP, Griffiths PR. Theoretical and experimental investigation of internal reflection at thin copper films exposed to aqueous solutions. Anal Chem 1994;66:522–30.

[49] Zippel E, Kellner R, Breiter MW. Instability of sputtered platinum films on germanium reflection elements in aqueous solutions. J Electroanal Chem 1990;289:297–8.

[50] Osawa M. Surface-enhanced infrared absorption. In: Kawata S, ed. Near-field optics and surface plasmon polaritons. Berlin (Germany): Springer; 2001:163–87.

[51] Osawa M. Surface-enhanced infrared absorption spectroscopy. New York (NY): Wiley; 2006.

[52] Hartstein A, Kirtley JR, Tsang JC. Enhancement of the infrared absorption from molecular monolayers with thin metal overlayers. Phys Rev Lett 1980;45:201.

[53] Garnett JCM. Colours in metal glasses and in metallic films. Philos Trans R Soc London A 1904;203:385–420.

[54] Bruggeman DA. Berechnung verschidener physikalischer konstanten von heterogenen substanzen. Ann Phys 1935;24:636–79.

[55] Niklasson GA, Granqvist CG. Optical properties and solar selectivity of coevaporated Co-Al_2O_3 composite films. J Appl Phys 1984;55:3382–410.

[56] Ishiguro K, Kuwahara G. Determination of the optical constants of ag films from the measurements of intensity and phase change [I]. J Phys Soc Jpn 1951;6:71–5.

[57] Fahsold G, Bartel A, Krauth O, Magg N, Pucci A. Infrared optical properties of ultrathin Fe films on MgO(001) beyond the percolation threshold. Phys Rev B 2000;61:14108.

[58] Ordal MA, Bell RJ, Alexander RW, Newquist LA, Querry MR. Optical properties of Al, Fe, Ti, Ta W, and Mo at submillimiter wavelengths. Appl Opt 1988;27:1203–9.

[59] Bockris JOM, Khan SUM. Surface electrochemistry a molecular level approach. New York (NY): Plenum Press; 1993.

[60] Boerio FJ. Measurements of the chemical characteristics of polymers and rubbers by vibrational spectroscopy. In: Chalmers JM, Griffiths PR, eds. Handbook of vibrational spectroscopy. Chichester (UK): Wiley; 2002:2419–536.

[61] Bower DI, Maddams WF. The vibrational spectroscopy of polymers. Cambridge (UK): Cambridge University Press; 1989.

[62] Chalmers JM. Infrared spectroscopy in analysis of polymers and rubbers. In: Meyers RA, ed. Encyclopedia of analytical chemistry. Chichester (UK): Wiley; 2000:7702–59.

[63] Chalmers JM, Everall NJ. Qualitative and quantitative analysis of polymers and rubbers by vibrational spectroscopy. In: Chalmers JM, Griffiths PR, eds. Handbook of vibrational spectroscopy. Chichester (UK): Wiley; 2002:2389–418.

[64] Chalmers JM, Hannah RW, Mayo DW. Spectra-structure correlations: polymer spectra. In: Chalmers JM, Griffiths PR, eds. Handbook of vibrational spectroscopy. Chichester (UK): Wiley; 2002:1893–918.

[65] Koenig JL. Spectroscopy of polymers. 2nd edn. Amsterdam (The Netherlands): Elsevier; 1999.

[66] Siesler HW, Holland-Moritz K. Infrared and Raman spectroscopy of polymers. New York (NY): Marcel Dekker; 1980.

[67] Stuart BH. Polymer analysis. Chichester (UK): Wiley; 2002.

[68] Dothee D, Vigoureux J-M, Camelot M. Application of attenuated total reflection (ATR) infrared spectroscopy to the study of microstructural changes in high density polyethylene during sliding friction. Polym Degrad Stab 1988;22:161–74.
[69] Mirabella FM. Surface orientation of polypropylene. I. Theoretical considerations for the application of internal reflection spectroscopy. J Polym Sci Polym Phys Ed 1984;22:1283–91.
[70] Mirabella FM. Surface orientation of polypropylene. II. Determination for uniaxially and biaxially oriented films using internal reflection spectroscopy. J Polym Sci Polym Phys Ed 1984;22:1293–304.
[71] Nishio E, Morimoto M, Nishikida K. Nondestructive quantitative analysis of ethylene-propylene blend polymer by the NIR/ATR/FT-IR method. Appl Spectrosc 1990;44:1639–40.
[72] Walls DJ. Application of ATR-IR to the analysis of surface structure and orientation in uniaxially drawn poly(ethyleneterephthalate). Appl Spectrosc 1991;45:1193–8.
[73] Lovinger A, Reed D. Inhomogeneous thermal degradation of poly(vinylidene fluoride) crystallized from the melt. Macromolecules 1980;13:989–94.
[74] Wentink T, Willwerth LJ, Phaneuf JP. Properties of polyvinylidene fluoride. Part II. Infrared transmission of normal and thermally decomposed polymer. J Polym Sci 1961;55:551–62.
[75] Kuhn KJ, Hahn B, Percec V, Urban MW. Structural and quantitative analysis of surface modified poly(vinylidene fluoride) films using ATR FT-IR spectroscopy. Appl Spectrosc 1987;41:843–7.
[76] Urban MW, Salazar-Rojas EM. Ultrasonic PTC modification of poly(vinylidene fluoride) surfaces and their characterization. Macromolecules 1988;21:372–8.
[77] Brandrup J, Immergut EH. Polymer handbook. New York (NY): Wiley; 1989.
[78] Nguyen T, Byrd E, Lin C. A spectroscopic technique for in situ measurement of water at the coating/metal interface. J Adhes Sci Technol 1991;5:697–709.
[79] Urban MW, Stewart MT. DMA and ATR FT-IR studies of gas plasma modified silicone elastomer surfaces. J Appl Polym Sci 1990;39:265–83.
[80] Flournoy PA. Attenuated total reflection from oriented polypropylene films. Spectrochim Acta 1966;22:15–20.
[81] Ishida H, Scott C. Fast polymerization and crystallization kinetic studies of nylon 6 by combined use of computerized micro-rim machine and FT-IR. J Polym Eng 1986:201.
[82] Snyder RW, Fuerniss SJ. ATR/IR spectroscopic method for following photo-polymer curing. Appl Spectrosc 1992;46:1113–6.
[83] Gellman AJ, Spencer ND. Surface chemistry in tribology. J Eng Tribol 2002;216:443–61.
[84] Hutchings IM. Tribology, friction and wear of engineering materials. Oxford (UK): Butterworth-Heinemann; 1992.
[85] Mang T, Dresel W, eds. Lubricants and lubrication. 2nd edn. New York (NY): Wiley; 2007.
[86] Mortier RM, Orszulik ST, eds. Chemistry and technology of lubricants. 2nd edn. London (UK): Blakie Academic & Professional; 1997.
[87] Pawlak Z. Tribochemistry of lubricating oils. Amsterdam (The Netherlands): Elsevier; 2003.
[88] Rudnick LR, ed. Lubricant additives: chemistry and applications. New York (NY): Marcel Dekker; 2003.
[89] Nicholls MA, Do T, Norton PR, Kasrai M, Bancroft GM. Review of the lubrication of metallic surfaces by zinc dialkyl-dithiophosphates. Tribology International 2005;38:15–39.
[90] Spikes H. The history and mechanisms of ZDDP. Tribol Lett 2004;V17:469–89.
[91] Kubsh J. Three-way catalyst deactivation associated with oil-derived poisons. In: Bode H, ed. Materials aspects in automotive catalytic converters. Weinheim (Germany): Wiley-VCH Verlag; 2003:215–22.
92] Spikes H. Low- and zero-sulphated ash, phosphorus and sulphur anti-wear additives for engine oils. Lubric Sci 2008;20:103–36.
[93] Donnet C. Problem-solving methods in tribology with surface-specific techniques. In: Rivière JC, Myhra S, eds. Handbook of surface and interface analysis: methods and problem-solving. New York (NY): Marcel Dekker; 1998:968.

[94] Spikes HA. In situ methods for tribology research. Tribol Lett 2003;14:1–1.
[95] Sawyer WG, Wahl KJ. Accessing inaccessible interfaces: in situ approaches to materials tribology. MRS Bull 2008;33:1145–50.
[96] Cann PM. In-contact molecular spectroscopy of liquid lubricant films. MRS Bull 2008;33:1151–8.
[97] Cann PM, Spikes HA. In-contact IR spectroscopy of hydrocarbon lubricants. Tribol Lett 2005;19:289–97.
[98] Olsen JE, Fischer TE, Gallois B. In situ analysis of the tribochemical reactions of diamond-like carbon by internal reflection spectroscopy. Wear 1996;200:233–7.
[99] Sasaki K, Inayoshi N, Tashiro K. Development of new in situ observation system for dynamic study of lubricant molecules on metal friction surfaces by two-dimensional fast-imaging Fourier-transform infrared-attenuated total reflection spectrometer. Rev Sci Instrum 2008;79:123702–7.
[100] Sasaki K, Inayoshi N, Tashiro K. In situ FTIR-ATR observation of phase transition behavior of n-alkane molecules induced by friction motion on a metal interface. J Phys Chem C 2009;113:3287–91.
[101] Gao F, Furlong O, Kotvis PV, Tysoe WT. Reaction of tributyl phosphite with oxidized iron: surface and tribological chemistry. Langmuir 2004;20:7557–68.
[102] Gao F, Kotvis PV, Stacchiola D, Tysoe WT. Reaction of tributyl phosphate with oxidized iron: surface chemistry and tribological significance. Tribol Lett 2005;18:377–84.
[103] Holbert AW, Batteas JD, Wong-Foy A, Rufael TS, Friend CM. Passivation of Fe(110) via phosphorus deposition: the reactions of trimethylphosphite. Surf Sci 1998;401:L437–43.
[104] Ren D, Gellman A. Initial steps in the surface chemistry of vapor phase lubrication by organophosphorus compounds. Tribol Lett 1999;6:191–4.
[105] Ren D, Gellman AJ. Reaction mechanisms in organophosphate vapor phase lubrication of metal surfaces. Tribol Int 2001;34:353–65.
[106] Mangolini F, Rossi A, Spencer ND. Reactivity of triphenyl phosphorothionate in lubricant oil solution. Tribol Lett 2009;35:31–43.
[107] Roeges NPG. A guide to the complete interpretation of infrared spectra of organic structures. Chichester (UK): Wiley; 1994.
[108] Socrates G. Infrared and raman characteristic group frequencies. 3rd edn. Chichester (UK): Wiley; 2001.
[109] Bargar JR, Kubicki JD, Reitmeyer R, Davis JA. ATR-FTIR spectroscopic characterization of coexisting carbonate surface complexes on hematite. Geochim Cosmochim Acta 2005;69:1527–42.
[110] Rèmazeilles C, Refait P. Fe(II) hydroxycarbonate Fe2(OH)2CO3 (chukanovite) as iron corrosion product: synthesis and study by Fourier transform infrared spectroscopy. Polyhedron 2009;28:749–56.
[111] Roonasi P, Holmgren A. An ATR-FTIR study of carbonate sorption onto magnetite. Surf Interface Anal 2010;42:1118–21.
[112] Wijnja H, Schulthess CP. Carbonate adsorption mechanism on goethite studied with ATR-FTIR, DRIFT, and proton coadsorption measurements. Soil Sci Soc Am J 2001;65:324–30.
[113] Efimov AM. IR fundamental spectra and structure of pyrophosphate glasses along the $2ZnO\cdot P_2O^{5-}2Me_2O\cdot P_2O_5$ join (Me being Na and Li). J Non-Cryst Solids 1997;209:209–26.
[114] Ross SD. Inorganic infrared and Raman spectra. London (UK): McGraw-Hill; 1972.
[115] Lide DR, ed. The CRC handbook of chemistry and physics. 72nd edn. Cleveland (OH): CRC Press; 2007;1991–1992.
[116] Oligomers I. Durasyn® polyalphaolefins. League City (TX): INEOS Oligomers; 2007.
[117] Silverstein RM, Webster FX, Kiemle DJ. Spectroscopic identification of organic compounds. 5th edn. New York (NY): Wiley; 2005.

Dalia Yablon

5 Scanning probe microscopy of polymers

5.1 Introduction

Scanning probe microscopy (SPM) is a field that was invented in the early 1980s with the invention of the scanning tunneling microscopy by Binnig and Rohrer [1]. This field has spawned a powerful set of characterization microscopies that today have been incorporated into the standard toolkit for nanoscale characterization. Atomic force microscopy (AFM) was invented in 1986 [2] and is perhaps the most versatile and widely used member of this family.

The power of AFM lies in its high-resolution capability to map a variety of material properties on the nanoscale. A schematic of AFM operation is shown in Fig. 5.1 and described briefly. AFM operation is based on tracking the motion of a very small cantilever/tip assembly as it interacts with a sample. The cantilever is typically very small (length of a few hundred micrometers, width of a tens of micrometers, and thickness of a few micrometers), and a very sharp tip is grown off the end of the cantilever. The motion of the cantilever/tip is tracked through an optical detection system comprised of a laser that is bounced of the back of the lever and then directed toward a position-sensitive detector (PSD). The AFM tip can then sensitively interact with the substrate through a variety of mechanisms to probe a variety of substrate properties. A simplified scheme of the apparatus is reported in Fig. 5.1. The series of properties that can be probed by AFM include mechanical, magnetic, electrical, optical/spectroscopic, and then other capabilities such as lithography and pulling experiments. For polymer applications, the most useful of these properties is mechanical.

For polymer applications, the AFM now resides alongside optical microscopy and electron microscopy [scanning electron microscopy (SEM) and transmission electron microscopy (TEM)] as essential tools for characterization. The AFM provides specific advantages over photon- and electron-based microscopies to probe materials. Because the AFM provides mechanical interaction between the tip and the sample, it often provides contrast in situations where electron- or photon-based microscopies struggle or even fail. For example, polyolefin materials are composed of primarily carbon and hydrogen so that most chemically sensitive microscopies cannot differentiate between them. However, as long as there is some mechanical difference between the materials (e.g. stiffness, energy dissipation), the AFM will be able to contrast the materials with ease. In addition, materials such as elastomers are very quickly and easily differentiated from thermoplastics because of the extreme difference in mechanical property in this case.

There are many AFM imaging modes available for mechanical contrast and specifically to measure polymer materials. The aim of the AFM, like any microscopy, is

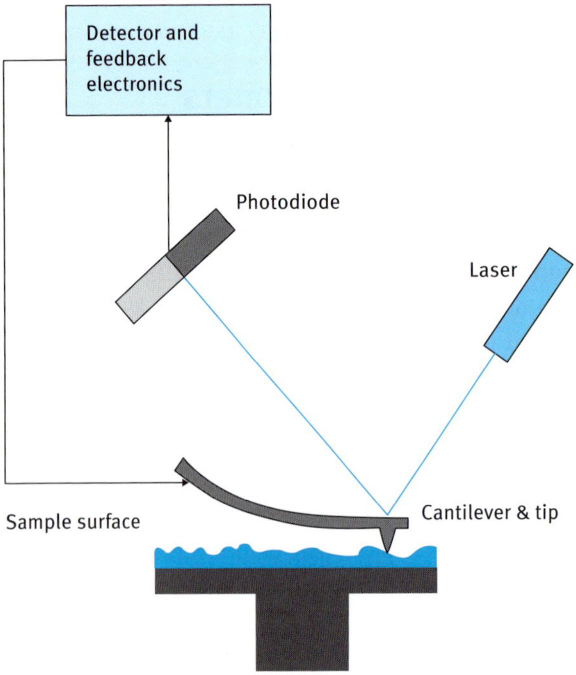

Fig. 5.1: A simplified scheme of an AFM apparatus showing a cantilever/tip assembly in contact with the sample. The cantilever motion is tracked through an optical detection mechanism, whose signal is fed through to the controller, allowing for active feedback control on the tip-sample interaction.

multifold: mapping, discrimination, and identification. In a multicomponent material, the AFM can discriminate among all the different components that vary in size, morphology, and material properties. Unambiguous identification of the actual mechanical properties of the nanoscale component is sometimes required together with discrimination. The latter relies on the AFM's ability to make quantitative measurements of the parameters of interest (e.g. storage modulus, loss modulus, loss tangent).

There are several reasons why one might want to have a quantitative nanomechanical measurement in polymer studies. The first reason is to aid in the discrimination capability described earlier to help unambiguously identify components in the AFM image. The second reason is independent of discrimination and is based on the need to measure mechanical properties on the nanoscale. Often, there is no access to a bulk analogue of the material probed with the AFM, and so a thin film or a domain of 10 nm or 1 μm may be all the available material to test. Bulk instruments are obviously not suitable for such measurements, and so a nanoscale method is required. Finally, AFM plays a critical role in establishing structure-property relationships in polymer-based materials. In many polymer applications, the product is designed around a mechanical property requirement – e.g. tensile pulling, impact strength, tear strength. Often, nanoscale and microscale structures play an important role in determining macroscopic properties, and so the ability of the AFM to discriminate and identify the various structural properties of the material on this length scale (e.g. phase segregation, continuity of phases, dispersion) is important to its ultimate performance.

This chapter will describe the main available AFM capabilities for imaging polymers, illustrated by applications taken from the literature. Sample preparation, where all AFM measurements begin, is elaborated in Section 5.2. Then, the most popular method of imaging polymer materials, phase imaging, is described in Section 5.3, along with its benefits and shortcomings. Section 5.4 describes advanced, state-of-the-art multifrequency-based imaging. Section 5.5 covers methods for quantitative nanorheological mapping. Section 5.6 follows with some "alternative" (not strictly mechanical characterization-based) AFM-based methods that are very useful to polymers including nanothermal analysis and spectroscopic-based methods. Finally, Section 5.7 concludes with environmental-based measurements of polymer systems as a function of temperature and deformation, an area in which the AFM provides unique capabilities.

5.2 Sample preparation

AFM imaging requires smooth and flat samples. An AFM can image at most approximately a 100×100-μm field of view, depending on the instrument and the scanner used. This field of view is actually quite small, especially compared with its electron- and photon-based microscopy counterparts. Vertically, the range is confined to features or tilt of less than 5 μm (this z limit is instrument specific) over the 100×100-μm region; these limitations are a function of the instrument hardware.

Because of these limitations, great care has to be taken to prepare smooth and flat samples appropriate for AFM imaging. There are many methods available from the general microscopy community to prepare samples. Preparing smooth thin films through spin coating is a popular method, although not always practical. For materials that are extruded or molded, cryomicrotomy, specifically cryofacing, is a preferred method. In cryomicrotomy, the sample is cooled down below the glass transition temperature (Tg) and then cut with a very sharp edge of a glass knife, followed by a sharp edge of a diamond knife. Several commercial vendors make suitable instruments, and the method is described in more detail elsewhere [3]. Cryomicrotomy can yield two kinds of surfaces: a smooth face or a thin section. For AFM imaging, cryofacing – where a pellet or component is simply smoothed out by the knives – is sufficient. Thin sections are essential for SEM and TEM imaging but are not required for AFM imaging. Once a sample has been cryomicrotomed, it can be immediately imaged by the AFM without further preparation (e.g. staining, again a common method required for SEM and TEM imaging is not needed for AFM imaging). Samples can age rapidly after cryomicrotoming, especially materials where oxidation and free radical formation upon microtoming can damage the samples. Imaging as soon as possible after cryomicrotoming is generally suggested. The generally minimal sample preparation required by AFM is another one of its many attractive features.

5.3 Phase imaging

5.3.1 Background on phase imaging

Phase imaging is perhaps one of the best known and most widely used modes in AFM. It is a simple method in that it is a channel collected in tapping mode (also called amplitude modulation mode, intermittent contact mode) and requires no postprocessing. There has been a significant amount of research activity trying to relate the phase signal to actual material properties. However, phase imaging is rather complicated, and interpretation of its meaning is still widely debated and researched [4–7]. Fundamentally, the phase shift measured in a phase image is a measure of the energy dissipated by the cantilever into the material. It is a measure of both conservative and nonconservative tip-sample interaction forces that depends on a number of factors including viscoelasticity, adhesion, capillary forces, and tip contact area.

Phase shift is always measured with respect to a reference and, in this case, to the driven cantilever. The cantilever is driven at a resonant frequency, and interacts with the sample at a given oscillation amplitude set by the user. The phase shift (ϕ) is then induced by interacting with the sample and is mapped as a channel while the tip raster scans over the surface.

At resonance, the phase lag of a driven harmonic oscillator with damping is 90° with respect to the driving or excitation force; a phase lag of 90° is easily observed at the frequency resonance peak. To understand the concept of phase shift and a 90° shift at resonance, one can try this experiment at home: shake a flexible ruler. Once the ruler is at resonance, watch the motion of the wrist with respect to the peak of the ruler or string. The peak of one's wrist will be 90° out of phase with the peak of the ruler or string's motion at resonance. (Note that different vendors define phase at resonance differently – e.g. phase is defined as "zero" at resonance by some vendors.)

5.3.2 Applications of phase imaging to polymer materials

A phase image of an 80/20 by weight blend of two plastics, polypropylene (PP) and polystyrene (PS), is shown in Fig. 5.2, where the PS is in the spherical domain in the image, whereas the PP is in the matrix. Note that the phase contrast is not very strongly differentiated for these two materials. This is due to a similarity in storage moduli (unlike, for example, in a plastic/elastomer blend) as well as both materials having very small and similar loss tangents of <0.1.

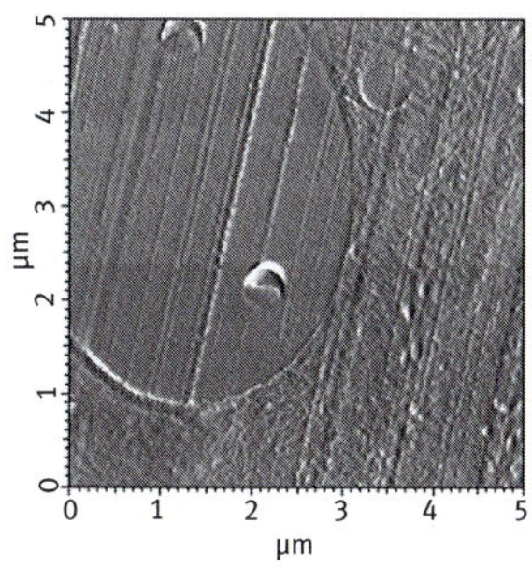

Fig. 5.2: The 5 × 5-µm phase image of a blend of PP and PS. The PS is in the spherical domain, differentiated by a smooth topography. The matrix contains the PP with a more textured topography. There is a pronounced curved border between the two materials, primarily due to the delamination between the two materials, which induces a topographical depression, and hence artifact, in the phase image.

Phase imaging is most useful to measure the morphology, dispersion, and, on occasion, high-resolution structure of multicomponent polymer samples that contain components with widely varying loss tangents and/or moduli. A good example of such a material – and a type of material that AFM phase imaging is extensively used to characterize – is a category of thermoplastic olefins called impact copolymer (ICP) (or impact-modified polymer) that contains a stiff thermoplastic matrix and elastomeric rubber domains. These materials are toughened plastics with little elasticity but with excellent impact toughness properties.

The example shown in Fig. 5.3 is an AFM phase image of a PP-based ICP with a-PP matrix and ethylene-propylene rubber (EPR) domains. The EPR domains are easily identified as the bright, circular regions in the image. The morphology, size, and dispersion of the EPR domains can be quickly ascertained and also quantitatively analyzed, if so desired. These characteristics of the EPR within ICP are essential to determining the material's ultimate mechanical properties such as stiffness, toughness, and strength, which are critical to understanding structure-property relationships as new materials in this class are being developed. High-resolution structures within the ICP are visible in Fig. 5.3b, both within the PP and the EPR domains. The lamellar structure within the PP is observed while the EPR shows inclusions, indicating a further level of complexity within the EPR. Finally, the interface between the PP and the EPR can be explored for mechanical strength, as discussed in Section 5.7.

Another category of polymer materials for which AFM phase imaging has been very popular and useful is block copolymers, where small nanoscale domains of various polymers are assembled together. Beautiful phase images of materials such as poly(styrene-*b*-ethylene-*co*-butylene-*b*-styrene) are now routinely used in differentiating the hard and soft blocks [8–11]. Phase imaging was also used to image PS-poly(acrylic acid) amphiphilic block copolymer films to visualize the different components [12]. However, the use of phase to discriminate the different components in block copolymers is very extensive, and these two areas are only offered as examples.

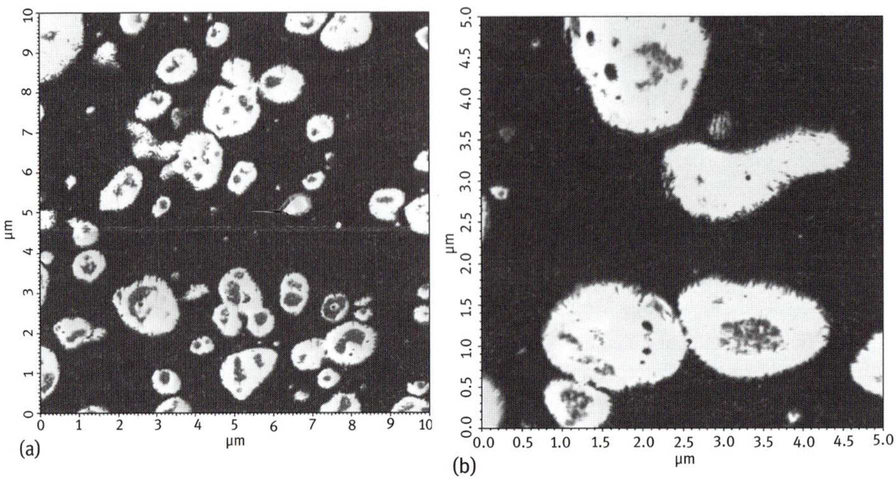

Fig. 5.3: (a) The 10 × 10-μm phase image of an ICP showing the surrounding PP matrix (dark) and the bright patches of EPR. Bright inclusions are visible inside the EPR. (b) High-resolution 5 × 5-μm image of the ICP highlighting lamellar structure within the PP and inclusions in the EPR.

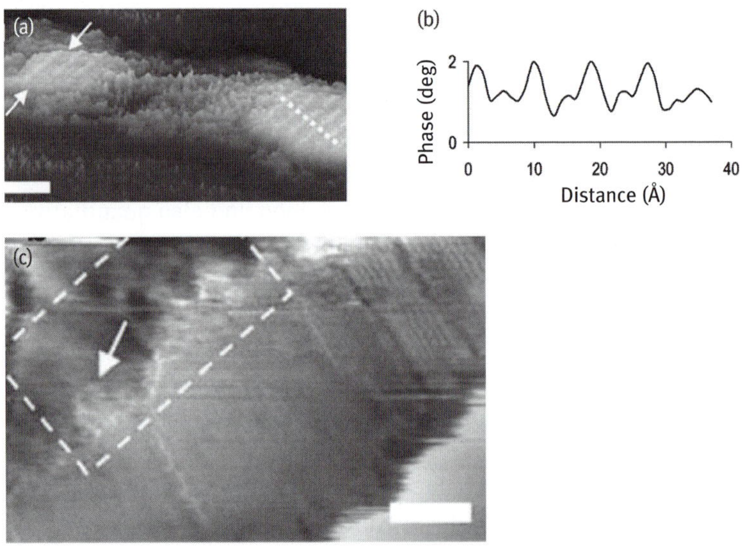

Fig. 5.4: (a) High-resolution phase image of a lamella of PE-resolving single chains in the (010) surface accompanied by (b) a line profile across these chains showing a spacing of 3.69 Å between chains. (c) Phase image of PE film showing the interface between the edge of a crystalline lamella and the amorphous region, revealing free loops from the crystalline area into the amorphous area. Scale bars: (a) 100 nm and (c) 10 nm. Reprinted with permission from Mullin N, Hobbs JK, *Phys Rev Lett* 107, 197801, 2011 [13].

The limits of resolution continue to be pushed on a regular basis. Recently, Hobbs and colleagues used torsional resonance imaging – where the cantilever's twisting (as opposed to up and down flexural) behavior is measured, providing exquisite sensitivity [13]. In this mode, single chains, buried folds, and chain ends within a polyethylene (PE) film are visualized with sub-nanometer resolution as shown in Fig. 5.4. Figure 5.4a shows a high-resolution torsional resonance phase image resolving single chains in the (010) surface accompanied by a line profile (Fig. 5.4b) across these chains showing a spacing of 3.69 Å between chains. The image in Fig. 5.4c reveals free loops from the crystalline area into the amorphous region as indicated by the white arrow.

5.4 Multifrequency imaging

Typically, an AFM cantilever is excited at its first or fundamental eigenmode for conventional tapping or intermittent contact mode imaging. The phase imaging discussed in Section 5.3 operates based on oscillation of the fundamental mode as well. Recently, with advances in hardware and software, it has become possible to oscillate the AFM cantilever at higher frequencies and even simultaneously multiple frequencies if the instrument has access to multiple lock-in amplifiers. Simultaneous oscillation at multiple frequencies can either occur at higher-order harmonics (integer multiple frequencies of the first eigenmode, see e.g. MAC mode by Agilent) or at higher-order eigenmodes (see e.g. bimodal AC mode by Asylum Research). Typically, in these modes of operation, excitation and feedback occur at the fundamental eigenmode. The higher-order mode or harmonics amplitude and phase signal can then be collected and has no feedback. More information about multifrequency operation, and theory can be found elsewhere [14–16].

The advantage conferred by multifrequency imaging is still a greatly researched topic; however, there have been noticeable improvements in discrimination capabilities with its operation. Multifrequency imaging has shown improvements in imaging a variety of materials including polymers [17, 18] and biological samples [19]. The ability to derive meaningful material property information from the higher-order mode or harmonic channel is still under investigation, but recent results reveal bistability in the higher-order mode with multiple regimes of operation depending on selection of operating parameters [20, 21].

Figure 5.5 shows the ability of multifrequency imaging to improve upon the contrast discrimination in complicated polymer samples. Figure 5.5a shows the typical phase image of a three component 60/20/20 wt% polymer blend of PE, PS, and PP obtained from the conventional phase image. The three different components cannot be differentiated by contrast in the phase image, although the topography does facilitate some differentiation showing the PE as the matrix material. An image of second eigen mode phase, collected in bimodal dual AC (where the cantilever is oscillated

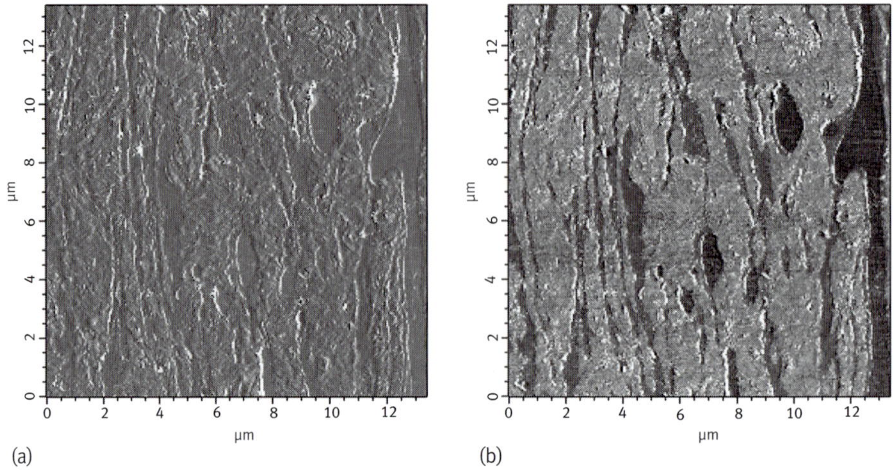

Fig. 5.5: (a) The 13 × 13-μm mode 1 phase image of a blend of 60% PE, 20% PP, and 20% PS showing little contrast between the three materials. (b) Mode 2 phase of the same location showing good contrast between the three materials where the PS is darkest and PE (matrix background) is lightest.

simultaneously at two eigenmodes with feedback occurring off the first eigenmode) is shown in Fig. 5.5b. Now, the three components of PP, PS, and PE are easily differentiated by contrast where the PE appears bright; PP, dark vertical "rivers"; and the PS, the darkest contrast of patches on the right and in the middle of the image. Clearly, the higher-order eigenmode clearly differentiates the three materials based on the phase contrast, thereby facilitating interpretation of dispersion and morphology of the various components.

5.5 Nanorheological mapping

Quantitative measurements of viscoelastic properties are important to the polymer community. First, quantitative measurements can facilitate unambiguous identification of components in various complex materials. Second, very often, there is either not enough material available for a bulk viscoelastic measurement or the material is made in a reactor so that its final form, e.g. micrometer-sized domains, is the only form that is available. Thus, being able to quantify properties such as storage modulus, loss modulus, and loss tangent on the nanoscale and use them to develop structure-property relationships can be very useful.

An important note of caution when making viscoelastic measurements with the AFM with regard to frequency space: AFM measurements are typically made at resonance frequencies of the AFM cantilevers, which occur at tens of kilohertz to megahertz. These frequencies are substantially higher than corresponding bulk viscoelastic measurements conducted by, for example, dynamic mechanical analysis (DMA), which occur at a few hertz up to a few hundred hertz. Time-temperature superposition can be used to convert viscoelastic measurements between time and frequency domains in order to extrapolate appropriately to the high frequency of AFM measurements.

Dynamic contact methods have emerged as a useful potential tool for quantitative viscoelastic measurements. Specifically, recent progress has been made with contact resonance to quantitatively probe viscoelastic properties of thermoplastics. In contact resonance, a tip oscillates with very small amplitudes while in contact with a sample, resulting in a contact resonance that is at a frequency that is typically substantially higher than the frequency of the freely oscillating frequency. Operation in the contact regime under a constant force set by the user results in a linear force displacement relationship between the tip and the sample that is simple to model (as opposed to nonlinear dynamic in tapping mode), resulting in the ability to extract viscoelastic parameters. The contact resonance frequency and lineshape are then measured and are a function of the substrate's material properties.

Contact resonance has thus been used to measure the storage modulus, loss modulus [22] and loss tangent [23] of a blend of PP, PE, and PS, as shown in Fig. 5.6. Corresponding topography (a) with the maps of storage modulus (b) and loss

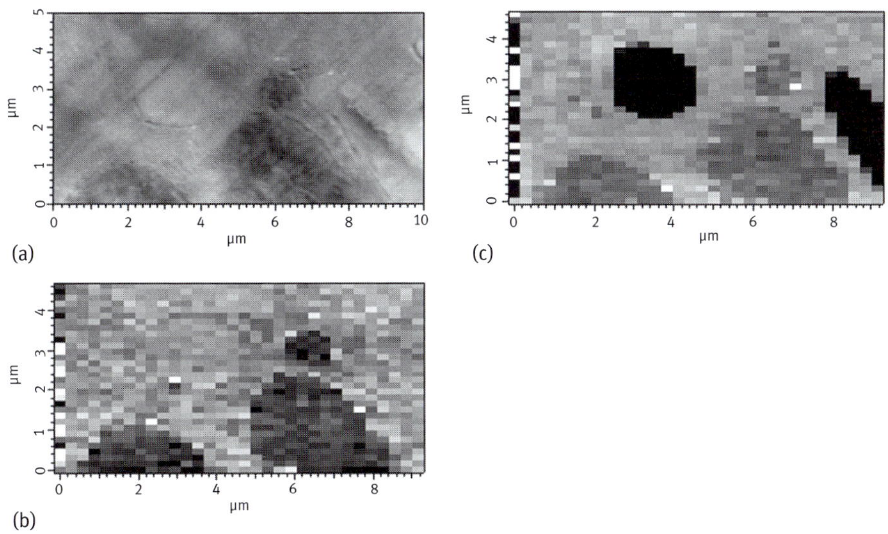

Fig. 5.6: (a) Topography of a blend of PP (matrix background), PS (large, smooth, round domains), and PE (smaller, rougher round domains). (b) Contact resonance storage modulus image showing the PE with lower (darker) storage modulus than the other two materials. The PP and PS are not differentiated here. (c) Contact resonance loss modulus image showing the PS with significantly lower (very dark) loss modulus with the PE having a slighter lower (slightly darker) loss modulus than the PP background. Reprinted with permission from Yablon DG, Grabowski J, Killgore JP, Hurley DC, Proksch R, Tsou AH, *Macromolecules*, 45, 4363–70, 2012 [22].

modulus (c) show that contact resonance effectively can measure the moduli of these materials and discriminate between them.

The values measured by contact resonance were in good agreement with the values measured for bulk individual components as measured by DMA. An important distinction between the storage/loss modulus measurement and the loss tangent measurement is that the former require a reference for absolute quantification, whereas the latter loss tangent measurement is reference free. A current limitation of the contact resonance technique is in its application to measuring elastomers. As shown in [22], the adhesion of the elastomer, coupled with a most certainly higher tip contact area due to the tip sinking into the elastomer, results in incorrect measurements on such materials.

Force-distance curves, which are single-point measurements, are a common source of nanorheological mapping as well. In a force-distance curve, the AFM tip is lowered onto the surface and then retracted while the force-*vs*-*z*-piezo position (which can be ultimately converted to a force vs tip-sample distance curve) is monitored. Force-distance curves exist in two forms: static and dynamic. Static curves are the force-distance curves described earlier. In dynamic force curves, the cantilever is oscillated at its resonance frequency as it is lowered onto the surface, and so the amplitude and phase are measured as a function of *z*-piezo position. More details about these single-point measurements can be found in elsewhere (see specifically chapter 3 of [24]). Force-distance curves can also be acquired in mapping mode called "force volume imaging." There are several approaches for rapid acquisition of force-distance curves so that it approaches realtime imaging speeds. The most popular rapid force curve acquisition methods are pulsed force mode (from Witec) and peak force QNM (from Bruker).

Once a force-distance curve has been measured, it then needs to be analyzed to extract viscoelastic properties. Several models exist, including Hertz model (no adhesion), Derjaguin-Muller-Toporov (DMT) model (adhesion outside the contact), and Johnson-Kendall-Roberts (JKR) model (adhesion within the contact); a complete treatment of these models is beyond the scope of this work but can be found elsewhere (see specifically chapter 2 of [24]).

Using force curves, the elastic modulus of a nanocomposite based on natural rubber and multi-walled carbon nanotubes (CNTs) was thus measured [25]. These studies used a JKR model in combination with a "two-point method" [26] to reveal a decrease in stiffness on the CNT and an intermediate modulus region around the CNT. This method was also used to investigate the interfacial properties and compatibilization effect in a blend of polyolefin elastomer and polyamide [27]. A similar method, but now with a superimposed oscillation while the cantilever was maintained on the specimen surface, was used to map the frequency dependence of the viscoelastic properties of various rubbers [28].

The DMT model has also been used to extract viscoelastic properties from force curves. The current commercial implementation of peak force QNM relies on the DMT model to extract a modulus value and has been used to measure moduli of a variety of polymers including low-density PE, acrylonitrile butadiene styrene, PS and polymethyl methacrylate (PMMA) [29], and polyurethanes and PS [30]. However, great care with calibrating various experimental parameters and matching the cantilever spring constant to the material need to be exercised to obtain accurate values with this method.

5.6 Thermal/spectroscopic measurements

In addition to topography- and mechanical-based measurements, AFM methods have been developed to probe thermal properties (scanning thermal microscopy) and even spectroscopic properties of polymers. Measurements of localized thermal properties take advantage of the cantilever to detect thermal transitions by several different methods including using an oscillating cantilever to sense temperature changes and using heated probes that can heat very small areas under the AFM tip. In this latter method, the cantilever deflection can then be measured as a function of localized heating, so that when the material under the tip softens or melts, the tip will drop or sink into the material, thereby producing a large deflection change at the thermal transition point. Localized thermal properties and transitions of a variety of polymers have been measured [31–33].

More recently, hybrid AFM-spectroscopic methods are being developed to measure chemical spectra with very-high-nanoscale resolution that can be correlated with simultaneously obtained topographic images. Near-field optical techniques to measure Raman and fluorescence on the nanoscale have been around for over almost two decades [34]. For the polymer community, however, infrared (IR) measurements to probe various organic stretches of interest have started to be the focus of much recent research and commercial activity. Typical IR measurements are limited by the diffraction of light, resulting in a best-case resolution of several micrometers (for attenuated total reflection measurements). Early work successfully measured IR spectra on the submicron scale of a thin film of PS and polyethyl acrylate blend [35]; however, these measurements required the use of a very expensive and complicated laser source and detectors. There are now two main current approaches to measure IR spectra on the nanoscale *via* either a thermo-mechanical approach or a scattering near-field scanning optical measurement. The thermo-mechanical approach has been successful at IR measurements of PE nanostructures [36, 37], with the latest results reporting measurements of 15-nm structures and a multilayer film of nylon and ethylene acrylic acid [38]. An example of this capability is given in Fig. 5.7, beside the topography image of overlapping PE and PS (a), AFM-IR single-wavenumber absorption images of PS and PE nanowires, where the 3026-cm^{-1} absorption (b) shows only PS nanowires, but absorption image at 2,920 cm^{-1} (c) reveals absorption of both PS and PE wires [37]. Scattering near-field optical microscopy measurements are also being developed with polymer specific applications reporting IR spectroscopic measurements of a PMMA/PS blend resolving several-hundred-nanometer domains of PS [39].

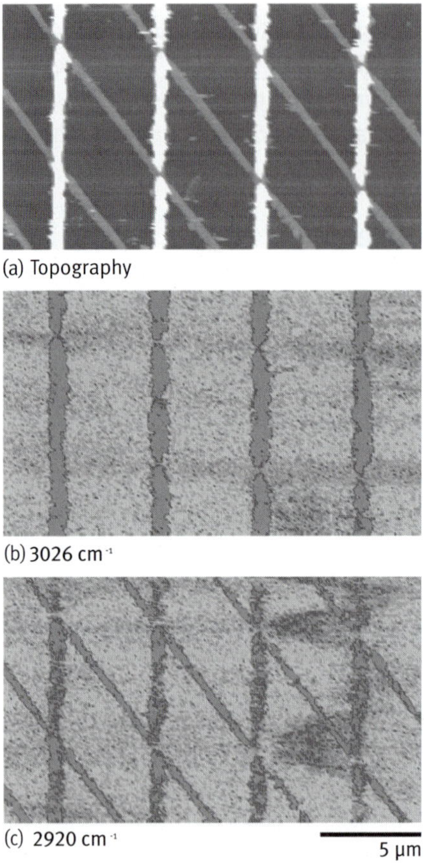

Fig. 5.7: (a) Topography image of overlapping PE and PS nanowires. (b) Single-wavenumber absorption AFM-IR image at 3,026 cm^{-1} image of PS nanowires (vertical features) showing only PS absorption. (c) Single-wavenumber absorption image at 2,920 cm^{-1} showing absorption of both PS (vertical) and PE (diagonal) wires. Reprinted with permission from Felts JR, Kjoller K, Lo M, Prater CB, King WP, *ACS Nano*, 6, 8015–21, 2012 [37].

5.7 Environmental measurements

The advantage of the SPM is that it can operate in a variety of environments including vacuum, fluid, and ambient (which are the conditions used in most of the measurements described in this chapter) and under a variety of conditions including heating/cooling and stretching. By taking advantage of heating stages, AFM has imaged *in situ* crystallization of a variety of polymers including PE, polyhydroxybutyrate, PP, and PMMA [40] following real-time processes such as nucleation, growth, and crystal thickening. The effect of environmental factors of humidity and temperature on polymer has been explored for polymer-based biomaterials of polybutyl methacrylate and polylauryl methacrylate (see specifically chapter 15 of [24]).

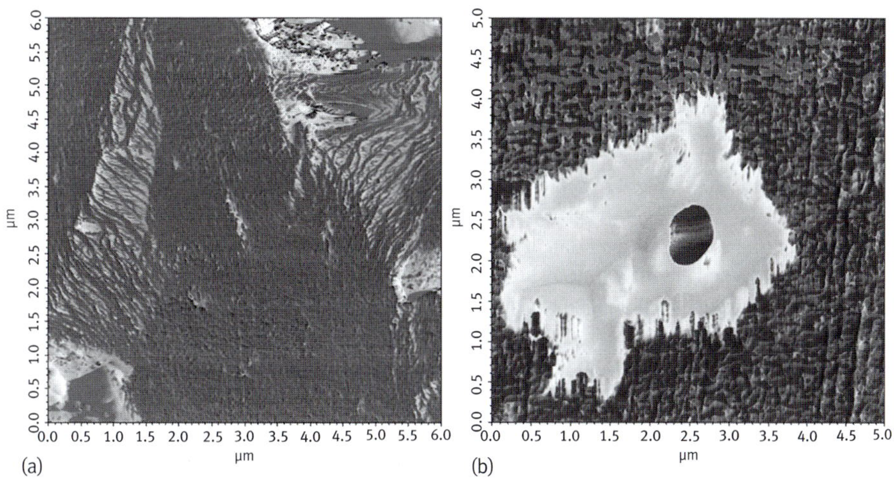

Fig. 5.8: (a) The 6 × 6-µm phase image of an ICP film under tensile stress showing crazing crack formation in the PP. (b) Stretch marks between the EPR and the PP matrix as the material is being pulled along the x-axis.

These polymers coated with drugs showed changes in topography and adhesion with temperature, important behaviors for performance testing of its application as a drug-eluting stent in the human body. Similar studies were conducted to study the effect of humidity and immersion in water of these coatings. Finally, AFM imaging of an ICP of PP with an EPR was studied as a function of tensile stress. As the material was stretched, AFM imaged crazing in the PP and equatorial stretching between the PP and rubber domains, as shown in Fig. 5.8a and b, respectively. *In situ* AFM measurements have been made on other polymer systems, including studying the effect on spherulite morphology in polybutene [41].

5.8 Conclusions

AFM has emerged as a powerful tool for nanoscale characterization of polymers. AFM can probe many material properties, where mechanical properties are the most useful for polymer applications with the ultimate goal of nanoscale discrimination and quantitative rheological measurements to guide, among other goals, the formation of structure-property relationships. Several key AFM-based methods to image *via* mechanical contrast are described here including the popular phase imaging, advanced multifrequency methods, and quantitative dynamic contact methods with applications to plastics blends, plastic/elastomer blends, nanocomposites, and thin films. Finally, AFM studies under environmental conditions such as heating and deformation as well as other AFM-based methods to measure nanoscale thermal and chemical properties of materials are also described.

References

[1] Binnig G, Rohrer H. Scanning tunneling microscopy. Surface Science Reports 1983;126:236–44.

[2] Binnig G, Quate CF, Gerber C. Atomic force microscopy. Phys Rev Lett 1986;56:930–3.

[3] Sawyer LC, Grubb DT, Meyers GF. Polymer microscopy. 3rd edn. New York: Springer; 2010.

[4] Cleveland JP, Anczykowski B, Schmid AE, Elings VB. Energy dissipation in tapping mode atomic force microscopy. Appl Phys Lett 1998;72:2613–5.

[5] Garcia R, Tamayo J, Calleja M, Garcia F. Phase contrast in tapping-mode scanning force microscopy. Appl Phys A 1998;66:s309–12.

[6] Garcia R, Tamayo J, Paulo AS. Phase contrast and surface energy hysteresis in tapping mode scanning force microscopy. Surf Interface Anal 1999;27:312–6.

[7] Martinez NF, Garcia R. Measuring phase shifts and energy dissipation with amplitude modulation atomic force microscopy. Nanotechnology 2006;17:S167–72.

[8] Ganguly A, Sarkar MD, Bhowmick AK. Morphological mapping and analysis of poly[styrene-b-(ethylene-co-butylene)-b-styrene] and its clay nanocomposites by atomic force microscopy. J Polym Sci B 2007;45:52–66.

[9] Han X, Hu J, Liu HL, Hu Y. A new approach to thick films of a block. Copolymer with ordered surface structures. Langmuir 2006;22:3428–33.

[10] Ganguly A, Bhowmick AK. Insights into montmorillonite nanoclay based ex situ nanocomposites from SEBS and modified SEBS by small-angle X-ray scattering and modeulated DSC studies. Macromolecules 2008;41:6246–53.

[11] Wang L, Hong S, Hu HQ, Zhao J, Han CC. The surface morphology evolution of an ultrathin film of poly[styrene-b-(ethylene-co-butylene)-b-styrene] during its dewetting process. Langmuir 2007;23(5):2304–7.

[12] Park JH, Sun Y, Goldman YE, Composto RJ. Amphiphilic block copolymer films: phase transition, stabilization, and nanoscale templates. Macromolecules 2009;42:1017–23.

[13] Mullin N, Hobbs JK. Direct imaging of polyethylene films at single-chain resolution with torsional tapping atomic force microscopy. Phys Rev Lett 2011;107.

[14] Lozano JR, Garcia R. Theory of multifrequency AFM. Phys Rev Lett 2008;100:076102.

[15] Rodriguez TR, Garcia R. Compositional mapping of surfaces in atomic force microscopy by excitation of the second normal mode of the microcantilever. Appl Phys Lett 2004;84:449.

[16] Proksch R. Multifrequency repulsive mode amplitude modulated atomic force microscopy. Appl Phys Lett 2006;89:113121.

[17] Gigler AM, Dietz C, Baumann M, Martinez NF, Garcia R, Stark RW. Repulsive bimodal atomic force microscopy on polymers. Beilstein J Nanotechnol 2012;3:456–63.

[18] Dietz C, Zerson M, Riesch C, et al. Nanotomography with enhanced resolution using bimodal atomic force microscopy. Appl Phys Lett 2008;92:143107.

[19] Martinez NF, Lozano JR, Herruzo ET, et al. Bimodal AFM imaging of isolated antibodies in air and liquids. Nanotechnology 2008;19:384001.

[20] Kiracofe D, Raman A, Yablon DG. Multiple regimes of operation in bimodal AFM: understanding the energy of cantilever eigenmodes. Beilstein J Nanotechnol 2013;4:385–93.

[21] Chakraborty I, Yablon DG. Cantilever energy effects on bimodal AFM: phase and amplitude contrast of multicomponent samples. Nanotechnology 2013.

[22] Yablon DG, Grabowski J, Killgore JP, Hurley DC, Proksch R, Tsou AH. Quantitative mapping of viscoelastic properties of polyolefin blends with contact resonance atomic force microscopy. Macromolecules 2012;45:4363–70.

[23] Yablon DG, Grabowski J, Chakraborty I. Measuring the loss tangent of polymer materials with atomic force microscopy based methods. Polymer 2014. http://dx.doi.org/10.1016/j.polymer.2014.02.027.

[24] Yablon DG, ed. Scanning probe microscopy in industrial applications: nanomechanical characterization. 1st edn. New York (NY): Wiley; 2013.
[25] Wang D, Fujinami S, Nakajima K, et al. Visualization of nanomechanical mapping on polymer nanocomposites by AFM force measurement. Polymer 2010;51:2455–9.
[26] Sun Y, Walker GC. Using the adhesive interaction between atomic force microscopy tips and polymer surfaces to measure the elastic modulus of compliant samples. Langmuir 2004;20:5837.
[27] Wang D, Fujinami S, Liu H, Nakajima K, Nishi T. Investigation of reactive polymer-polymer interface using nanomechanical mapping. Macromolecules 2010;43:5521–3.
[28] Igarashi T, Fujinami S, Nishi T, Asao N, Nakajima AK. Nanorheological mapping of rubbers by atomic force microscopy. Macromolecules 2013;46:1916–22.
[29] Young TJ, Monclus MA, Burnett TL, Broughton WR, Ogin SL, Smith PA. The use of the PeakForce™ quantitative nanomechanical mapping AFM-based method for high-resolution Young's modulus measurement of polymers. Meas Sci Technol 2011;22:125703.
[30] Dokukin ME, Sokolov I. Quantitative mapping of the elastic modulus of soft materials with HarmoniX and PeakForce QNM AFM modes. Langmuir 2012;28:16060–71.
[31] Fryer DS, Nealey PF, Pablo JJD. Thermal probe measurements of the glass transition temperature for ultrathin polymer films as a function of thickness. Macromolecules 2000;33:6439–47.
[32] Ihalainen P, Backfolk K, Sirvio P, Peltonen J. Thermal analysis and topographical characterization of latex films by scanning probe microscopy. J Appl Phys 2007;101:043505/043501.
[33] Zhou J, Berry B, Douglas JF, Karim A, Snyder CR, Soles C. Nanoscale thermal-mechanical probe determination of softening transitions in thin polymer films. Nanotechnology 2008;19;495703/495701.
[34] Lewis A, Lieberman K. Near field optical imaging with a non-evanescently excited high brightness light source of sub-wavelength dimensions. Nature 1991;354:214–6.
[35] Michaels CA, Gu X, Chase B, Stranick SJ. Appl Spectrosc 2004;58:257.
[36] Felts JR, Cho H, Yu MF, Bergman LA, Vakakis AF, King WP. Atomic force microscope infrared spectroscopy on 15 nm scale polymer nanostructures. Rev Sci Instrum 2013;84:023709.
[37] Felts JR, Kjoller K, Lo M, Prater CB, King WP. Nanometer scale infrared spectroscopy of heterogeneous polymer nanostructures fabricated by tip-based nanofabrication. ACS Nano 2012;6:8015–21.
[38] Kjoller K, Felts JR, Cook D, Prater CB, King WP. High sensitivity nanometer scale infrared spectroscopy using a contact mode microcantilever with an internal resonator paddle. Nanotechnology 2010;21:18705.
[39] Taubner T, Hillenbrand R, Keilmann F. Nanoscale polymer recognition by spectral signature in scattering infrared near-field microscopy. Appl Phys Lett 2004;85:5064–6.
[40] Hobbs JK, Farrance OE, Kailas L. How atomic force microscopy has contributed to our understanding of polymer crystallization. Polymer 2009;50:4281–92.
[41] Thomas C, Seguela R, Detrez F, Miri V, Vanmansart C. Plastic deformation of spherulitic semi-crystalline polymers: an in situ AFM study of polybutene under tensile drawing. Polymer 2009;50:3714–23.

Miroslav Šlouf, Taťana Vacková, František Lednický,
and Petr Wandrol

6 Polymer surface morphology: Characterization by electron microscopies

6.1 Introduction

This chapter deals with electron microscopy (EM) and its applications on polymer surface characterization. We show that the surface morphology of a polymer system is always closely connected with its composition (types of polymers/additives/fillers and overall composition), the processing conditions (thermal and mechanical history of the sample), and the sample preparation (we can observe polymer surfaces either "as received" or, more often, polymer surfaces intentionally prepared by fracturing, cutting, smoothing, etching, etc.). Moreover, we demonstrate how to use EM in characterization of subsurface and internal structure of polymer materials.

There are two basic types of EM: scanning electron microscopy (SEM) and transmission electron microscopy (TEM). In SEM, we usually work with *bulk* specimens, investigate the surface morphology, and collect signal with detectors placed *above* the sample. In TEM, we almost exclusively work with *ultrathin* specimens, investigate the internal morphology, and collect transmitted electrons using detectors localized *below* the sample. Therefore, SEM is the dominant EM technique for characterization of polymer surfaces. This chapter focuses on SEM (Section 6.2), briefly mentions TEM (Section 6.3), discusses the key step of polymer morphology studies, i.e. sample preparation (Section 6.4), and shows the typical applications of EM in polymer science (Section 6.5).

6.2 Scanning electron microscopy

This section summarizes the basic principles of SEM (Section 6.2.1), classical modes of SEM (Section 6.2.2), and new developments in SEM connected with characterization of polymers (Section 6.2.3). Other possibilities of SEM are briefly mentioned (Section 6.2.4). We explain the elements of SEM theory, but more details are to be found in specialized textbooks [1–4].

6.2.1 SEM: Principles

6.2.1.1 Microscope and image formation

SEM consists of a column and a chamber (Fig. 6.1). In the top of the column, primary electrons (PEs) (electron beam) are emitted from the electron gun (cathode). The com-

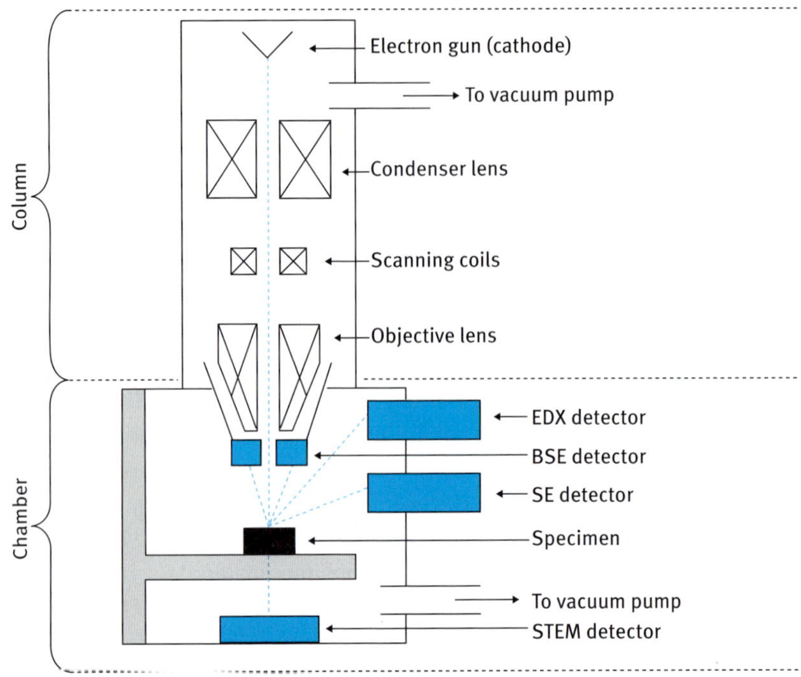

Fig. 6.1: Scheme of the SEM. PEs are emitted from the electron gun and interact with the specimens, generating secondary electrons (SE), backscattered electrons (BSE), characteristic X-rays (collected with EDX detector) and transmitted electrons (collected with STEM detector).

monly used cathodes include tungsten filament (thermionic gun), lanthanum hexaboride (LaB_6), or field-emission gun (FEG). The quality (and the price) of the electron guns is as follows: tungsten < LaB_6 < FEG. The FEG sources produce the electron beam with the highest brightness, the lowest energy spread, and the smallest final probe size, which all lead to the highest resolution of FEG-equipped SEM microscopes (which are occasionally called FEGSEM or FESEM).

The negatively charged electrons (charge $Q = e$) are emitted from the cathode and move to the positively charged anode due to accelerating voltage (U); the potential energy of the electron at the anode ($E_1 = eU$) changes to the kinetic energy of the electron ($E_2 = 1/2m_e v^2$; m_e = electron mass; v = velocity of the accelerated electron). Supposing that the potential energy is completely transferred to the kinetic energy ($E_1 = E_2$) and neglecting relativistic effects, we calculate the velocity, v, of the accelerated electrons by a straightforward combination of the above relations:

$$v = \sqrt{2eU/m_e} \tag{6.1}$$

In the column, the accelerated electrons go through the set of electromagnetic lenses/coils (from top to bottom: condenser lenses, scanning coils, and objective lens) and are focused on a small spot (typical spot size ranges from 1 µm to 1 nm). The scanning coils deflect the electron beam in the x and y directions so that the beam scans on the surface of the sample and the micrograph is collected pixel by pixel. Typically, both the column and the chamber are kept under high vacuum ($p < 10^{-3}$ Pa) because any gas molecules will spread and attenuate the electron beam. The *magnification* of a SEM microscope is given by the ratio of the scanned area size to the final micrograph size (typical magnifications range from 10× to 100,000×). The *resolution* of an SEM microscope depends on a number of parameters as discussed in Section 6.2.1.2.

6.2.1.2 Interaction of the electron beam with the specimen

The PEs in the electron beam travel at a very high speed. Typical accelerating voltages in SEM range from 30 to 1 kV. This corresponds, according to Eq. 6.1, to velocities of 0.06c and 0.34c, respectively, where c is the velocity of light. Despite these extreme speeds, the PEs do not penetrate too deep into the sample. A characteristic length scale of the PE penetration depth, called the maximum electron range R, can be estimated using the Kanaya-Okayama formula [5]:

$$R = 27.6\frac{M}{\rho Z^{8/9}}U^{5/3} \tag{6.2}$$

where R, M, ρ, Z, and U are the electron range (in nm), the molecular weight (in g/mol), the density (in g/cm^3), the sum of atomic numbers of consisting atoms, and the accelerating voltage of PEs (in kV), respectively. Calculation of R for polyethylene (M = 28.0 g/mol, ρ = 0.96 g/cm^3, Z = 16) gives the result ~20,000 and ~1,000 nm for 30 and 5 kV, respectively. Analogous calculation for gold (M = 197.0 g/mol, ρ = 19.32 g/cm^3, Z = 79) yields ~1,600 and ~80 nm, respectively. Several important conclusions follow from these simple calculations: (1) certain parts of SEM signal may come from subsurface layers in light-element polymer systems, (2) lower accelerating voltage reduces penetration depth, and thus it enhances signal from surface, and (3) very thin polymer films, fibers, and particles may appear translucent in SEM because a part of the electrons penetrates through.

The trajectories of PEs penetrating into a sample can be also modeled with Monte Carlo simulation programs [6]. The simulations for polyethylene and Au are shown in Fig. 6.2. We note that (1) the overall size of PEs interaction volume correlates with the predictions based on Eq. 6.2, (2) the shape of interaction volume changes from pear shape, in the case of light-element materials, to apple shape for heavy-element materials, and (3) the interaction volume decreases dramatically with the decreasing accelerating voltage U.

A scheme of electron beam interaction with the specimen is given in Fig. 6.3a. When the electron beam hits the specimen surface, numerous processes take place, the first four of which are of particular importance in polymer science:

- emission of low-energy (<50 eV), secondary electrons (SE);
- emission of high-energy (>50 eV), backscattered electrons (BSE);
- emission of X-radiation (characteristic X-rays);

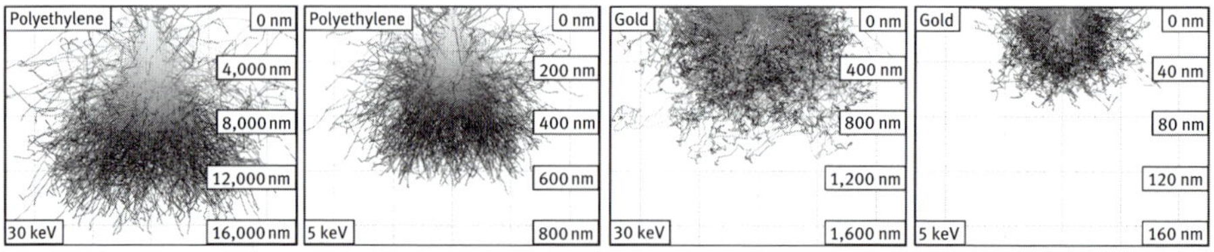

Fig. 6.2: Monte-Carlo simulations for 1,000 electron trajectories through polyethylene and Au, using accelerating voltage 30 and 5 kV and probe diameter 10 nm in all cases. (left to right) polyethylene, 30 kV; polyethylene, 5 kV; Au, 30 kV; Au, 5 kV. The thick trajectories going back to surface represent BSEs; their number increases for heavier elements. All other trajectories represent electrons that are absorbed in the sample. Note the scale change in the images.

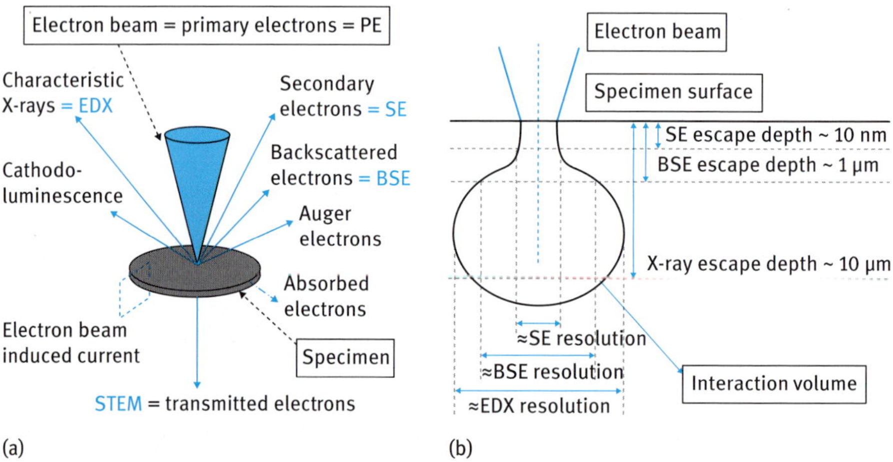

Fig. 6.3: Electron beam interaction with the specimen in SEM: (a) overview of signals generated during the interaction and (b) typical escape depths of the three most important signals – SE, BSE, and X-rays. The signals appear simultaneously, but we usually detect one of them at the time. Real escape depths depend strongly on the accelerating voltage and specimen (Eq. 6.2, Fig. 6.2).

- penetration of electrons through an ultrathin sample and their detection in the form of transmitted electrons (TE);
- emission of Auger electrons (AE);
- emission of light – cathodoluminiscence (CL);
- absorption of electrons.

SEM resolution is closely connected with interaction volume and escape depth, which differ for different signals as illustrated in Fig. 6.3b. The highest resolution (~1 nm in top microscopes) is achieved in SE imaging because the SE are emitted from the very subsurface region. Lower resolution (usually units to tens of nanometers) is obtained in BSE imaging because the BSE electrons have higher escape depth, which results in wider lateral spread. The lowest resolution (usually units to tens of micrometers) is obtained in energy-dispersive analysis of X-rays (EDX, also called energy dispersive spectroscopy, or EDS) because the characteristic X-rays are able to escape from the highest depths, where the beam is widely spread. Scanning transmission electron microscopy (STEM) resolution is comparable to that of SE, as the samples for STEM must be ultrathin (<100 nm) so that they were electron-transparent, and as a result, the electrons leave the sample before being spread laterally.

6.2.2 SEM: Classical modes

The above-listed four main SEM signals (SE, BSE, X-ray, and TE) correspond to four classical modes of SEM, which are applied for polymer morphology investigations: SE imaging, BSE imaging, EDX spectroscopy, and imaging with transmitted electrons (STEM). SE micrographs (Fig. 6.4a) exhibit topographic contrast, i.e. the morphology

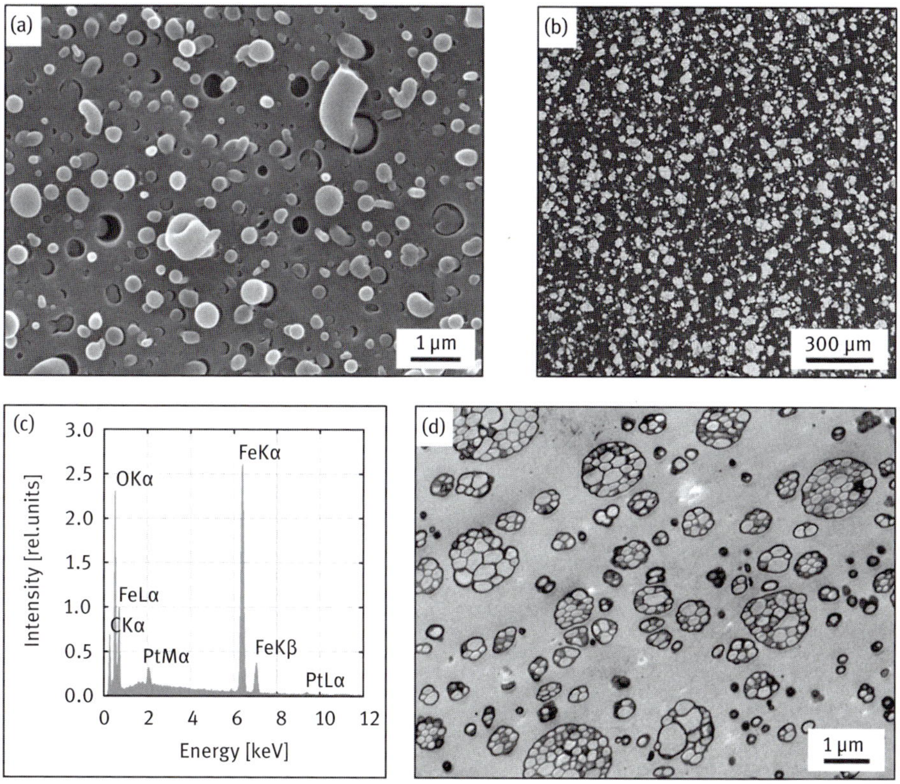

Fig. 6.4: Examples of the four main modes of SEM in polymer science. (a) SE imaging: fracture surface of PP/COC blend – topographic contrast. (b) BSE imaging: smooth surface of epoxy composite filled with aluminum oxide particles – material contrast. (c) EDX spectrum: microanalysis of polymer microspheres filled with magnetic nanoparticles – elemental analysis. (d) STEM micrograph: OsO_4-stained ultrathin section of high impact PS – internal morphology.

of the surface. BSE micrographs (Fig. 6.4b) display material contrast, i.e. the difference between atomic numbers of the components. EDX spectra (Fig. 6.4c) yield information about elemental composition of the sample. STEM micrographs (Fig. 6.4d) visualize the internal morphology of samples.

6.2.2.1 SE imaging

Secondary electron imaging (Fig. 6.4a) represents the most common mode of SEM. SE micrographs show *topographic contrast*. This is because SE have low energies (<50 eV), and so they can escape only from the thin surface layer (~10 nm). The geometry of the surface layer is closely associated with the number of emitted electrons (Fig. 6.5). Surfaces perpendicular to electron beam yield average SE signal (Fig. 6.5a), which increases with the inclination of the surface (Fig. 6.5b), and achieves maximum at surfaces parallel with the electron beam (Fig. 6.5c) – this extreme intensity of SE signal on sharp edges of the specimen is called *edge effect*. Holes in the sample surface exhibit a decrease in SE intensity due to SE electron absorption (Fig. 6.5d), whereas hills may show an intensity increase if their dimension is comparable to probe size (Fig. 6.5e).

The SE micrograph in Fig. 6.4a exhibits all types of contrast shown in Fig. 6.5: flat surfaces have medium intensity (effect in Fig. 6.5a); inclined areas and edges exhibit increased intensity (effects in Fig. 6.5b and c); the hills above the surface and the holes below the surface display higher and lower intensity, respectively (effects in Fig. 6.5d and e). Three additional pieces of information are worth mentioning here: (1) there are yet another, not-so-frequent types of SE contrast [7], but those shown in Fig. 6.5 suffice for interpretation of most SE micrographs of polymer materials; (2) Lower accelerating voltage increases surface detail at the expense of resolution; (3) edge effect weakens with decreasing accelerating voltage.

6.2.2.2 BSE imaging

Backscattered electron imaging (Fig. 6.4b) is applied mostly for multicomponent systems because the number of backscattered electrons depends on chemical composition – this is called *material contrast*. Let us first consider the difference between SE and

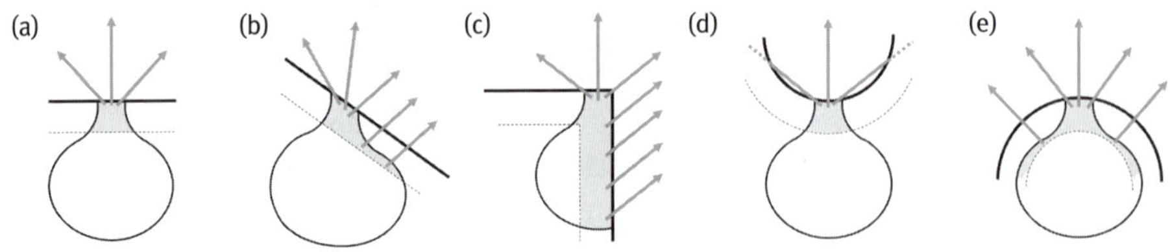

Fig. 6.5: Origin of topographic contrast in SE imaging. Thick black line, sample surface; thinner black dashed line, SE escape depth; arrows, SE emitted from the surface.

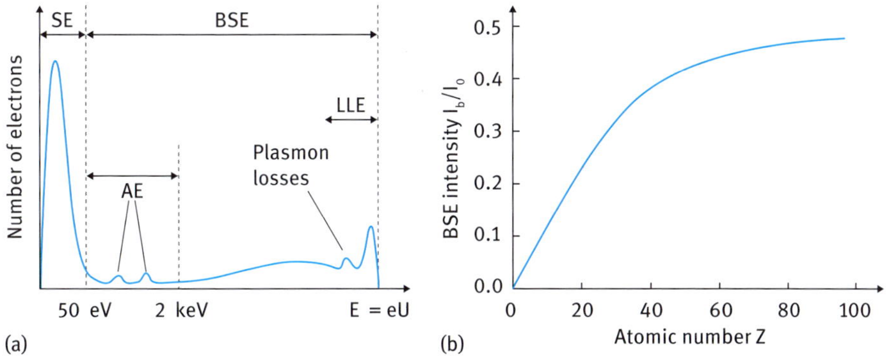

Fig. 6.6: Properties of BSEs. (a) The energy of BSE ranges from 50 eV to energy of PEs ($E = eU$, Section 6.2.1.1). (b) The intensity of BSE signal increases with atomic number Z. Redrawn according to Reimer L, *Scanning electron microscopy*, Springer, Berlin (Germany), 1998 [2].

BSE: SEs are regarded as the electrons detached from specimen atoms during collisions of PE with the specimen, whereas BSEs are supposed to be the original PE that were very deflected by the collisions with atoms that they were turned back out of the specimen.

SEs exhibit low energies because most of the PE energy is consumed for the detachment from the atom, whereas BSE electrons have higher energies, depending on their collision history. By convention, the limit between SE and BSE is set at 50 eV (Fig. 6.6a). It has been shown experimentally [8] and later also calculated theoretically that the intensity of backscattering increases with increasing atomic number Z (Fig. 6.6b). The fact that number of BSE electrons increases with Z is evident also from Monte Carlo simulations shown in Fig. 6.2.

Material contrast in BSE imaging is demonstrated in Fig. 6.4b: we clearly differentiate aluminum oxide filler (bright) from the polymer matrix (dark). At low accelerating voltages (~1 kV), the relationship in Fig. 6.6b ceases to hold, and increased backscattering may not indicate higher Z [8]. If BSE detector is composed of two semi-annular segments A and B, it can show either material contrast (signal from A + B) or topographic contrast (signal from A–B) [7]; however, the topographic contrast (and higher resolution; see Fig. 6.3b) is usually obtained with SE imaging.

6.2.2.3 EDX spectra

EDX (Fig. 6.4c) is a spectroscopic method that yields elemental composition in microscopic scale. Typical EDX resolution in polymer materials ranges from 10 to 20 μm in both horizontal and vertical directions (Fig. 6.2 and Fig. 6.3). A somewhat lower resolution can be achieved with lower accelerating voltages (Fig. 6.2 and Fig. 6.3), but the energy of the electrons (in keV), which equals the accelerating voltage (in kilovolts), must be at least three times higher than the rightmost peak in the EDX spectrum to get the correct intensities for quantitative analysis [9].

The origin of characteristic X-rays, which are detected in EDX, is shown schematically in Fig. 6.7a; the image illustrates that the energy of emitted X-rays corresponds to the energy difference between the electron shells. In a one-electron atom (only a rough approximation but it is sufficient for showing the principle), the energy of the electron, E, depends only on the principal quantum number, n (Eq. 6.3 [9]):

$$E(n) = -\mathrm{Ry}\frac{Z^2}{n^2} \tag{6.3}$$

where Ry = 13.60 eV is the Rydberg unit of energy and Z is the atomic number. The transitions in a one-electron atom are shown in Fig. 6.7b. The energy corresponding to those transitions can be estimated by Eq. 6.3. For instance, the energy of FeKα transition (main peak in Fig. 6.4d) is calculated as $E = E(2) - E(1) = -13.60(26^2/2^2 - 26^2/1^2) =$ 6.9 keV, which reasonably agrees with the experiment (E = 6.4 keV) (Fig. 6.4d). More precise energies of X-ray emission lines can be calculated using Moseley's laws [10], and precise values are to be found in specialized software, databases, and tables [10].

Modern EDX detectors can detect the lightest elements starting from lithium. *Qualitative* EDX analyses are usually reliable, but experienced user should always check the computer-generated results. *Quantitative* analysis can be done without standards (standardless analysis; results are only semiquantitative) or with standards (more

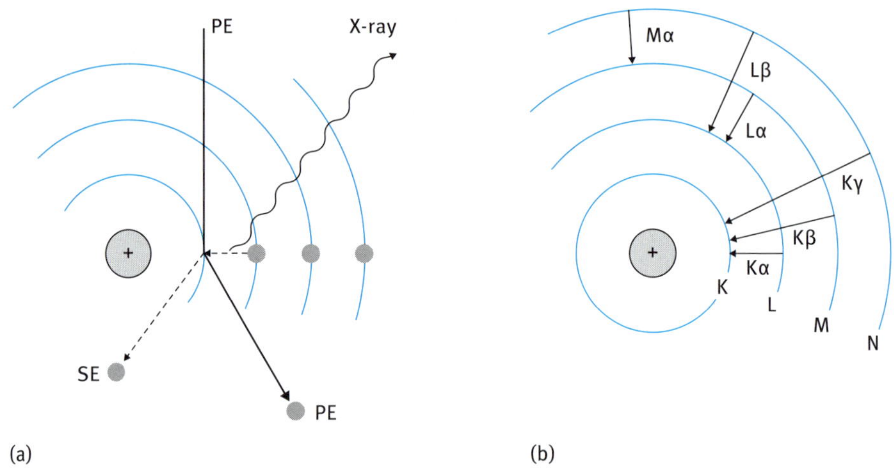

Fig. 6.7: (a) Origin of the EDX signal. PE ejects the electron from the inner shell, the electron from outer shell fills the hole, and the energy difference is emitted as a characteristic X-ray. (b) Simplified nomenclature of X-ray emission lines. The lines are named according to the shell in which the initial vacancy occurs (K, L, M, etc. shells correspond to principal quantum number n = 1, 2, 3, etc.) and the shell from which the electrons fills the hole (α, β, γ, etc. correspond to the first, second, third, etc. adjacent shell). The emission/transition shown in (a) is Kα; in real atoms, the nomenclature becomes more complex due to the complexity of their electronic structure, but the main emissions/peaks in EDX spectra are Kα, Kβ, Lα, Lβ, and Mα (Fig. 6.4d [9]).

precise, but suitable standards are rarely available for newly developed polymer systems). Commercial EDX systems perform an analysis from a selected point or user-defined area on a previously saved SE/BSE/STEM micrograph. It is also possible to measure line scans or two-dimensional (2D) maps (i.e. concentration profiles/maps of selected elements); these are rarely applied to polymer systems because longer data collection times cause sample damage.

6.2.2.4 STEM imaging

STEM imaging (Fig. 6.4d) in SEM is regarded as an additional technique of minor interest in most fields of materials science [2], but it is one of the useful microscopic methods in polymer science [3, 4]. STEM requires electron-transparent, ultrathin samples (thickness <100 nm) with sufficient contrast between the components (contrast of one component is frequently increased by *staining*). Preparation of these ultrathin and stained specimens is discussed in Section 6.4.

Thicker, heavier-atom-containing or diffracting areas of the sample appear dark on a bright background if we collect transmitted electrons (bright-field imaging, or BF) or bright on a dark background if we detect scattered electrons (dark-field imaging, or DF). Segmented STEM detectors make it possible to record both BF and DF images from the same location. STEM/BF micrographs are more common, but STEM/DF micrographs may exhibit higher contrast at the expense of total intensity (higher data collection times, increased risk of sample damage).

In the field of polymer science, STEM can be applied to the (1) visualization of internal morphology using ultrathin sections from the interior of the material (Fig. 6.8a), (2) visualization of surface and sub-surface morphology using ultrathin cross sections from the surface of the material (Fig. 6.8b), and (3) direct visualization surface/internal morphology of thin nanoparticles, nanofibers, or nanofilms deposited on electron-transparent carbon films (Fig. 6.8c).

6.2.3 SEM: Modern trends

6.2.3.1 Variable-pressure SEM

Variable-pressure SEM (VP-SEM) can be used for the observation of nonconductive, frozen, and wet specimens. In VP-SEM, the electron column is kept under high vacuum ($p \leq 10^{-3}$ Pa), whereas the pressure in the specimen chamber is higher ($p > 10$ Pa). The gas in the specimen chamber (usually H_2O or N_2) removes the charge from a nonconductive specimen (typical chamber pressure $p \approx 100$ Pa; the method is called low-vacuum SEM, or LV-SEM) or allows the observation of wet/frozen samples (typical pressures $p \approx 1{,}000$ Pa; the method is called environmental SEM, or ESEM).

In standard, high-vacuum SEM (HV-SEM), the negative charge delivered by the PEs to the surface of the insulator is not conducted off. This effect, called *charging*,

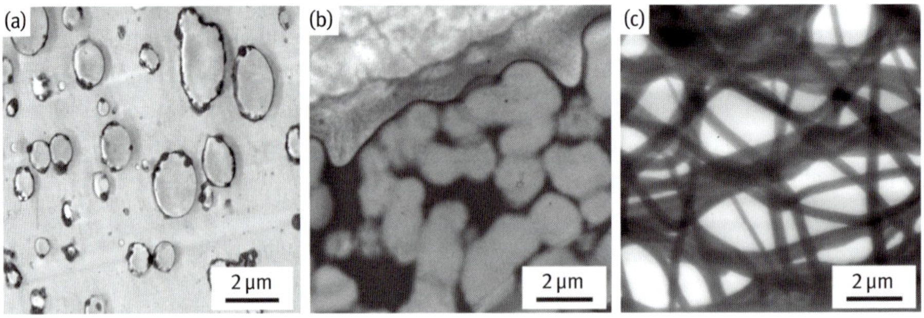

Fig. 6.8: Typical applications of STEM in polymer science. (a) OsO_4-stained ultrathin section from the interior of the sample, (b) RuO_4-stained ultrathin cross section showing the surface and subsurface regions of the sample, and (c) polymer nanofibers on an electron transparent carbon film. The first micrograph (a) shows the localization of the styrene-butadiene block copolymer on the surface of PE microparticles in the PS matrix [11]. The second micrograph (b) displays the surface and subsurface phase structure of PP/PE copolymer microparticle embedded in epoxy resin [12]. The third micrograph (c) visualizes the surface and overall morphology of PS nanofibers [13]. All micrographs are BF images.

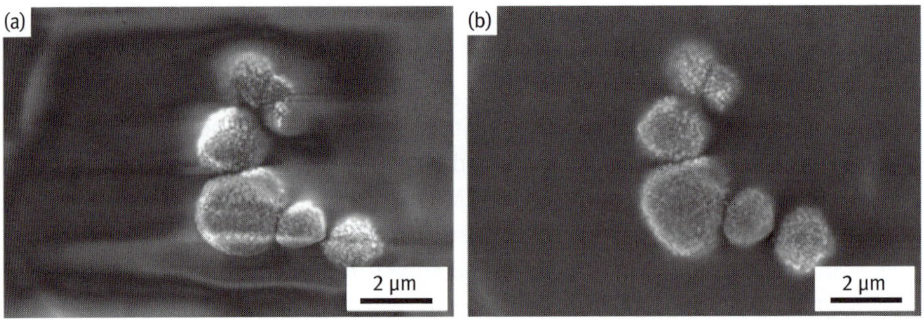

Fig. 6.9: Polymer microcomposite, uncoated, accelerating voltage 5 kV: (a) HV-SEM with oversaturated and undersaturated areas caused by sample charging and (b) the same sample observed in LV-SEM at $p = 90$ Pa.

influences both primary and signal electrons and causes artifacts in the image (Fig. 6.9a). The sample can be covered with a thin conductive layer, which usually solves the charging problem (Section 6.4.2.1). If covering the sample is impossible or impractical, the surface can be visualized by LV-SEM (Fig. 6.9b). The negative charge on the specimen surface in LV-SEM is removed, thanks to *cascade ionization*: signal electrons ionize gas molecules in the specimen chamber and produce positive ions and further electrons for ionization – *ionization avalanche*. The positive ions from the avalanche are attracted toward the negatively charged specimen surface and neutralize the surface charge, whereas the negative electrons are usually used as the imaging signal. Also, PEs in LV-SEM interact with the gas molecules, being scattered due to collisions with the gas molecules in the experimental chamber (*skirt effect*). The skirt effect negatively influences resolution and noise but can be minimized by very short sample-objective distance (minimal PE gas path length).

VP-SEM is also applied on the observation of samples containing liquids. The technique is usually called ESEM, as outlined in the first paragraph of this section. A relative humidity of ~100% at the sample area must be maintained to keep the sample hydrated. Humidity depends on the temperature (T) of the sample and pressure (p) in the specimen chamber. At room temperature (T = 22°C), we need high pressure (p = 2,700 Pa) to keep the sample 100% hydrated. Such a pressure prevents high-resolution, low-noise imaging. Therefore, the sample is usually cooled to lower temperatures ($T \leq 5$°C; using a cooling stage inside the specimen chamber), at which lower pressures (600–900 Pa) are sufficient in maintaining 100% humidity and the quality of the image improves significantly. Thanks to pressure and temperature control, ESEM is suitable for both static experiments such as visualization of morphology of polymer hydrogels and dynamic experiments such as hydration and contact angle measurements [14].

In VP-SEM, we may use SE and BSE imaging as well as EDX analysis. Special SE and BSE detectors exist for both LV-SEM ($p \approx 100$ Pa) and ESEM ($p \approx 1{,}000$ Pa), whereas EDX detectors work at all pressures (HV-SEM, LV-SEM, and ESEM). Finally, the VP-SEM can be applied on frozen samples (in combination with cooling/freezing stage; Section 6.2.3.2, Fig. 6.10b and c) or even on polymer nanoparticles in solution (cryo-SEM [15] or STEM-in-ESEM [16, 17]). Detailed explanation of VP-SEM principles and practice can be found in [14]; a few polymer-related applications are shown in Section 6.5.5.1.

6.2.3.2 Variable-temperature SEM

Samples can be cooled, freezed, or heated inside the SEM chamber. Sample cooling to temperatures around 0°C (typically by a simple cooling stage inside the microscope chamber) is usually used in ESEM to control sample humidity (Section 6.2.3.1, Fig. 6.10a). Fast sample freezing at a very low temperature (typically by liquid N_2, using special equipment both outside and inside the microscope [15]) can be applied to solidify samples containing volatile components, such as polymer hydrogels (Fig. 6.10b) or polymer self-assemblies (Fig. 6.10c), which may contain >90% of water, or self-assembled polymer nanoparticles, micelles, and vesicles in solution [15–17]. More detailed description of the cryogenic SEM can be found in [18]. Sample heating enables us to study temperature-induced transformations of sample morphology such as crystallization or phase transitions [19, 20]. Various heating modules capable of heating the sample up to 1,500°C are commercially available. The heating experiments can be performed in the high vacuum or in a gas atmosphere (H_2O, N_2, air, He) influencing the observed processes, such as corrosion [21] or reduction and oxidation reactions [22]. The monitoring of processes at high temperatures and/or pressures is called *in situ* SEM [14].

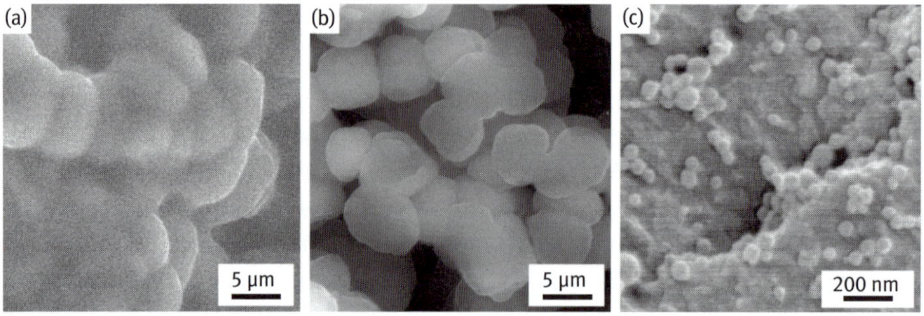

Fig. 6.10: SEM micrographs of (a,b) poly(2-hydroxyethyl methacrylate) hydrogel and (c) poly(4-hydroxystyrene)-block-poly(ethylene oxide) block copolymer nanoparticles. The first two micrographs show the same sample imaged by ESEM at 2°C (a) and LV-SEM at –10°C (b) [23]. The last micrograph is a cryo-HV-SEM image at –135°C, displaying the spherical nanoparticles in flash-frozen solution/ice [15].

6.2.3.3 Low-voltage SEM

The observation of the sample at accelerating voltages <5 kV [24] is called low-voltage SEM. In principle, it is possible to perform low-voltage SEM with a standard/thermionic SEM, but the resolution and image quality rapidly decrease with decreasing kilovolts. High-resolution low-voltage SEM investigations require instruments equipped with a FEG (Section 6.2.1.1 [25]) and immersion final lens (ultrahigh-resolution SEM). Further improvements in the electron optics [26] together with the deceleration of electrons just above the sample [27] completely equalize the resolution at high and low kilovolts. The main benefit of low-voltage SEM is *detailed information that can be obtained from the sample surface* due to lower penetration depths at low kilovolts, which means that the signal comes from the topmost surface layer (Section 6.2.1.2, Fig. 6.2). The second important advantage of low-voltage SEM is the *lower charging and sample damage*, because the difference between PEs reaching the sample surface and the signal electrons (SE, BSE, etc.) leaving the sample surface decreases at lower kilovolts. The main disadvantages of low-voltage SEM, in comparison with high-voltage SEM, is its somewhat lower resolution and signal, although with the latest microscopes, these drawbacks are minimized. State-of-the-art low-voltage SEM provides both SE and BSE information at accelerating voltages down to hundreds of volts [28]. Recently, low-voltage EDX analysis, which leads to higher lateral resolution, became available for materials research [29]. Low-voltage SEM is of particular importance for nonconductive, electron-beam-sensitive materials such as synthetic polymers, as exemplified in Section 6.5.5.2.

6.2.3.4 Multidimensional SEM

SEM yields 2D information of sample topography and composition. However, real samples are three-dimensional (3D), and the depth information is often important. As we focus our attention on polymer surface characterization, 3D SEM methods will be briefly discussed.

The 3D *visualization of surface morphology* can be obtained by stereomicroscopy. Two images of the sample at two different angles/tilts (*stereo images*) are taken and processed by a specialized software [30]. Quantitative depth information can be extracted from the stereoimages using sophisticated mathematical algorithms [31]. Another way to obtain 3D information of the sample topography is to combine images from two differently located SE detectors [32] or images from SE and BSE detectors [33].

The *3D visualization of subsurface morphology* is achieved by methods that incorporate either mechanical sectioning or focused ion beam sputtering. Mechanical sectioning (or slicing) is made using ultramicrotomy (Section 6.4.3.4). The ultramicrotome [34] is located either outside the microscope (ultrathin sections in a given order are observed using TEM or SEM) or even inside the specimen chamber (such a method is called serial block face SEM, and the surfaces are observed with SE or BSE). Focused ion beam SEM (also called FIB-SEM or dual-beam SEM [35]) uses a focused beam of heavy ions (ion beam; the first beam) to etch off a thin layer from a selected area on the specimen surface and an electron beam (the second beam) to observe the subsurface morphology. In the final step of all 3D methods, a set of 2D images is reconstructed into a final 3D model.

There are *further methods of multidimensional microscopy* that deserve at least a brief note. Nondestructive 3D SEM was introduced recently: the area of interest is scanned repetitively by PEs of different energies, and hence different depths of origin of detected BSE, which is used in the final 3D reconstruction [36]. 3D TEM is used mostly in biology: set of 2D micrographs from one (ultrathin) sample at various tilt angles (typically ±70°, step 1°) is reconstructed into the final 3D model. Time-resolved SEM/TEM (where the third dimension is time) can monitor *in situ* processes in liquid or gas environment in both SEM/TEM [37].

6.2.4 SEM: Further possibilities

The remaining modes of SEM that have not been mentioned in the previous sections are rarely used in polymer science but are included in this work for the sake of completeness. The methods can be divided into three groups.

6.2.4.1 Spectroscopic methods

EDX (or EDS, Section 6.2.2.3) belongs to the classical modes of SEM. Wavelength-dispersive analysis of X-rays (WDX, or WDS) is analogous to EDX (the same signal is detected), but WDX data collection times are longer (the main disadvantage for electron-beam-sensitive polymers) and the results are more precise (with better signal-to-noise ratio and narrower peaks). Auger electron spectroscopy (AES) detects emitted electrons with characteristic energies, which are used analogously with EDX and WDX for microanalysis. Cathodoluminiscence (CL) detects electron-beam-generated light (luminescence) and is quite often applied to analysis of luminescent minerals.

6.2.4.2 Methods connected with crystal structure of samples

These methods connected with crystal structure of samples are hardly applicable on polymer materials because synthetic polymers are almost never completely crystalline. The methods are electron backscattered diffraction (EBSD), electron channeling contrast imaging (ECCI), and electron channeling pattern (ECP); in special microscopes, it is possible to use low-energy electron diffraction (LEED) and reflection high-energy electron diffraction (RHEED) [1, 2].

6.2.4.3 Methods using sample-specific interactions with electron beam

In semiconductors, PE can generate electron-hole pairs, which results in measurable electron-beam-induced current (EBIC mode of SEM). In magnetic specimens, the PE and SE are deflected due to the magnetic fields, which make even the observation of magnetic fields possible. Other special SEM methods are discussed elsewhere [2].

6.3 Transmission electron microscopy

TEM is applied on polymer characterization analogously with STEM (Section 6.2.2.4). Polymer sample preparation is the same for both methods. If we observe the same sample in STEM and TEM, STEM micrographs usually exhibit higher contrast (which is related with the lower accelerating voltages of SEM microscopes in comparison with conventional TEMs) and lower resolution (which is limited by spot size in SEM, but not in conventional TEM [1]). The typical applications of STEM (Fig. 6.8) and TEM (Fig. 6.11) in polymer science are analogous: TEM can be applied to study the detailed internal morphology using ultrathin sections from the interior of samples, such as

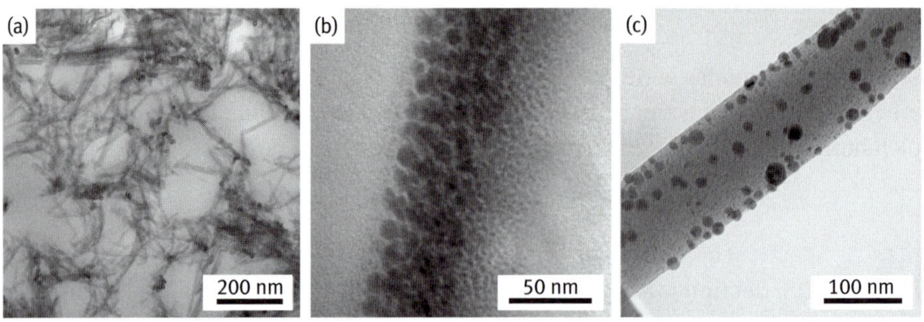

Fig. 6.11: Typical applications of TEM in polymer science: (a) ultrathin section from the interior of a sample, (b) ultrathin cross section showing surface and subsurface regions, and (c) polymer nanoparticles on an electron transparent carbon film. The first micrograph (a) shows the dispersion of titanate nanotubes [38] in the PP matrix. The second micrograph (b) displays the cross section of poly(ethylene terephthalate) embedded in epoxy resin; the polymer surface contains implanted nickel nanoparticles [39]. The third micrograph (c) visualizes polyurethane nanofiber covered with Ag nanoparticles [40]. All micrographs are BF images.

polymer nanocomposites (Fig. 6.11a), surface and subsurface morphology using ultrathin cross sections perpendicular to sample surface (Fig. 6.11b), and morphology of polymer nanoparticles and nanofibers, which are tiny enough to be directly deposited and visualized on electron-transparent carbon film (Fig. 6.11c).

As for polymer morphology, BF imaging (shown in Fig. 6.11) is by far the most important mode of TEM. EDX in TEM is based on the same principles as in SEM (Section 6.2.2.3), but a higher resolution can be achieved [1]. Other modes of TEM (electron diffraction, DF imaging, scanning transmission mode of TEM in combination with electron spectroscopy, high-resolution TEM, etc.) are used mostly for analysis of inorganic crystalline materials [41].

As for polymer surface morphology, SEM is the dominant technique, and TEM plays only a supplementary role. One specific TEM sample preparation method for the study of polymer surfaces – replication – is mentioned in Section 6.4. Other details concerning TEM (such as scheme of the microscope, principle of imaging, electron diffraction, etc.), which are of minor interest for polymer surface characterization, can be found elsewhere [1, 3, 41–43].

6.4 Sample preparation

6.4.1 Overview of polymer materials

From the point of view of morphology and sample preparation, polymer materials can be divided into several groups (Fig. 6.12). Single-phase systems (amorphous polymers, miscible polymer blends) are morphologically homogeneous, and thus, the microscopic studies are limited to phenomena such as orientation, deformation, and fracture. Multiphase polymer materials containing only polymer components (semicrystalline polymers, block copolymers, and immiscible blends) usually require specific sample preparations due to low contrast between the phases. Multiphase systems with non-polymer components (filled systems and composites) must be prepared by different techniques, taking into account the different mechanical properties of the polymer matrix and the filler.

6.4.2 Specific features of polymer materials

6.4.2.1 Charging and electron-beam damage

In a typical SEM experiment (nonconductive materials such as polymer, bulk sample, and accelerating voltage >5 kV), the amount of PE penetrating into the sample is higher than the amount of SE and BSE emitted from the surface. Consequently, the sample tends to *charge*, which results in lower-quality micrographs or in image artifacts (Section 6.2.3.1 and Section 6.2.3.3). Moreover, the nonconductive polymer

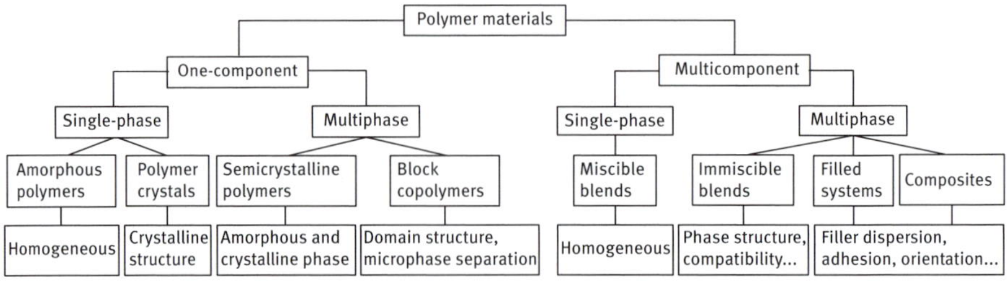

Fig. 6.12: Types of polymer materials according to their initial composition and resulting morphology.

specimens are more sensitive to electron-beam-induced damage (surface deformation, cracking, and/or contamination). Charging and electron-beam damage can be avoided by means of several methods. The first one is coating the sample with a thin layer of a (semi)conductive material using commercially available sputter-coating or thermal evaporation devices [44]. In SEM, we usually coat the samples with heavy elements such as Au (average grain size ~10 nm), Pt (~4 nm), or Cr (~2 nm), which also increase the signal, i.e. the number of emitted SE and BSE electrons due to higher Z (Fig. 6.2). In both SEM and TEM, we use a carbon (semiconductive light element, amorphous layer), which does not hide the observed subsurface structures and increases sample stability under electron beam. The second method, which can be used for the elimination of charging is VP-SEM (Section 6.2.3.1). The third method is the use of low-voltage SEM, which usually leads to lower sample damage and to an increase in surface detail at the expense of resolution (Section 6.2.3.3.). In TEM, a lower acceleration voltage (lower than ~5 kV) usually means higher contrast but also higher sample damage because electrons are absorbed in the sample.

6.4.2.2 Skin-core effect

Most bulk synthetic polymer products are made from the melt using processes such as mixing, molding, compression, extrusion, etc. During these processes, the polymer melts are subjected to flow (different flow rates at the surface and inside) and thermal treatment (different cooling rates at the surface and inside). Consequently, the morphology close to the surface (skin) frequently differs from the morphology of the interior (core). Morphological changes include size, concentration and/or orientation of lamellae in semicrystalline polymers (Fig. 6.13a), particles in polymer blends (Fig. 6.13b), or fillers in polymer composites (Fig. 6.13c). The skin-core effects are also observed in solution-prepared polymer materials (Fig. 6.13d) and natural polymers.

Two important conclusions concerning sample preparation follow from the skin-core effects: (1) If we study the external surface, we should be aware of the fact that the internal morphology may differ significantly. (2) If we analyze cross sections (such as fracture surfaces, cut surfaces, or sections), we should take micrographs from at least two regions – close to the original surface (skin morphology) and from the middle of the specimen (core morphology).

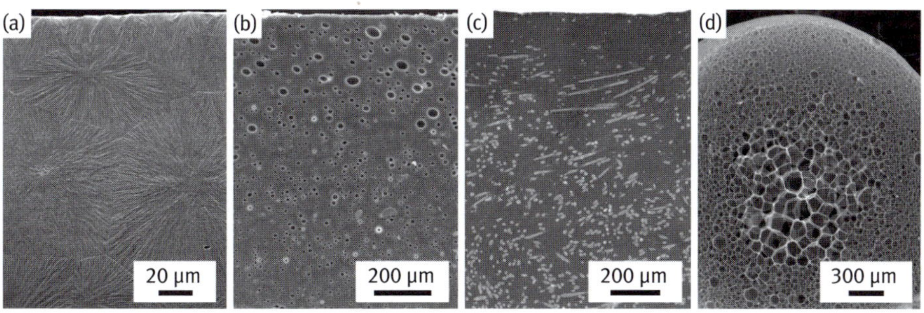

Fig. 6.13: Typical skin-core effects in polymers: (a) crystalline lamellae in PP crystallized on highly oriented pyrolytic graphite, (b) smooth surface of PS with styrene maleic anhydride, (c) short glass fibers embedded in polyamide polymer matrix, and (d) cross section of PS foam bead.

6.4.2.3 Low contrast between components

From the point of view of elemental composition, synthetic polymers are frequently quite similar. Consequently, the yield of SE, BSE, TE, and characteristic X-rays from the individual components of multiphase polymer systems is almost the same. Therefore, the polymer systems are rarely observed "as received." Instead, their morphology has to be made visible by suitable preparation techniques (fracturing, etching, staining, etc.) as described in Section 6.4.3.

6.4.3 Preparation techniques for polymer materials

The polymer materials can be observed "as received" (Fig. 6.14a) or their morphology is revealed by preparation techniques, the most important of which are fracturing, etching, or cutting (Fig. 6.14b–d). Direct observation of polymer surfaces (Fig. 6.14a) is usually limited to the observation of overall shape of polymer micro/nanoparticles (Section 6.4.3.1). Often, we "open" the structure, preparing fracture surfaces (Fig. 6.14b), etched surfaces (Fig. 6.14c), and cut surfaces or sections (Fig. 6.14d). On those surfaces, we usually achieve higher topographic contrast in SE (fracture surfaces; Section 6.4.3.2) or material contrast in BSE (cut and stained surfaces; Section 6.4.3.3 and Section 6.4.3.4).

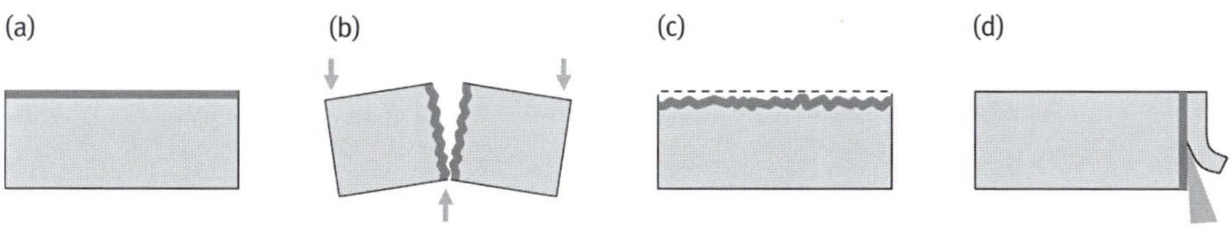

Fig. 6.14: Main types of polymer surfaces from the point of view of sample preparation: (a) original surface, observed directly; (b–d) surfaces prepared by fracturing (b), etching (c), or cutting (d), which often show more details of surface, subsurface, and/or interior morphology. The observed surfaces are denoted with thick line.

6.4.3.1 Direct observation of polymer surface

The morphology and surface of polymer microparticles (Fig. 6.15a), nanoparticles (Fig. 6.15b), or fibers (Fig. 6.15c) are usually observed without complicated sample preparation. The sample is placed onto a flat surface, fixed with a paste or glue, surface-coated and observed in the SE mode of SEM. Nanoparticles do not have to be glued because the surface coating fixes them sufficiently. If we want to observe a sample without surface coating, conductive support (such as mica pre-sputtered with platinum), conductive glue (such as commercially available silver paste), lower accelerating voltage (lower penetration depth and charging), and/or low-vacuum microscopy have to be applied. Direct observation of bulk polymer surfaces is less frequent. The surfaces of bulk polymers only reflect the processing-associated features (roughness of the molds, scratches due to subsequent machining or polishing). If these features are of interest, the sample preparation is analogous to that of polymer micro/nanoparticles.

6.4.3.2 Fracturing

The fracture of any material is controlled by its mechanical properties and applied forces. In polymer sample preparation, we differentiate between brittle fractures (smooth fracture surfaces, sharp fracture lines, no deformation textures; Fig. 6.16a) and ductile fractures (fracture surface with visible plastic deformations/textures, no sharp fracture lines; Fig. 6.16c). If the fractured material contains heterogeneities, these frequently become centers of secondary fractures (Fig. 6.16b), and typical conic fracture lines are formed on the brittle fracture surface [45].

Because (micro)plastic deformations hide the original morphology of the sample, the brittle fracture surfaces are usually preferred. Fracturing at low temperature (mostly in liquid nitrogen) is used with those polymer materials, which are not sufficiently brittle at room temperature. Care should be taken to let the sample cool thoroughly (samples completely submerged in liquid nitrogen bath, cooling time ≥ 5 min, tools for fracturing cooled in the same way as the sample). After fracturing, the polymer sample is fixed on a conductive support, sputter-coated, and observed in SEM.

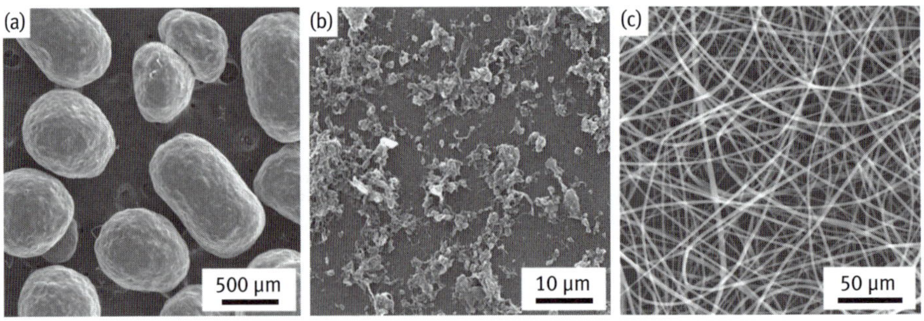

Fig. 6.15: Typical examples of the direct observation of surface morphology: (a) polymer microparticles for drug delivery, (b) isolated ultrahigh-molecular-weight PE wear microparticles and nanoparticles from periprosthetic tissues, and (c) poly-γ-benzyl-L-glutamate nanofibers.

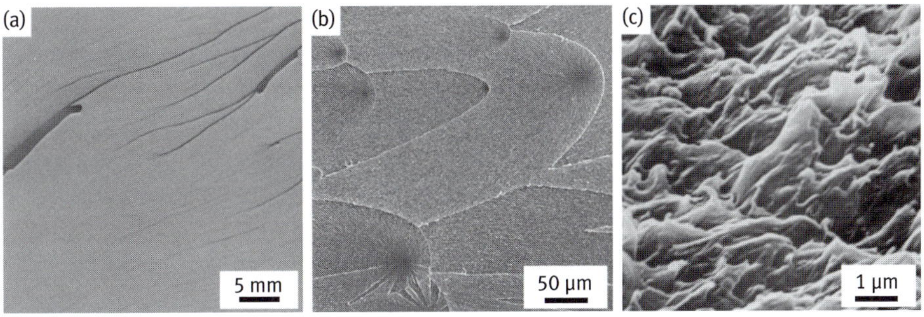

Fig. 6.16: Fracture surfaces: (a) brittle fracture in epoxy resin, (b) brittle fracture in epoxy resin/rubber polymer with centers of secondary fracture, (c) ductile fracture with microplastic deformations in acrylonitrile-butadiene-styrene copolymer [46].

Fracturing is a quite universal and widely used technique in polymer microscopy. It reveals the morphology of polymer blends (Fig. 6.4a [47]) and composites (Fig. 6.18c [48]). It is also a unique method in finding minute amounts of heterogeneities or defects present in material, as the defects are easily localized on fracture surface because they are the centers of secondary fracture (Fig. 6.17a [45]). Special fracturing conditions can be used when classical preparations fail: for instance, fracturing with elastomeric matrix (at elevated temperatures) is advantageous for amorphous polymers with low amount of inorganic filler (Fig. 6.17b [49]).

6.4.3.3 Etching

Etching consists of selective removal of component(s) from a multiphase polymer system (Fig. 6.12) by chemical or physical treatment. The surface for etching is obtained by any of the methods given in Fig. 6.14, i.e. "as received," fractured, etched, or cut surfaces. Chemical etching means treating the polymer surface with a

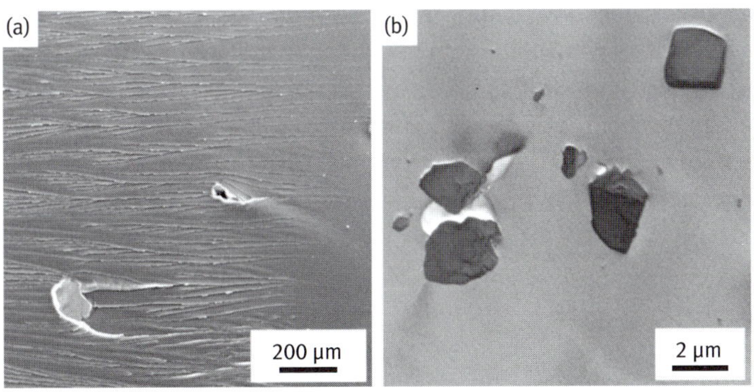

Fig. 6.17: (a) Brittle fracture in an epoxy resin, with inorganic inhomogeneities acting as sources of secondary fracture. (b) Soft matrix fracture surface of ethylene-propylene-diene-monomer rubber with a low content of $CaCO_3$ particles.

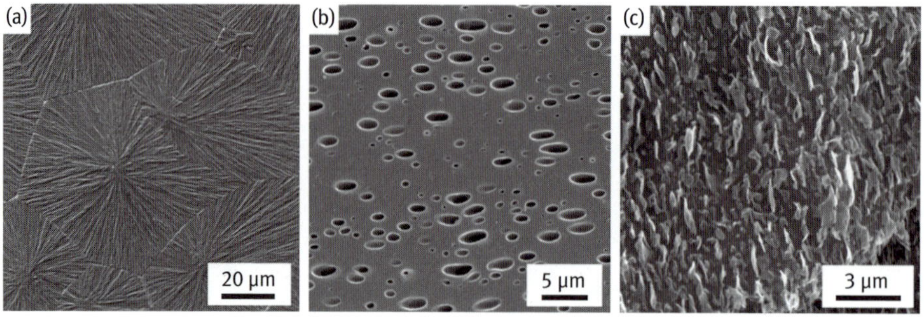

Fig. 6.18: SEM/SE micrographs of (a) PP spherulites etched with permanganic mixture, (b) PP/EPR polymer blend etched with *n*-hexane, and (c) PP/MMT composite fracture surface etched with plasma.

sample-specific chemical, called *etching agent*. The physical etching of the surface, called ion beam etching or plasma etching, is carried out in special devices, where the surface is exposed to a high-energy beam.

Figure 6.18 shows three typical examples of etching, applied on multiphase polymer systems: semicrystalline polymer (Fig. 6.18a), immiscible polymer blend (Fig. 6.18b), and polymer composite (Fig. 6.18c). The polypropylene (PP) in Fig. 6.18a was crystallized on mica decorated with Au nanoparticles, the mica was peeled off, and the unmodified surface was etched with permanganic mixture [50]. The PP/ethylene propylene rubber (PP/EPR) polymer blend in Fig. 6.18b was cut with a microtome and the cut surface was smoothed under liquid nitrogen [51] and etched with permanganic mixture [52]. The PP/montmorillonite (PP/MMT) composite in Fig. 6.18c was fractured and etched by plasma [53].

The permanganic mixture is a universal etching agent that etches almost all polymers, and their different rates of etching lead to the visualization of their morphology. In semicrystalline polymers (Fig. 6.18a), the amorphous phase is etched faster, and the crystalline lamellae emerge [54]. In immiscible polymer blends (Fig. 6.18b), the different etching rate of the components reveals the phase structure [55]. Other etching agents applied on specific polymers include toluene [56], *n*-hexane [57], etc. Plasma etching gives the best result if one component is much more resistant to ion bombardment – a typical example is a polymer composite with hard inorganic filler (Fig. 6.18c [53]).

6.4.3.4 Cutting and staining

Cutting is a frequently used technique that yields *cut surfaces* for SEM and *ultrathin sections* for (S)TEM. In most cases, cutting is combined with *etching* (described in Section 6.4.3.3) or *staining* to achieve sufficient contrast. The two most

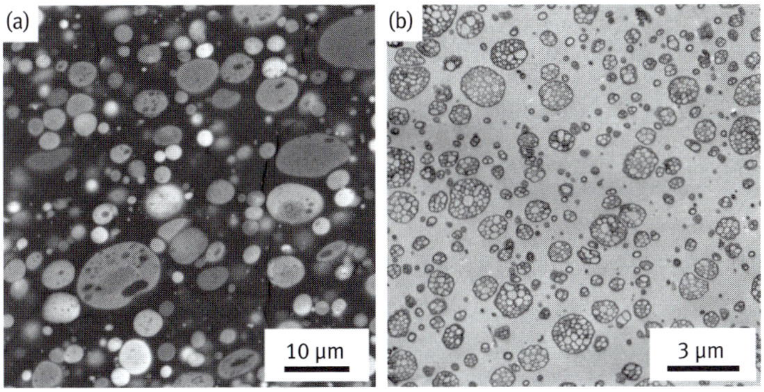

Fig. 6.19: SEM micrographs of high-impact PS prepared by ultramicrotomy followed by OsO_4 staining: (a) cut surface, BSE imaging; (b) ultrathin section, TE micrograph, BF. The stained particles appear darker than unstained matrix in TE micrograph and brighter in BSE micrograph due to the higher atomic number of the staining agent.

important staining agents for polymer materials are OsO_4 and RuO_4. During staining, the heavy atoms (Os or Ru) selectively bind to one phase, which then yields higher signal in SEM/BSE (Fig. 6.19a) or (S)TEM (Fig. 6.19b) due to a higher atomic number Z.

OsO_4 staining is more specific because osmium attacks selectively the components containing double bonds [58]. RuO_4 staining is more universal because ruthenium stains almost all common polymers (an important exception is polymethyl methacrylate and its derivatives) [47]. Other staining agents containing heavy elements (uranyl acetate, phosphorotungstenic acids) are used less often [59, 60].

Cut surfaces and/or thin sections can be prepared by three different techniques: (a) microtomy, (b) ultramicrotomy, and (c) smoothing in liquid nitrogen. Other methods are unsuitable for polymer systems because they produce rough and/or deformed surfaces. Microtomy yields cut surfaces and thin sections (thickness ~10 µm) that are thin enough for light microscopy. Ultramicrotomy gives even smoother cut surfaces and ultrathin sections (thickness ~50 nm), which are transparent for (S)TEM. Smoothing consists in polishing the sample submerged in liquid nitrogen using a freshly broken glass. This method yields only cut surfaces, but it is simpler and faster than (ultra)microtomy and yields high-quality surfaces for morphology studies of polymer blends (Fig. 6.18b and Fig. 6.19a [51, 52]).

6.4.3.5 Special techniques

In addition to the classical sample preparation techniques mentioned above (Sections 6.4.3.2–6.4.3.4), any other physical or chemical procedures can be applied to reveal the surface or subsurface morphology. The techniques can be divided into several groups: (1) combined techniques, (2) less frequently used techniques, and (3) specific techniques for wet specimens. Typical examples follow.

Combinations of various preparation techniques

Combined techniques may be advantageous, if not necessary. The combinations are not limited to the typical examples illustrated in Fig. 6.18 and Fig. 6.19 (cutting/etching, Fig. 6.18b; fracturing/etching, Fig. 6.18c; cutting/staining, Fig. 6.19). Other examples of combined techniques are cutting/annealing [61] or heating/fracturing [49]. Possible combinations and other procedures depend solely on the investigated systems, their structure, and properties.

Less frequently used techniques

These techniques may improve the study of specific samples. Interfacial adhesion between polymer matrix and filler is usually examined on fracture surfaces (Section 6.5.3), but it can be also monitored by scratching the surface and SE imaging of the scratch morphology [62, 63]. The phase structure of multicomponent systems is revealed by fracturing, cutting, staining, etching, and their combinations, as described above (Sections 6.4.3.2–6.4.3.4, and 6.4.3.5.1). However, the surface relief on the cut surface can also be induced by selective swelling [64]. The heterogeneous phase structure of polymer composites can be revealed by fracturing at the temperature, at which the polymer matrix is soft (above its glass transition temperature T_g) while the dispersed phase is hard (below its T_g) [49]. In the cases when neither cutting nor fracturing is applicable (hard and resistant polymer microcomposites), sample polishing in combination with BSE imaging can solve the problem (Fig. 6.13c). Abrasive papers are to be used with successively finer graininess, similar to metallographic sample preparation procedures [65]. As for STEM and TEM, two specific, less common, but useful techniques are worth mentioning: (1) negative staining facilitates the visualization of polymer micro/nanoparticles [59] and (2) surface replication makes it possible to visualize surface at high-resolution using TE imaging [66].

Specific preparation techniques for wet specimens

These techniques have already been discussed in Section 6.2.3.1 and Section 6.2.3.2 because of their close connection with variable-pressure and variable-temperature SEM. To summarize, water-containing specimens in electron microscopy can be observed in dried, frozen, or hydrated form. The *drying* of water-containing specimens must be carried out in such a way that their morphology did not collapse [67]. *Freezing* must be performed quickly (liquid N_2, liquid propane) so that the ice crystals will be negligible and will not destroy the internal structure [15, 68]. *Direct observation of hydrated/wet samples* requires a specialized SEM microscope that allows controlling pressure and temperature in the specimen chamber (Section 6.2.3.1 [69]).

6.5 Applications

6.5.1 Homopolymers

Homopolymers contain one type of repeating unit in the polymer chain. Very definite differences in morphology and properties occur between amorphous homopolymers (which cannot crystallize) and semicrystalline homopolymers (which crystallize to a significant extent). Fully crystalline homopolymers (or polymer single crystals) are exceptional, and their morphology is discussed elsewhere [70]. The ability of a polymer to form a crystal depends on the regularity, symmetry, and chirality of the monomer units forming the homopolymer chain [71].

Amorphous polymers (such as polystyrene, or PS) are morphologically homogeneous (Fig. 6.12). They do not exhibit any ordered structures in wide-angle X-ray scattering and do not have a first-order melting transition in differential scanning calorimetry (DSC). Fig. 6.20 summarizes the proposed supermolecular structure models of amorphous homopolymers: random chains (Fig. 6.20a), partially ordered chains, resembling highly irregular crystalline lamellae (Fig. 6.20b and c), and meander model (Fig. 6.20d). Small-angle neutron scattering measurements demonstrated that amorphous polymers contain randomly oriented molecules (Fig. 6.20a). Meanwhile, aging (change of macroscopic properties as a function of time) of amorphous polymers indicates that some structural changes and chain rearrangements take place [73]. Consequently, the main morphological features of amorphous polymers are density fluctuations and entanglements between the polymer chains, which exhibit very low contrast and cannot be observed by SEM or TEM [74, 75]. Therefore, the morphology of amorphous polymers limits microscopic research to studies of deformation and fracture (for the type of fracture, see Fig. 6.16; for deformation zones, crazes, deformation, and shear bands, see [76]).

Fig. 6.20: Possible morphologies of amorphous homopolymers [72]: (a) statistical or random-coil model, (b) aligned chains in fringed micellar model, (c) bundle or micelle model, and (d) meander model.

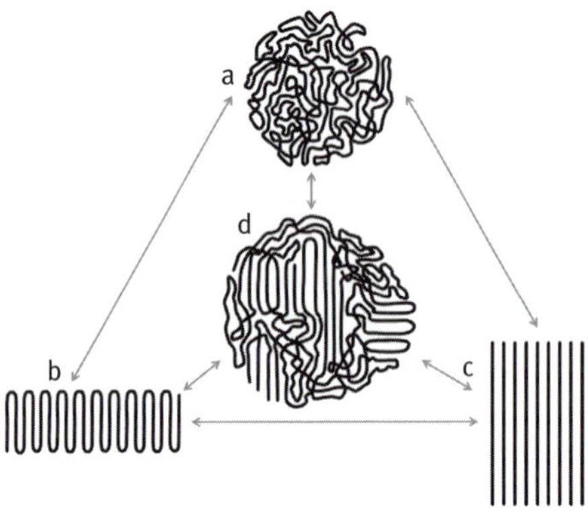

Fig. 6.21: Chain conformations in semicrystalline polymers: (a) random coils, (b) folded and (c) extended chain lamella, and (d) mixture of (*a* + *b* + *c*).

Semicrystalline polymers (such as polyethylene) contain amorphous and crystalline regions/phases (Fig. 6.12). The polymer chains are either randomly oriented (Fig. 6.21a) or packed in lamellae (Fig. 6.21b and c). Real conformations of the polymer chains are usually a mixture of all three cases (Fig. 6.21d) and depend on processing history (temperatures, pressures, nucleating agents/fillers, etc.). Typical bulk semicrystalline polymer consists of an amorphous phase (Fig. 6.21a) and crystalline lamellae (Fig. 6.21b) in the form of flat crystallites organized into polymer spherulites (scheme in Fig. 6.22). An example of polymer spherulites, observed in SEM microscope, was given in Fig. 6.13a. Other common morphologies of semicrystalline polymers include randomly oriented lamellae in amorphous matrix, oriented lamellae and/or fibrous structures, epitaxial lamellar overgrowth on microfibrils, hedrites or axialities, dendrites, and single lamellae [77].

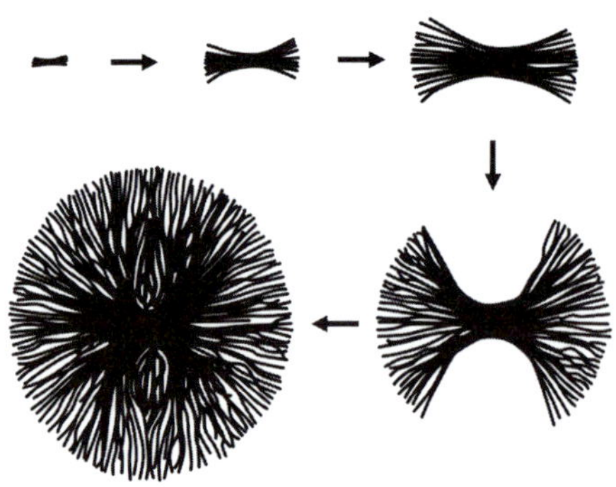

Fig. 6.22: Flat crystalline lamella (that corresponds to Fig. 6.21b) and its growth with branching, which results in formation of polymer spherulite [78].

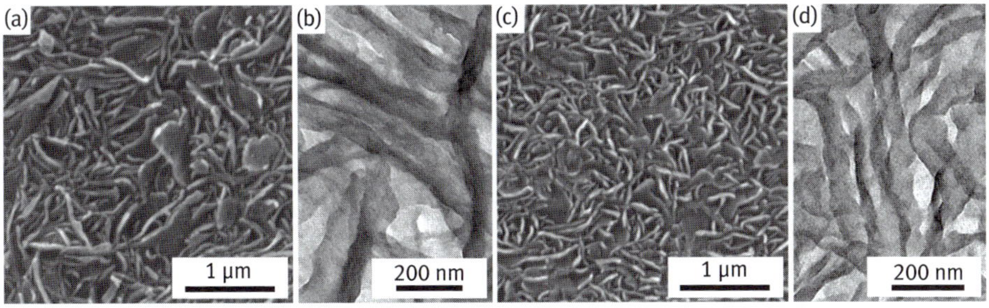

Fig. 6.23: SEM/SE (a,c) and TEM/BF (b,d) micrographs of UHMWPE crystalline lamellae in amorphous matrix: (a,b) virgin UHMWPE and (c,d) remelted UHMWPE (sample thermally treated at 150°C).

Ultrahigh-molecular-weight PE (UHMWPE) is a semicrystalline polymer used as bearing material in total joint replacements [79]. It contains crystalline lamellae randomly oriented in the amorphous matrix (Fig. 6.23). The weight fraction and thickness of the lamellae depend on the processing, and they influence the mechanical properties of the polymer [80]. The lamellae can be visualized by EM: SEM micrographs (Fig. 6.23a and c) show cut surfaces (prepared by microtomy) etched by permanganic mixture, whereas TEM micrographs (Fig. 6.23b and d) display ultrathin sections (prepared by ultramicrotomy) stained with oleum (details on sample preparation are given in [80]). The difference between virgin polymer (Fig. 6.23a and b) and remelted polymer (Fig. 6.23c and d) is clearly visible.

6.5.2 Copolymers

Copolymers contain two or more repeating units in the polymer chain. *Statistical/random copolymers* have statistically/randomly linked repeating units in the linear chain, whereas in *alternating copolymers*, the repeating units alternate regularly. *Block copolymers* contain two (or more) long blocks of the same repeating units. Other geometries include *branched* and *star copolymers* [81].

From the point of view of morphology, statistical, random, branched, and star copolymers are usually amorphous because their irregular molecular structure prevents crystallization. Alternating copolymers can be both amorphous and semicrystalline, depending on the symmetry and regularity of their chains. Block copolymers, composed of immiscible blocks, exhibit microphase separation (Fig. 6.24 [81]) and a wide range of morphologies (Fig. 6.25 [81]).

Figure 6.26 shows three typical examples and applications of block copolymers. The pentablock copolymer of PS and polybutadiene (PS-b-PB; Fig. 6.26a) belongs to an important class of polymers called thermoplastic elastomers, i.e. materials exhibiting both thermoplastic and elastomeric properties [74]. The bi-continuous lamellar structure of the block copolymer in Fig. 6.26a was obtained from 50% PS and 50% PB;

Fig. 6.24: Microphase separation in block copolymers composed of two immiscible blocks A and B. Blocks A and B tend to minimize mutual contacts and separate from each other, forming various structures shown in Fig. 6.25 [81].

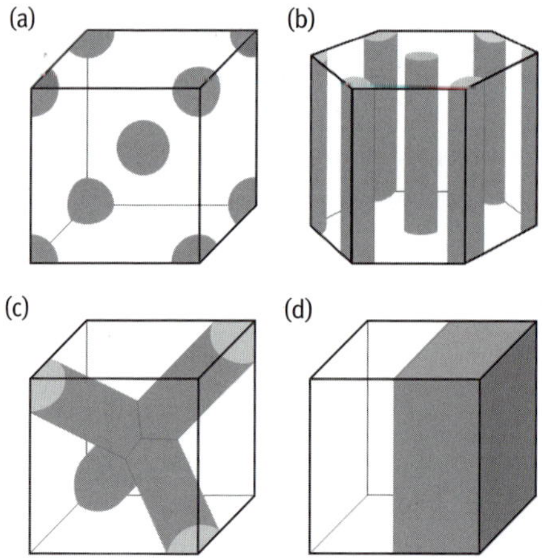

Fig. 6.25: Morphology of block copolymers A-block-B, composed of two immiscible blocks A and B. As the concentration of block A increases, the structure changes from cubic (a) to hexagonal (b), gyroid (c), and lamellar (d) [81].

the sample was prepared by ultramicrotomy, stained with OsO_4, and observed in TEM using BF imaging.

The copolymer of PS and poly(acrylic acid) (PS-b-PAA; Fig. 6.26b) is self-assembling in the solution and forms polymer vesicles [16]. Block copolymer self-assemblies, such as micelles and vesicles, are studied due to their possible applications in drug delivery. Self-assembled PS-b-PAA vesicles can be observed in dried state using standard SEM (as illustrated in Fig. 6.26b) or in hydrated state using wet STEM (as shown in our study [16]).

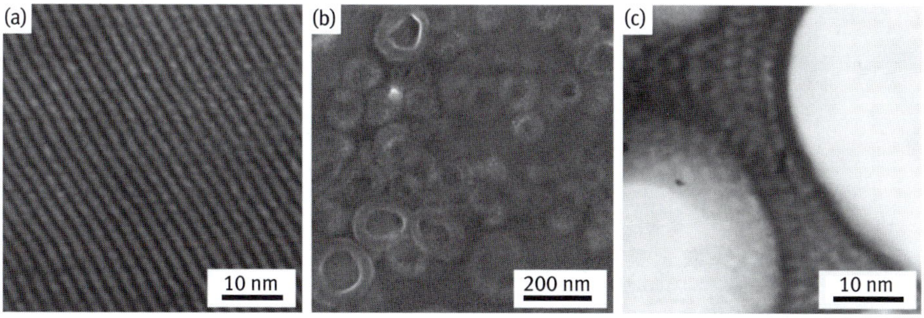

Fig. 6.26: Samples showing the variability of copolymer structures: (a) TEM micrograph of lamellar pentablock PS and PB copolymer with lamellar morphology (Fig. 6.25d); (b) SEM micrograph of dried polymer vesicles, which are formed by copolymer of PS-b-PAA; (c) TEM micrograph of PB particles with PS-block-PB compatibilizer in PS/PB polymer blend; the PS-b-PB compatibilizer between the particles exhibits hexagonal morphology (Fig. 6.25b).

A di-block copolymer of PS-b-PB (Fig. 6.26c) was used as a *compatibilizer* in the PS/PB polymer blend. The block copolymer compatibilizers are used to increase interphase adhesion between the components of polymer blends, stabilize the morphology, and improve final mechanical properties [82]. Fig. 6.26c shows a TEM micrograph of PS-b-PS copolymer with hexagonal morphology (Fig. 6.25b) between the two PS particles. The bulk PS/PB blend sample was prepared by melt mixing, and the specimen for TEM was prepared by ultramicrotomy in combination with OsO_4 staining (Section 6.4.3.4).

6.5.3 Polymer blends

Polymer blends consist of two or more different polymers. Most polymers are mutually incompatible [82], and thus, the polymer blends are multiphase systems (Fig. 6.12). The phase morphology of the immiscible polymer blends (particulate, fibrous, co-continuous) is closely connected with their properties (modulus, toughness, gas permeability) [82]. Therefore, blends morphology control (by optimizing composition, processing conditions, and/or compatibilization) and visualization by electron microscopy have been attracting the attention of numerous researchers [11, 75].

Compatibilization of polymer blends increases the adhesion between the blend components (interphase adhesion) and improves mechanical performance. Block copolymers (Section 6.5.2) with blocks that are identical or miscible, or that adhere to the related components of a blend, are frequently used as compatibilizers [11]. SEM micrographs in Fig. 6.27 illustrate the typical changes of blend morphology as a function of compatibilizer concentration: a non-compatibilized PS/low-density PE blend (PS/LDPE) exhibits strong phase separation with the formation of large particles and fracture running along the interface (Fig. 6.27a). The addition of 5% of PS-b-PB

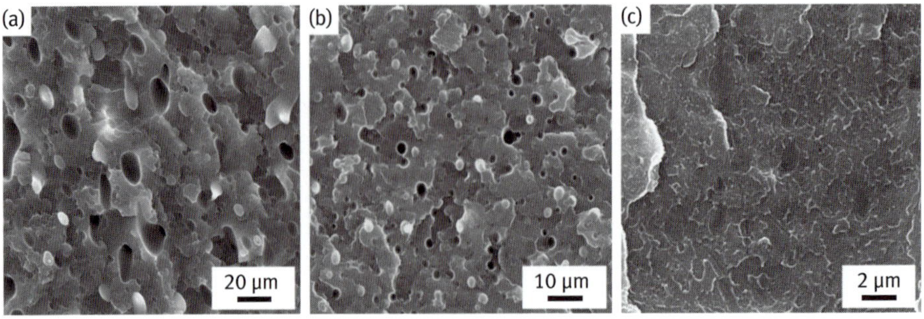

Fig. 6.27: SEM micrographs (SE imaging) of fracture surfaces of PS/LDPE (80/20) blends: non-compatibilized (a) and compatibilized with 5% (b) or 10% (c) of PS-b-PB copolymer [11].

decreases the average particle size (Fig. 6.27b), but the fracture still propagates along the interface. Further increase in compatibilizer concentration to 10% leads to even smaller average particle size and improvement of interphase adhesion, which is evidenced by the fact that the fracture runs directly through the LDPE particles and not along the interface (Fig. 6.27c).

Phase morphology is another important parameter influencing the final properties of polymer blends [75, 82]. SEM of fracture surfaces, cut surfaces, and/or ultrathin sections has become a key tool for the investigation of polymer blend morphology [83]. Figure 6.28 shows selected results from a morphological study of PP/cyclo-olefin copolymer blend (PP/COC) [47]. The morphology of the blend was optimized by careful selection of processing conditions so that the minority component (COC) could form reinforcing fibers, which imparted to the PP matrix higher modulus and yield strength [84]. The fibrous morphology of the PP/COC blends was clearly evidenced in SEM micrographs of fracture surfaces parallel with injection molding direction (Fig. 6.28a).

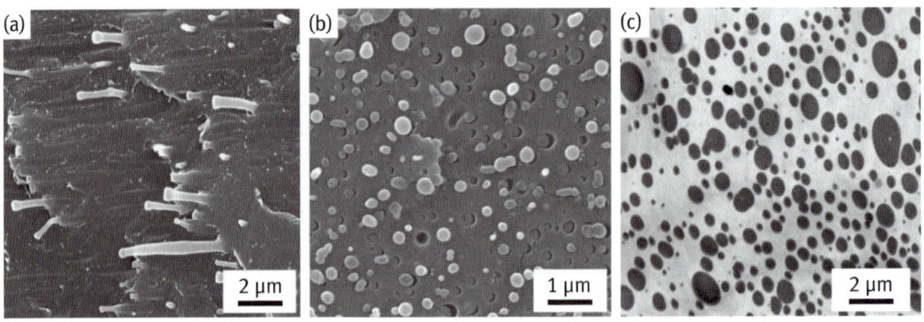

Fig. 6.28: SEM micrographs of PP/COC (70/30) polymer blends: (a) SE imaging of fracture surface parallel to the injection molding direction, (b) SE imaging of fracture surface perpendicular to the injection direction, and (c) STEM micrograph of RuO_4-stained ultrathin section perpendicular to the injection molding direction.

SEM micrographs of fracture surfaces perpendicular to injection direction confirmed the uniaxial orientation of the fibers (Fig. 6.28b). Finally, STEM micrographs of RuO_4-stained ultrathin sections perpendicular to the injection direction yielded detailed information on the average diameters of the COC fibers in the PP matrix. Compatibilization of PP/COC blends (and analogous PE/COC blends, where PE is high-density PE [56, 83]) was not necessary because the SEM micrographs, tensile testing results, and comparison with theoretical predictions confirmed very high interphase adhesion [56].

6.5.4 Polymer composites

Polymer composites represent a popular group of polymer materials, in which isometric particles, fibers, or plates dispersed in the polymer matrix act as reinforcing elements. In a typical polymer composite, a stiff filler (glass fibers, clay platelets, etc.) is embedded in a tough polymer matrix (either thermoplastic or thermosetting). The final polymer composite properties can be adjusted to the final application by suitable combination of the components and processing [74, 75].

Polymer microcomposites contain filler particles with average size >1 μm. Homogeneous microfiller dispersion and good matrix-filler interphase adhesion are required to achieve good mechanical performance. Both dispersion and interphase adhesion can be visualized by SE imaging of liquid-nitrogen fracture surfaces of the composite, as illustrated in Fig. 6.29, which shows recycled PET filled with basalt fibers (Fig. 6.29a and b; sample 1) and glass fibers with coupling agent (Fig. 6.29c and d; sample 2). In both samples 1 and 2, the fibers are dispersed homogeneously in the polymer matrix. In sample 1, the fracture is brittle (sharp fracture lines in Fig. 6.29a) and the interphase adhesion is low (distinct interface between fibers and matrix in Fig. 6.29b). In sample 2, the fracture is semibrittle (microplastic deformations in Fig. 6.29c) and the interphase adhesion is high (matrix adhering to the fiber in Fig. 6.29d). The improved interphase adhesion in sample 2 resulted in higher modulus, tensile, and impact strength [48].

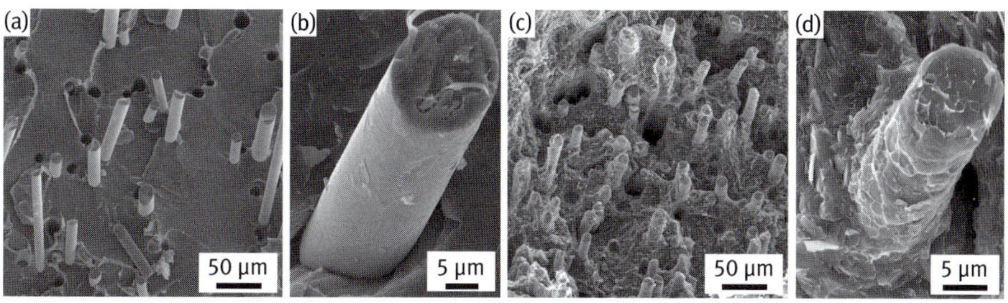

Fig. 6.29: SEM micrographs of fracture surfaces of recycled PET polymer composites with embedded: (a,b) basalt fibers [85] and (c,d) glass fibers with coupling agent [48].

Polymer nanocomposites contain particles with at least one dimension <1 μm. If the adhesion between the embedded nanofiller and the polymer matrix is strong, further improvement of mechanical properties can be achieved because nanoparticles have a higher specific surface than microparticles. Polymer nanocomposites are studied mostly by TEM of ultrathin sections from the interior of the sample, as briefly mentioned in Section 6.3. This can be attributed to the fact that nanofillers are best observed at higher resolutions in TEM microscopes.

It is worth noting that some researchers [86] differentiate between polymer composites and filled systems (see also Fig. 6.12). The basic difference between polymer composites (or reinforced systems) and filled systems is that fillers are used in composites/reinforced systems to improve properties, whereas in filled systems, the fillers are used to reduce cost. However, the distinction between reinforced and filled systems is sometimes unclear [87].

6.5.5 Special applications

Synthetic polymers have several specific features (Section 6.4.2), which are addressed in the following three subsections (6.5.5.1–3.). Low-voltage SEM helps to overcome problems with nonconductivity and beam sensitivity of the polymers and enhances surface structures (Section 6.5.5.1). VP-SEM makes it possible to visualize wet, hydrated, or frozen polymer specimens close to their natural state (Section 6.5.5.2). Other special applications include microscopy at elevated temperatures, under deformation, morphology changes associated with processing, etc. (Section 6.5.5.3).

6.5.5.1 Low-voltage SEM in polymer science

Most of the samples in the polymer science are insulators, which are very difficult to observe in the SEM because of their charging and electron-beam-induced damaging. Typical way of observation is covering their surface by a thin conductive layer. However, such a layer can cover details on the surface, and therefore, coating should be minimized if the details on the surface are to be investigated.

The low-voltage SEM enhances details on the surfaces of uncoated or mildly coated insulators as explained in Section 6.2.3.3. The surface sensitivity of low-voltage SEM is demonstrated in the images of silica particles embedded in a polyurethane matrix captured at the accelerating voltage of 20 kV (Fig. 6.30a) and 1 kV (Fig. 6.30b). The morphology of the silica particles is less clear, and particles embedded below the surface are recognizable due to a high penetration depth of 20 kV electrons (see also Eq. 6.2), whereas at a lower accelerating voltage of 1 kV, the surface detail is increased.

The sensitivity of the polymers to electron beam is also associated with sample contamination and/or electron-beam-induced sample damage. Both effects are best visualized if we observe the specimen at higher magnification and then (without

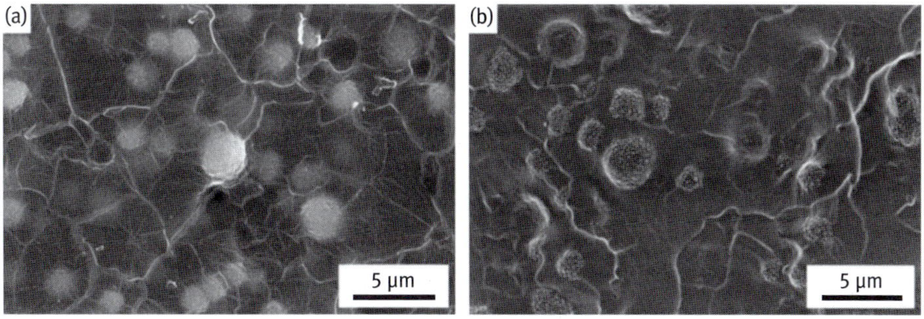

Fig. 6.30: Influence of the penetration depth on the visualization of a polyurethane sample filled with silica microparticles observed at accelerating voltage of 20 kV (a) and at 1 kV (b). The sample was uncoated for the 1-kV imaging and carbon coated for the 20-kV imaging.

moving the stage) at lower magnification. If the sample is contaminated or damaged, the lower-magnification SEM micrograph shows dark rectangle in the center of the image. The sample damage can be minimized using lower accelerating voltages in combination with low doses (small spot size and/or probe current) [88].

Some polymer specimens may charge so extensively that low accelerating voltage and electron dose do not help it if the standard way of image acquisition is used. The SEM image is usually acquired by *sequential scanning*: the frame is scanned pixel by pixel with the electron beam staying at each pixel long enough to obtain low noise image; each pixel of the image is irradiated once but using a relatively high dose. Another image acquisition strategy is the *integration method*: the beam dwells at each pixel a few hundred nanoseconds at most, a series of quickly scanned frames is collected, and in the last step, they are integrated together. Image quality improvement owing to the integration method is demonstrated in Fig. 6.31. Although the total electron dose necessary to acquire images in Fig. 6.31 is the same, the integration method (Fig. 6.31b) distributes the total dose over the whole acquisition time, and hence, it decreases the charging. Similar, but even more advanced, scanning strategies such as *scan interlacing* and *drift correction* have recently been described (Fig. 6.31) [89].

6.5.5.2 Wet specimens in polymer science

Wet polymer specimens, such as polymer hydrogels (for tissue engineering), polymer nanoparticles (micelles and vesicles for drug delivery), or polymer nanofibers (scaffolds for antimicrobial photodynamic therapy), represent important and intensively studied materials. These samples frequently contain up to 90 wt% of water, and thus, they could not be observed in their natural state in standard HV-SEM. Several examples have already been shown in Section 6.2.3. Here, we summarize how the morphology of the samples can be visualized with electron microscopy. In the first step, we

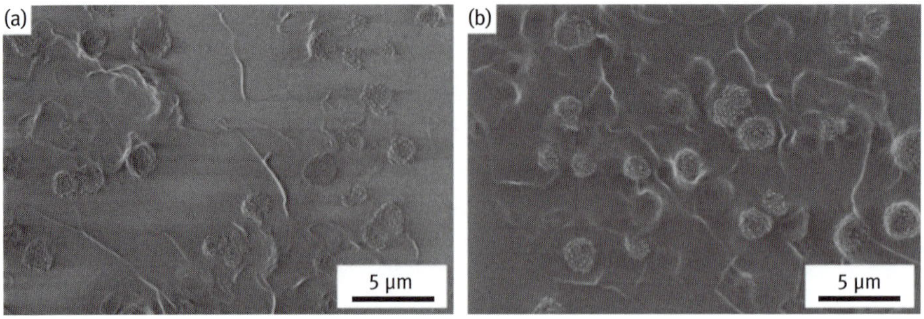

Fig. 6.31: Image of a polymer composite (polyurethane/silica microparticles; the same sample as in Fig. 6.30) acquired by sequential scanning (a) and by integration of 64 frames (b). Total acquisition time, accelerating voltage and electron dose was identical.

should verify if the samples can be dried in such a way that their structure does not break down (air drying, freeze drying, critical point drying [67, 90]) and observed using standard HV-SEM. In the second step, if the drying destroys the structure, the samples can be flash-frozen and observed at low temperatures ($T < 0°C$; cryomicroscopy) using either HV-SEM or LV-SEM. HV-SEM can be applied if the samples are frozen at very low temperatures, where the rate of sublimation of ice is negligible ($T < -100°C$ [15]), whereas LV-SEM can be applied at samples frozen a few degrees below zero ($T \sim -10°C$ [68]). If even the fast freezing of the samples results in artifacts, the samples can be observed in the hydrated/wet state at close-to-room temperature ($T > 0°C$) using either SE/BSE imaging (ESEM [23]) or TE imaging (STEM-in-ESEM, or wet-STEM [16, 17]). The degree of difficulty of sample preparation and observation techniques is as follows: HV-SEM < LV-SEM < ESEM < wet-STEM. Technical details about the methods are given in Section 6.2.3.1 and Section 6.2.3.2); sample micrographs of wet polymer specimens are shown in Fig. 6.10.

6.5.5.3 Further applications

This chapter is focused on the typical applications of EM on polymer surface and/or subsurface morphology characterization. A few other, more specialized applications are mentioned in this section for the sake of completeness. Polymer morphologies associated with processing, deformation, and fracture are frequently studied, being focused on craze development, shear band formation, and fracture mechanics [76]. 3D microscopy of subsurface structures has become quite popular recently but has not been discussed here because this chapter concentrates on polymer surfaces (basic information was given in Section 6.2.3.4). Polymer liquid crystals are usually studied by polarized light microscopy in combination with DSC and X-ray scattering, but EM methods can be applied as well [91].

Acknowledgments

This work was supported by the Technology Agency of the Czech Republic (project TE01020118) and Czech Science Foundation (project P205/10/0348).

References

[1] Brandon D, Kaplan WD. Microstructural characterization of material. Chippenham (UK): Wiley; 2008.
[2] Reimer L. Scanning electron microscopy. Berlin (Germany): Springer; 1998.
[3] Watt IM. The principles and practice of electron microscopy. 2nd edn. Cambridge (UK): Cambridge University Press; 1997.
[4] Sawyer LC, Grubb DT. Polymer microscopy. 2nd edn. London (UK): Chapman & Hall; 1996.
[5] Kanaya K, Okayama S. Penetration and energy-loss theory of electrons in solid targets. J Phys D Appl Phys 1972;5:43–8.
[6] Drouin D, Couture AR, Gauvin R, et al. Program CASINO Monte Carlo simulation of electron trajectory in solids. Version 2.48; 2011 (Accessed on 25 Aug 2013 at http://www.gel.usherbrooke.ca/casino).
[7] Reimer L. Image contrast and signal processing. In: Scanning electron microscopy. Berlin (Germany): Springer; 1998:207–52.
[8] Watt IM. Electron-specimen interactions: processes and detectors. In: The principles and practice of electron microscopy. Cambridge (UK): Cambridge University Press; 1997:30–57.
[9] Brandon D, Kaplan WD. Microanalysis in electron microscopy. In: Microstructural characterization of material. 2nd edn. Chippenham (UK): Wiley; 2008:333–86.
[10] Fultz B, Howe JM. Characteristic radiation. In: Transmission electron microscopy and diffractometry of materials. Berlin (Germany): Springer; 2001:16–9.
[11] Fortelný I, Šlouf M, Sikora A, et al. The effect of the architecture and concentration of styrene-butadiene compatibilizers on the morphology of polystyrene/low-density polyethylene blends. J Appl Polym Sci 2006;100:2803–16.
[12] Lednický F, Hromádková J, Šlouf M, Dybal J. Uložení E/P kaučuku v houževnatých E/P kopolymerech. Research report. Brno (Czech Republic): Polymer Institute. Edition: 2008, Macro T-727 [in Czech].
[13] Henke P, Lang K, Kubát P, Sýkora J, Šlouf M, Mosinger J. Polystyrene nanofiber materials modified with an externally bound porphyrin photosensitizer. ACS Appl Mater Interfaces 2013;5:3776–83.
[14] Stokes DJ. Principles and practice of variable pressure/environmental scanning electron microscopy (VP-ESEM). Cornwall (UK): Wiley; 2008.
[15] Štěpánek M, Hajduová J, Procházka K. Association of poly(4-hydroxystyrene)-block-poly(ethylene oxide) in aqueous solutions: block copolymer nanoparticles with intermixed blocks. Langmuir 2012;28:307–13.
[16] Šlouf M, Lapčíková M, Štěpánek M. Imaging of block copolymer vesicles in solvated state by wet scanning transmission electron microscopy. Eur Polym J 2011;47:1273–8.
[17] Hajduova J, Prochazka K, Šlouf M et al. Polyelectrolyte-surfactant complexes of poly[3,5-bis(dimethylaminomethyl)-4-hydroxystyrene]-block-poly(ethylene oxide) and sodium dodecyl sulfate: anomalous self-assembly behavior. Langmuir 2013;29:5443–9.
[18] Goldstein J, Newbury D, Joy D, et al. Low-temperature specimen preparation. In: Scanning electron microscopy and X-ray microanalysis. 3rd edn. New York (NY): Kluwer Academic/Plenum Publishers; 2003:621–46.

[19] Seward GGE, Celotto S, Prior DJ, Wheeler J, Pond RC. In situ SEM-EBSD observations of the hcp to bcc phase transformation in commercially pure titanium. Acta Mater 2004;52:821–32.
[20] Srinivasan NS. Dynamic study of changes in structure and morphology during the heating and sintering of iron powder. Powder Technol 2002;124:40–4.
[21] Reichmann A. High-temperature corrosion of steel in an ESEM with subsequent scale characterisation by Raman microscopy. Oxid Met 2008;70:257–66.
[22] Appel CC, Rasmussen A-M, Ullmann S, Hansen PL. Experiments with a modified heating stage on an environmental scanning electron microscope. In: Engelbrecht J, ed. International Congress on Electron Microscopy. Proceedings of the 15th International Congress on Electron Microscopy, Sep 1–6, 2002, Durban, South Africa; 2002:227–8.
[23] Karpushkin E, Dušková-Smrčková M, Šlouf M, Dušek K. Rheology and porosity control of poly(2-hydroxyethyl methacrylate) hydrogels. Polymer 2013;54:661–72.
[24] Reimer L. Image formation in low-voltage scanning electron microscopy. Bellingham (WA): The International Society for Optical Engineering; 1993.
[25] Goldstein J, Newbury D, Joy D, et al. Low-voltage operation. In: Scanning electron microscopy and X-ray microanalysis. 3rd edn. New York (NY): Kluwer Academic/Plenum Publishers; 2003:55–6.
[26] Young R, Templeton T, Roussel L, et al. Extreme high resolution SEM: a paradigm shift. Microsc Today 2008;16:24–8.
[27] Müllerová I, Frank L. Scanning low-energy electron microscopy. Adv Imag Elect Phys 2003;128:309–443.
[28] Frank L, Zadražil M, Müllerová I. Low energy imaging of nonconductive surfaces in SEM. Mikrochim Acta 1996;13:289–98.
[29] Xiaobing L, Bhadare S, Statham P, Burgess S, Holland J, Rowlands N. Improving low energy sensitivity of EDS detectors towards lithium detection. In: Prince RL, ed. Instrumental Symposium. Proceedings of Microscopy and Microanalysis, Jul 29–Aug 2, 2012, Phoenix (AZ); 2012.
[30] Scandium software, module Stereo (electronic manual available at www.soft-imaging.net).
[31] Danzl R, Schroettner H, Helmli F, Scherer S. Coordinate measurement with nano-metric resolution from multiple SEM images. In: The European Microscopy Society Proceedings of the 15th European Microscopy Congress, Sep 16–21, 2012, Manchester (UK); 2012.
[32] Suganuma T. Measurement of surface topography using SEM with two secondary electron detectors. J Electron Microsc 1985;34:328–37.
[33] Kodama T, Li X, Nakahira K, Ito D. Evolutionary computation applied to the reconstruction of 3-D surface topography in the SEM. J Electron Microsc 2005;54:429–35.
[34] Denk W, Horstmann H. Serial block-face scanning electron microscopy to reconstruct three-dimensional tissue nanostructure. PLoS Biol 2004;2:1900–9.
[35] Yao N, Imanishi N, Kang HHC, et al. Focused ion beam systems: basics and applications. Cambridge (UK): Cambridge University Press; 2007.
[36] Boughorbel F, Zhuge X, Potocek P, Lich B. SEM 3D reconstruction of stained bulk samples using landing energy variation and deconvolution. In: Prince RL, ed. Instrumental Symposium. Proceedings of Microscopy and Microanalysis, Jul 29–Aug 2, 2012, Phoenix (AZ); 2012:560–1.
[37] Wandrol P, Unčovský M, Vystavěl T. In-situ observation of solutions by WetSTEM. In: Prince RL, ed. Instrumental Symposium. Proceedings of Microscopy and Microanalysis, Jul 29–Aug 2, 2012, Phoenix (AZ); 2012:1090–1.
[38] Kralova D, Šlouf M, Klementova M, Kuzel R, Kelnar I. Preparation of gram quantities of high-quality titanate nanotubes and their composites with polyamide 6. Mater Chem Phys 2010;124:652–7.
[39] Malinsky P, Mackova A, Hnatowicz V. Properties of polyimide, polyetheretherketone and polyethyleneterephthalate implanted by Ni ions to high fluences. Nucl Instrum Methods B 2012;272:396–9.

[40] Dolina J, Lederer T. Silver particles incorporation to nanofibre structure for surface membrane modification. In: TANGER Ltd, ed. NANOCON 2011. Proceedings of the Third International Conference on NANOCON, Sep 21–23, 2011Brno (Czech Republic); 2011:331–40.
[41] Fultz B, Howe JM. Transmission electron microscopy and diffractometry of materials. Berlin (Germany): Springer; 2001.
[42] Rossiter BW, Hamilton JF, eds. Physical methods of chemistry. Volume IV: microscopy. 2nd edn. New York (NY): Wiley; 1991.
[43] Reimer L, Kohl H. Transmission electron microscopy. New York (NY): Springer; 2008.
[44] Sawyer LC, Grubb DT, Meyers GF. Conductive coatings. In: Polymer microscopy. 2nd edn. London (UK): Chapman & Hall; 1996:201–11.
[45] Lednický F. Morphology of coalescence processes: secondary fracture, cracks and spherulites. Polym Bull 1984;11:579–84.
[46] Lednický F, Pelzbauer Z. Specific use of electron microscopic techniques in the characterization of ABS polymers. Polym Test 1987;7:91–107.
[47] Šlouf M, Kolařík J, Fambri L. Phase morphology of PP/COC blends. J Appl Polym Sci 2004;91:253–9.
[48] Kráčalík M, Pospíšil L, Šlouf M, et al. Effect of glass fibres on rheology, thermal and mechanical properties of recycled PET. Polym Compos 2008;29:915–21.
[49] Lednický F, Michler GH. Soft matrix fracture surface as a means to reveal the morphology of multi-phase polymer systems. J Mater Sci. 1990;25:4549–54.
[50] Jordan ND, Bassett DC, Olley RH, Smith NG. In-depth morphological changes and embrittlement near the wear surface of UHMWPE inserts from uncemented hip systems. J Biomed Mater Res 2001;55:158–63.
[51] Fortelný I, Jůza J, Dimzoski B. Coalescence in quiescent polymer blends with a high content of the dispersed phase. Eur Polym J 2012;48:1230–40.
[52] Šlouf M, Kolařík J, Kotek J. Rubber-toughened polypropylene/acrylonitrile-co-butadiene-co-styrene blends: morphology and mechanical properties. Polym Eng Sci 2007;47:582–92.
[53] Merinska D, Kovarova L, Kalendova A, et al. Polypropylene nanocomposites based on the montmorillonite modified by octadecylamine and stearic acid co-intercalation. J Polym Eng 2003;23:241–57.
[54] Galeski A, Bartczak Z, Kazmierczak T, Šlouf M. Morphology of undeformed and deformed polyethylene lamellar crystals. Polymer 2010;51:5780–7.
[55] Fortelný I, Jůza J, Vacková T, Šlouf M. The effect of anisometry of dispersed droplets on their coalescence during annealing of polymer blends. Colloid Polym Sci 2011;289:1895–903.
[56] Vacková T, Šlouf M, Nevoralová M, Kaprálková L. HDPE/COC blends with fibrous morphology and their properties. Eur Polym J 2012;48:2031–9.
[57] Dimzoski B, Fortelný I, Šlouf M, Nevoralová M, Michálková D, Mikešová J. Coalescence during annealing of quiescent immiscible polymer blends. e-Polymers 2011;11:115–26.
[58] Lednický F, Hromádková J, Pientka Z. Ultrathin sectioning of polymeric materials for low-voltage electron microscopy. Polymer 2001;42:4329–38.
[59] Harris JR, Roos C, Djalali R, Rheingans O, Maskos M, Schmidt M. Application of the negative staining technique to both aqueous and organic solvent solutions of polymer particles. Micron 1999;30:289–98.
[60] Stara H, Šlouf M, Lednický F, Pavlova E, Baldrian J, Stary Z. New and simple staining method for visualizing UHMWPE lamellar structure in TEM. J Macromol Sci Phys 2008;47:1148–60.
[61] Lednický F, Hromádková J, Kolařík J. Revealing the phase structure of polymer blends via volume relaxation. Polym Test 1992;11:205–13.
[62] Kučera J, Lednický F. An investigation of properties of filled polypropylene-elastomer blends. Plasty a kaučuk 1989;26:97–102 [in Czech].

[63] Lednický F, Kučera J. Possibilities of visualization of the structures of composites. Plasty a kaučuk 1989;26:289–92 [in Czech].
[64] Hromádková J, Lednický F, Kolařík J. Visualization of the phase structure of selected polycarbonate blends using various preparation techniques. Polym Test 1994;13:461–78.
[65] Haworth B, Hindle CS, Sandilands GJ, White JR. Assessment of internal stresses in injection-molded thermoplastics. Plast Rubber Proc Appl 1982;2:59–71.
[66] Clay CS, Peace GW. Ion beam sputtering: an improved method of metal coating SEM samples and shadowing CTEM samples. J Microsc 1981;123:25–34.
[67] Meredith P, Donald AM, Payne RS. Freeze-drying: in situ observations using cryoenvironmental scanning electron microscopy and differential scanning calorimetry. J Pharmaceut Sci 1996;85:631–7.
[68] Přádný M, Lesný P, Smetana K jr. Macroporous hydrogels based on 2-hydroxyethyl methacrylate Part II Copolymers with positive and negative charges, polyelectrolyte complexes. J Mater Sci Mater Med 2005;16:767–73.
[69] Nedela V. Methods for additive hydration allowing observation of fully hydrated state of wet samples in environmental SEM. Microsc Res Technol 2007;70:95–100.
[70] Wittmann JC, Lotz B. Polymer decoration: the orientation of polymer folds as revealed by the crystallization of polymer vapors. J Polym Sci Phys Ed; 1985;23:205–26.
[71] Woodward AE. Crystallizable polymers. In: Understanding polymer morphology. Munich (Germany): Carl Hanser Verlag; 1995:19–29.
[72] Sperling LH. The amorphous state. In: Introduction to physical polymer science. 4th edn. Hoboken (NJ): Wiley; 2008:197–238.
[73] Lednický F. Density measurements of annealed amorphous polymers. Collect Czech Chem Commun 1995;60:1935–40.
[74] Sperling LH. Introduction to physical polymer science. Hoboken (NJ): Wiley; 2008.
[75] Michler GH. Electron microscopy of polymers. Heidelberg (Germany): Springer-Verlag; 2008.
[76] Woodward AE. Morphologies associated with deformation and fracture. In: Understanding polymer morphology. Munich (Germany): Carl Hanser Verlag;1995:101–12.
[77] Woodward AE. Morphologies of crystallized polymers. In: Understanding polymer morphology. Munich (Germany): Carl Hanser Verlag; 1995:31–56.
[78] Bassett DC. Spherulites. In: Principles of polymer morphology. Cambridge (UK): Cambridge University Press; 1981:16–36.
[79] Kurtz SM, ed. UHMWPE biomaterials handbook. 2nd edn. Amsterdam (The Netherlands): Academic Press/Elsevier; 2009.
[80] Šlouf M, Kotek J, Baldrian J, et al. Comparison of one-step and sequentially irradiated ultrahigh-molecular-weight polyethylene for total joint replacement. J Biomed Mater Res Part B 2013;101B:414–22.
[81] Woodward AE. Block copolymers. In: Atlas of polymer morphology. Munich (Germany): Carl Hanser Verlag; 1989:189–197.
[82] Horák Z, Fortelný I, Kolařík J, Hlavatá D, Sikora A. Polymer blends. In: J. Kroschwitz, ed. Encyclopedia of polymer science and technology. Indianapolis (IN): Wiley; 2005:1–59.
[83] Kolařík J, Kruliš Z, Šlouf M. High-density polyethylene/cycloolefin copolymer blends. Part 1: phase structure, dynamic mechanical, tensile and impact properties. Polym Eng Sci 2005;45:817–26.
[84] Vacková T, Šlouf M, Nevoralová M, Kaprálková L. Processing-improved properties and morphology of PP/COC blends. J Appl Polym Sci 2011;122:1168–75.
[85] Kráčalík M, Pospíšil L, Šlouf M, et al. Recycled poly(ethylene terephthalate) reinforced with basalt fibres: rheology, structure, and utility properties. Polym Compos 2008;29:437–42.

[86] Balta-Calleja FJ, Fakirov S. Microhardness of polymer blends, copolymers and composites. In: Microhardness of polymers. Cambridge (UK): Cambridge University Press; 2000:127–77.
[87] Karger-Kocsis J. Composites. In: Salamone JC, ed. Polymeric materials encyclopedia. New York (NY): CRC Press; 1996:1378–1383.
[88] Stokes DJ. A lab in a chamber is situ methods VP-ESEM and other applications. In: Principles and practice of variable pressure/environmental scanning electron microscopy (VP-ESEM). Cornwall (UK): Wiley; 2008:169–214.
[89] Wall D, Bosch E, Sluyterman S, Wandrol P. Advanced technologies for charge mitigation on non-conductive samples. In: Prince RL, ed. Instrumental Symposium. Proceedings of Microscopy and Microanalysis, Jul 29–Aug 2, 2012, Phoenix (AZ); 2012:1284–5.
[90] Watt IM. Preparation of moist specimens. In: The principles and practice of electron microscopy. 2nd edn. Cambridge (UK): Cambridge University Press; 1997:169–74.
[91] Woodward AE. Liquid crystalline morphologies. In: Understanding polymer morphology. Munich (Germany): Carl Hanser Verlag; 1995:57–70.

Fabio Palumbo and Rosa Di Mundo

7 Wettability: Significance and measurement

7.1 Introduction

Many technological issues as well as natural phenomena are based on the interaction between solids and liquids. Wetting, the phenomenon of bringing a liquid in contact with a solid surface, considered both when two like substances are involved and when two different ones are forced to interact, has been investigated in the last 200 years by many researchers, often in a controversial way and yet still not fully understood. Situations where a good contact between a solid and a liquid is preferred are those such as gluing, painting, printing paper, dyeing, or washing [1–2]. Conversely, plant leaves resisting water uptake or even surviving underwater, waterproof clothes/plastics, and drag reduction of fluid in pipes are examples of anti-adhesive surfaces having very poor interaction with liquid [3–6]. However, even phenomena such as metal/polymer adhesion and fluid lubrication in an engine depend at least in part from the wettability characteristic of the solids involved [7, 8]. In this chapter, some highlights will be given on the meaning of wettability and its link with materials surface energy and liquid surface tension, the main properties involved in the surface contact related phenomena. Then, the importance of contact angle (CA) and its measurement for the evaluation of the wettability will be emphasized together with the discussion on hydrophilic and hydrophobic materials. Finally, some special applications of CA measurement will be introduced: acid-base titration of polymer surfaces and the kinetics of liquid drops on porous materials.

7.2 CA and surface energy

The process of wetting leads to the formation of an interface between a liquid and a solid, which are normally both in contact with a fluid environment. In fact, the most typical experiment is placing a drop of liquid on a solid in the presence of air, but the case of a solid in contact with oil in the presence of water, such as in the petroleum extraction field, is not uncommon. In the first case, the fluid environment is air, whereas in the latter, it is water. Hence, because an interface is formed between the liquid and the solid, an adhesion issue is implicated as well as surface tension of the various phases of interest.

When two semi-infinite bodies made of different materials, *a* and *b*, separated at such a distance that the mutual interaction energy is negligible, are brought in contact, the *a-b* interface is formed and the interfaces between each material and the fluid, supposing for simplicity being air, where the bodies are immersed in, disappear as shown in Fig. 7.1a. The variation of free energy per unit area of the system, ΔG_{ab}, due to the interfaces' energy variation is

$$\Delta G_{ab} = \gamma_{ab} - \gamma_a - \gamma_b, \tag{7.1}$$

where γ_{ab} is the interfacial energy between a and b and γ_a and γ_b are the surface energies of each material. As a consequence, the energy per unit area necessary to achieve the separation of the bodies, called work of adhesion, W_{ab}, is equal to the opposite of the free energy variation, reported in Eq. 7.1, and in turn,

$$W_{ab} = \gamma_a + \gamma_b - \gamma_{ab}. \tag{7.2}$$

It is interesting to observe that if the two materials are the same, we do not deal with adhesion force, which is defined as the force necessary to detach two different materials, but with cohesion, and the energy per unit area necessary to overcome the intermolecular (cohesive) forces of molecules of the same species is

$$W_c = 2\gamma_a. \tag{7.3}$$

Here, it is necessary to highlight that, as depicted in Fig. 7.1b, the surface energy can be considered as the work per unit area, dW, due to the force, F, necessary to stretch a surface, A, of the quantity dA (the units are typically mN/m or mJ/cm^2). When dealing with liquids, normally, the term *surface tension* is used.

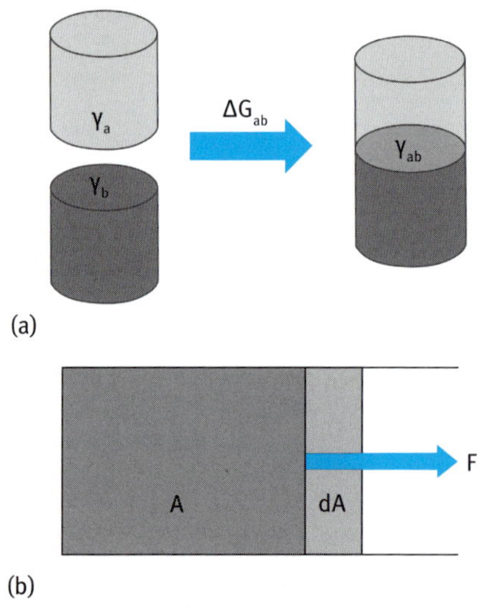

Fig. 7.1: (a) Gibbs energy variation due to the work of adhesion between bodies *a* and *b* and (b) sketch of surface area increase from *A* to *A* + *dA* due to force *F*.

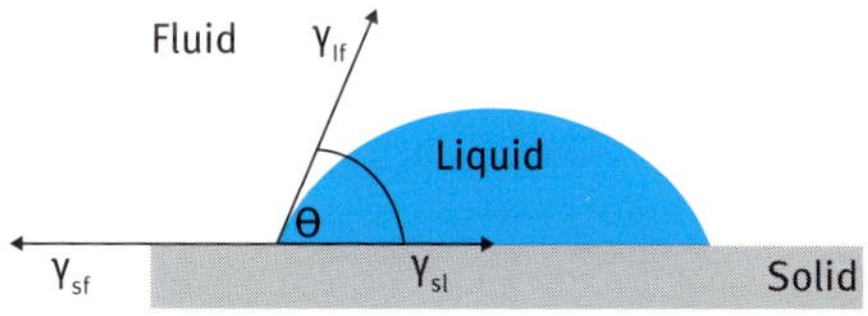

Fig. 7.2: *CA*, θ, between a liquid l and a solid s, immersed in a fluid f.

The shape of a liquid drop wetting a solid surface depends on the balance of the surface energies of the materials in contact with the fluid, as depicted in Fig. 7.2. More than 200 years ago, Thomas Young [9], in an essay on the cohesion of fluids, derived the following fundamental equation nowadays used by scientists to describe the phenomenon of wetting between a solid s and a liquid l immersed in a fluid f:

$$\frac{\gamma_{sf}-\gamma_{sl}}{\gamma_{lf}}=\cos\theta. \tag{7.4}$$

When the fluid corresponds to air, γ_{sf} and γ_{lf} can be replaced with γ_s and γ_l, respectively, indicating the surface energy of the solid and the liquid, Eq. 7.4, being transformed to the more familiar form

$$\frac{\gamma_{s}-\gamma_{sl}}{\gamma_{l}}=\cos\theta \tag{7.5}$$

The angle θ, called the contact angle, is defined as the angle between the tangent at the liquid-fluid interface and the tangent at the solid surface at the three-phase contact point. The Young equation has been derived for an ideal surface at equilibrium, which is smooth, rigid, chemically homogeneous, and not soluble in the liquid phase, where no external forces are applied.

To better understand the meaning of the Young equation, some qualitative comments can be drawn. The first observation is that, for a given solid material, a lower CA (hence higher spreading of the liquid) indicates a better wetting of the surface (Eq. 7.5 and Tab. 7.1), and this is obtained with apolar liquids, which have a lower surface tension. For example, Teflon, considered a typical hydrophobic material, can be wetted by a hydrocarbon solvent such as *n*-decane.

Water is well known as the most polar and common liquid solvent, with a surface tension as high as 72.8 mN/m, and thus influences a number of natural phenomena based on capillarity or water repellency [6, 10, 15]. Meanwhile, if one considers water as the wetting liquid, as illustrated in Tab. 7.2, surfaces with a higher surface energy will show a lower CA, being more wettable. Most of the organic polymeric surfaces rarely display water contact angle (WCA) lower than 70°, and this is the reason for the interest in developing methods for increasing the wettability of polymers. In fact, scarce wettability means poor adhesion of dyes or inks or of high-surface-energy metals, e.g. in metallization process for packaging [7].

Tab. 7.1: Surface tension of some common liquids

Liquid	γ(mN/m)
n-Hexane	18.4
Methanol	22.5
n-Decane	23.4
Tetrahydrofuran	27.4
Benzene	28.9
Dimethyl sulfoxide	44.0
Diiodomethane	50.8
Formamide	58.0
Glycerol	64.0
Water	72.8
Mercury	484.0

Tab. 7.2: Surface energy of some common polymers and their corresponding WCA

Polymer	Surface energy (mN/m)	WCA (°)
Polycarbonate (PC)	29	93[a] [10]
Polydimethylsiloxane (PDMS)	20	114 [11]
Polyethylene (PE)	27	93 [12]
Polyethyl methacrylate (PMMA)	37	65 [13]
Polypropylene (PP)	28	95 [14]
Polystyrene (PS)	29	90 [15]
Polytetrafluoroethylene (PTFE)	16	117 [16]
Polyvinylalcohol (PVA)	39	51 [17]
Polyvinyl chloride (PVC)	27	103 [18]
Polyvinylidene fluoride (PVDF)	32	85 [17]

[a] This has been determined advancing; the others have been determined in static mode.

The correlation of CA and adhesion of the liquid appears more evident when combining Eq. 7.2 and Eq. 7.5, obtaining

$$W_{sl} = \gamma_l (1 + \cos\theta). \tag{7.6}$$

This relation, known as Young-Dupree equation, highlights that a liquid drop sticks better on a solid surface where a low CA is formed; furthermore, polar liquid having higher surface tension will show higher work of adhesion.

7.2.1 Surface energy evaluation

Getting back to the Young equation, it is possible to observe that, in principle, it can drive scientists to evaluate the surface energy of polymers by measuring the CA.

However, serious concerns have been faced by investigators in the following two centuries: in fact, supposed that the CA, θ, can be measured (Section 7.4) and several methods exist to get the surface energy of the liquid, γ_l, two unknowns are present in Eq. 7.5, the surface energy of the solid, γ_s, and the liquid-solid interfacial energy, γ_{sl}, and none can be directly measured because solids cannot be deformed like liquids [19]. Hence, the Young equation, even if extremely appealing in describing wetting, requires a further equation relating γ_{sl} to γ_s to leave only one unknown in Eq. 7.5. From a historical point of view, the interpretation of the Young equation to get solid surface energy data started at the end of the 1950s with the work of Zisman [20], who identified a critical surface energy γ_c, as will be described in Section 7.2.1.1. Then, in the 1960s and the 1970s, two main approaches were developed to get solid surface energy from CA measurement [21–25]:

- Multicomponent approach: The total surface energy consists of a sum of different components, each of which arises due to a specific type of intermolecular forces, acid-base, dispersive, hydrogen bonding, and so on.
- Equation of state approach: Based on the thermodynamic definition of the work of adhesion and the London theory of dispersion forces, investigators derived a modified Young equation relating the CA to the surface energy of the liquid and the solid.

In the following, the Zisman method and the main models of the multicomponent approach will be reviewed leaving the equation of state to appropriate references [26, 27].

7.2.1.1 Critical surface energy method

Zisman and colleagues observed that for a given solid surface, the CA of different liquids did not vary randomly, and $\cos\theta$ rather changed quite linearly with the surface tension of the liquids, and they considered that a solid would be completely wetted by a liquid having the same surface tension of the solid. This is clear in a graph such as that in Fig. 7.3, where the cosine of CA of various liquids is plotted *vs* the liquid surface tension for a polyethylene (PE)-like solid. Furthermore, they defined the critical surface energy γ_c for the investigated solid as the surface tension of the liquid for which the CA is 0°, when the solid is completely wetted. The critical surface energy, hence, is obtained from similar plots built from CA measurements with different liquids, the obtained line is extrapolated at $\cos\theta = 1$ and the corresponding γ_l is γ_c. Obviously, γ_c is close, but it is not the surface energy of the solid.

This approach, even extremely important in its pioneer idea, has several problems, the most important being:

- the $\cos\theta / \gamma_l$ relation is not a straight line, but more hyperbolic, and the extrapolation to $\cos\theta$ is not straightforward;
- γ_c has no physical meaning;
- γ_s is not obtained.

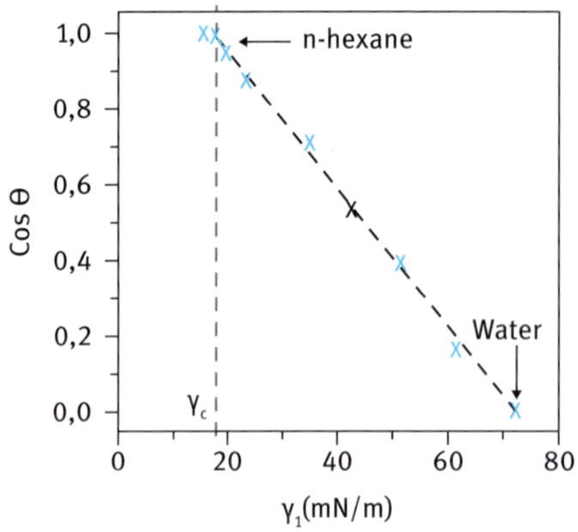

Fig. 7.3: Cosine of WCA *vs* surface tension of different liquids for a generic PE-like polymer. γ_c is the critical surface tension of the polymer.

However, the Zisman plot methods has been very useful for an applicative comparison of the wetting behavior of solids and still important for lubricant, ink, lacquer, and varnish applications: a liquid having a surface tension below the critical surface energy of the material will wet that surface completely.

7.2.1.2 Multicomponent approaches

Fowkes method

Fowkes and colleagues [23, 24, 28] introduced two main assumptions: the additivity of the surface tension components and the geometric mean relationship to describe solid-liquid interfacial interactions. According to this approach, the dispersive component of the surface energy commonly results in matter from the interaction between adjacent molecules arising from the electron dipole fluctuations, namely London forces. These interactions depend on the kind of mutually attracting elements of the matter, and what is more important, they act independent of other types of forces. The remaining kind of interactions, called polar, mostly those like van der Waals ones, induction (the Keesom and Debye ones), acid-base, also are supposed to be additive and operating independently.

With this assumption, one can write the surface energy in the following way

$$\gamma = \gamma^{d} + \gamma^{p}, \tag{7.8}$$

where

$$\gamma^{p} = \gamma^{vdw} + \gamma^{i} + \gamma^{h} + \gamma^{AB} + \ldots, \tag{7.9}$$

where d, p, vdW, i, h, and AB indicate the dispersive, polar, van der Waals, induction, hydrogen bonding, and acid-base interaction, respectively. Combining such additive hypothesis with the geometric mean description of the interfacial interaction energy applied to the work of adhesion for a force f, $W^f_{ab} = 2\left(\gamma^f_a \gamma^f_b\right)^{1/2}$, Eq. 7.2 can be written as follows:

$$\gamma_{sl} = \gamma_s + \gamma_l - 2\left[\left(\gamma^d_s \gamma^d_l\right)^{1/2} + \left(\gamma^p_s \gamma^p_l\right)^{1/2}\right], \tag{7.10}$$

where superscript p groups all the nondispersive interactions. Equation 7.10 is a generalized form coming from the pioneering contribution of Fowkes [23] and then further revised by Owens and Wendt [24]. It represents the desired relation between γ_{sl} and γ_s necessary to solve the problem of the excess number of unknowns in the Young equation.

The simplest case for taking advantage of Eq. 7.10 is a liquid that can interact with a polymer surface only through dispersive forces, such as diiodomethane and polytetrafluoroethylene (PTFE). In this case,

$$\gamma^p_s = \gamma^p_l = 0,$$

and Eq. 7.10, taking into account Eq. 7.2 and Eq. 7.6, can be rewritten as follows:

$$2(\gamma_s \gamma_l)^{1/2} = \gamma_l(1 + \cos\theta). \tag{7.11a}$$

Hence,

$$\gamma_s = \frac{\gamma_l(1 + \cos\theta)^2}{4}, \tag{7.11b}$$

from which it is evident that simply measuring the CA of a nonpolar liquid of well known surface tension, the surface energy of a purely dispersive polymer can be obtained.

In a more general case, the polymer surface will express one dispersive and n polar components of the surface energy. It can be demonstrated that if $n + 1$ liquids can be used, each with known corresponding surface tension components, the measurement of each CA will allow for the complete assessment of the polymer surface energy [29].

It should be mentioned here that Wu [25] introduced the possibility that the interaction forces between the liquid molecules and the polymer surface should be better described by a harmonic mean and not a geometric one, and Eq. 7.10 can be written with an analogous formalism.

Acid-base theory

This approach was developed by van Oss and colleagues [30, 31] based on the Lifschitz theory of the attraction between macroscopic bodies of condensed matter [32]. This model foresees that all the dipole-dipole and dipole-induced dipole interactions that contributed to the polar component of the surface energy are smaller than it can be supposed because in the condensed phase, orientation of the dipoles is more difficult. Furthermore, it is unlikely that these contributions, dispersive or Keesom/Debye-like, can be separated; rather, they cannot be distinguished, and the term γ^{LW} should be better considered to embrace such forces effects, with superscript LW denoting Lifschitz-van der Waals electromagnetic interactions. The term γ^{LW} does not include AB interactions, where acids and bases should be considered with the Lewis theory meaning. Remember that according to this theory, acid is an electron pair acceptor substance, and a base is an electron pair donor one. In this frame also a hydrogen bond is an acid-base interaction because an electronegative group (e.g. a carbonyl) donates an electron pair to a hydrogen bonded to another electronegative atom (e.g. an O-H group). Much more, even a non conventional acid substance such as chloroform, $CHCl_3$, should be considered an acid, and an ether, R-O-R, a base. For the sake of clarity of the following considerations, it is possible to distinguish between monopolar species, those having only an acid or a base character; bipolar, like water, exhibiting both acid and base character (even not of the same entity); and *apolar*, having no base or acid character.

Hence, since the two main kind of interactions, LW and AB, are additive, the surface energy for species *i* can be written as

$$\gamma_i = \gamma_i^{LW} + \gamma_i^{AB}, \tag{7.12a}$$

where γ_i^{AB} is the acid-base contribution to the surface energy of material *i*.

It is possible to define $\gamma^{(+)}$ and $\gamma^{(-)}$ as the acid and base contribution to the surface energy, respectively, and to derive the following relation:

$$\gamma_i^{AB} = 2\left(\gamma_i^{(+)}\gamma_i^{(-)}\right)^{1/2}. \tag{7.12b}$$

It should be considered that

- the acid component of the surface interacts only with the base counterpart of the liquid and *vice versa*;
- if the solid and the liquid are monopolar in the same direction, no interaction is possible;
- γ_i^{AB} is 0 if the acid or base contribution is null (see Eq. 7.12b)

Taking these elements into account, an equation analogous to Eq. 7.10 can be composed to express the interfacial surface tension:

$$\gamma_{sl} = \gamma_s + \gamma_l - 2\left(\gamma_s^{LW}\gamma_l^{LW}\right)^{1/2} + 2\left[\left(\gamma_s^{(+)}\gamma_l^{(-)}\right)^{1/2} + \left(\gamma_s^{(-)}\gamma_l^{(+)}\right)^{1/2}\right]. \tag{7.13}$$

With a rationale analogous to that of the Fowkes theory, one can couple the Young-Dupree equation of interfacial work of adhesion with Eq. 7.13, obtaining

$$2\left(\gamma_s^{LW}\gamma_l^{LW}\right)^{1/2} + 2\left(\gamma_s^{(+)}\gamma_l^{(-)}\right)^{1/2} + 2\left(\gamma_s^{(-)}\gamma_l^{(+)}\right)^{1/2} = \gamma_l(1 + \cos\theta). \tag{7.14}$$

One can measure the CA of a given solid surface with three different liquids, whose components of the surface tension (γ_l^{LW}, $\gamma_l^{(+)}$ and $\gamma_l^{(-)}$) are known and can write the corresponding Eq 7.14 for each probing liquid. Then, solving the system of the obtained three equations, the components of the solid surface energy, γ_s^{LW}, $\gamma_s^{(+)}$ and $\gamma_s^{(-)}$, can be determined. It is best to use at least one apolar liquid for which no acid and base components are present ($\gamma_l^{(+)} = \gamma_l^{(-)} = 0$); thus, from Eq. 7.12a and b $\gamma_l = \gamma_l^{LW}$, and the Lifschitz-van der Waals component of γ_s, of the solid under investigation, can be obtained from Eq. 7.14 by measuring the CA with that liquid.

It is important to highlight that database set $\gamma_l^{(+)}$ and $\gamma_l^{(-)}$ for suitable probing liquids can be found in literature, and they have been derived from eq 7.14 supposing that the corresponding values for water are equal ($\gamma_{H20}^{(+)} = \gamma_{H20}^{(-)}$). Considering that at 20°C, the surface tension of water is 72.8 mN/m and the apolar component is 21.8 mN/m it is possible to obtain that for water $\gamma_{H_2O}^{(+)} = \gamma_{H_2O}^{(-)} = 25.5$ mN/m (from Eq. 7.12a and Eq. 7.12b). In Tab. 7.3, a selection of probing liquids is reported with the corresponding components of the surface tension. It should be noted that the basic contribution of surface tension for methanol and dimethylsulfoxide is quite high (77 and 32 mN/m, respectively), but the total γ_l^{AB} is low, being the product of the acid and base. However, a strong acid-base interaction could develop with a polymer functionalized with acid groups, deeply affecting the wettability.

7.2.1.3 Particular cases: High-energy surfaces and granular materials

On high-surface-energy materials, the main part of liquids will spread (from Eq. 7.5); thus, it is not possible to directly evaluate γ_s using the methods described in Section 7.2.1.2. Schultz *et al.* [33] proposed an experimental approach where the solid is immersed

Tab. 7.3: Surface tension parameters (in mN/m) of polar and apolar liquids

Liquid	Total	γ^{LW}	γ^{AB}	$\gamma^{(+)}$	$\gamma^{(-)}$
1-Bromonaphthalene	44.4	44.4	0.0	0.0	0.0
Diiodomethane	50.8	50.8	0.0	0.0	0.0
Dimethyl sulfoxide	44	36	8	0.5	32
Ethanol	21.4	18.8	2.6	0.02	68
Ethylene glycol	48	29	19	3.0	30.1
Formamide	58	39	19	2.28	39.6
Glycerol	64	34	30	3.92	57.4
Methanol	22.5	18.2	4.3	0.06	77
Tetrahydrofuran	27.4	27.4	0.0	0.0	15
Water	72.8	21.8	51.0	25.5	25.5

in a liquid and a drop of another liquid immiscible with the first one is placed on the surface and its CA measured. Typically, the first liquid is a hydrocarbon one (e.g. *iso*-octane or *n*-decane) and the probing liquid is water. One can choose the more appropriate model among the ones discussed in Section 7.2.1.2 to describe the solid-liquid interfacial surface energy $\gamma_{sl} = f(\gamma_s \gamma_l)$ for each liquid. Then, the CA of liquid l_1 is measured on the solid surface immersed in the liquid l_2, and the Young equation for such system can be written as

$$\gamma_{sl_2} - \gamma_{sl_1} = \gamma_{l_1 l_2} \cos\theta, \tag{7.15}$$

where $\gamma_{l_1 l_2}$ is the interfacial tension between the two liquids, and Eq. 7.15 can be solved to obtain γ_s, using different immersion liquids.

In the case of granular materials, the measurement of the CA with a goniometer is severe because the probing liquid more or less penetrates in the porous structure, and because the surface is not flat and smooth, the CA is geometrically inaccessible. For granular materials, the Washnurn method is usually used, where the powder to be measured is placed into a glass tube closed to the bottom end with a filter. The tube is then hung at a balance, and when the vessel has contacted the liquid, the liquid rise rate through the bulk powder is measured.

The equation describing such phenomenon is

$$\cos\theta = \mathrm{A1A2}(m^2/t), \tag{7.16}$$

where m is the mass of the material, t is the probing time, and $\mathrm{A1} = \eta/\rho^2\gamma_l$, and $\mathrm{A2}=(2/\pi^2 r^5 n^2)$. Thus, parameter A1 contains information on the liquid, viscosity, η, density, and surface tension, whereas A2 is related to material properties supposed to be composed of n pores of radius r.

Parameter A2 is determined for the specific material with a measurement with a liquid that completely spreads on that surface ($\cos\theta = 1$). Then the experiment is carried out with the liquid of interest, and parameter A1 is calculated using the specific parameters (density, viscosity, and surface tension) for that liquid. It should be considered that no measurements can be obtained with any liquid exceeding 90° of CA with the granular material under investigation because in such a case, the liquid will not rise in the vessel.

7.2.2 Considerations on surface energy evaluation

Even if surface energy evaluation is a quite well-known subject and results interpretation lies on quite well-understood classic thermodynamic principles, debates and different opinions are still present in the scientific community. As an example, scientists are divided between those that follows the multicomponent approach and those searching for an equation of state describing the interfacial solid liquid tension, but others are still looking for a more unifying theory [32, 34]. Nevertheless, automatic CA apparatus (as described in Section 7.4) are often equipped with software tools that allow to approach the surface energy determination with all the discussed methods, thus leaving to the user the decision on the more feasible interpretation.

For a summary and hints on the complexities of the surface energy methods discussed in Section 7.2.1, see Tab. 7.4.

Tab. 7.4: Summary and applicability of surface energy determination methods

Method	Characteristics	Application
Critical surface energy (Zisman plot)	Provide γ_c, which is not the solid surface energy but is the surface tension of the liquid necessary to get complete wetting	Useful for easy polymer comparison: high γ_c are more wettable
Fowkes method and the like	– Surface tension is additively composed of independent contribution for each intermolecular interaction, $$\gamma = \gamma^{d} + \gamma^{vdW} + \gamma^{i} + \gamma^{h} + \gamma^{AB} + \ldots$$ – Each contribution was derived using the geometric mean equation – See Eq. 7.10, one probing liquid is necessary for each surface tension component – The equation can be slightly modified introducing the harmonic mean [25]	Mostly for moderately polar polymers
Acid-base method	– Similar to the Fowkes method in assuming the additivity and geometric mean validity for the components to the total surface energy – All dispersive and dipole-dipole interactions grouped in Liftschitz-van der Waals term, with the acid-base ones separated: $$\gamma = \gamma^{LW} + \gamma^{AB}$$ – The AB term of surface tension is divided into an acid, $\gamma^{(+)}$, and a base, $\gamma^{(-)}$, term such that $$\gamma_i^{AB} = 2\left(\gamma_i^{(+)}\,\gamma_i^{(-)}\right)^{1/2}$$ – Solid-surface energy obtained from Eq. 7.14 using one apolar liquid and two polar ones	With respect to the Fowkes method more suitable for polar solid that can engage acid-base interaction with Lewis acid or base liquids

(Continued)

Tab. 7.4: Summary and applicability of surface energy determination methods *(Continued)*

Method	Characteristics	Application
Schultz method	CA measured from a test liquid l_1 (water) and immersing the solid surface in a liquid l_2. All the aforementioned methods can be applied to describe this system and derive the surface tension of a solid with a high surface energy	Necessary for high-surface-energy substrates (protein layer, most metal/metal oxides, and ceramics), completely wetted by common testing liquids
Washburn method	Solid particles packed in a glass tube and brought into contact with the testing liquid. The mass of liquid absorbed into the solid is measured as a function of time. Once viscosity, density, and surface tension of the liquid are known, the material constant of the solid and the CA are derived from Eq. 7.16	Suitable for granular solid, powders, fibers, and the like, with a liquid CA lower than 90°

It should be noted that surface energy evaluation is highly dependent on the following conditions, which are often very hard to accomplish in a the real-laboratory situation:

- no dissolution or swelling of the solid material during the ca measurement;
- smoothness of the sample surface, allowing to measure the equilibrium CA;
- surface cleanliness;
- surface homogeneity, such that sessile drops of test liquids have a hemispherical shape.

However, if these parameters are not compliant with the solid under investigation, the CA measurement alone, even when not leading to complete surface energy characterization, will indicate the wettability in the field of interest of that specific material.

7.3 CA hysteresis

In Section 7.2 the liquid-solid work of adhesion has been derived as a function of the CA θ (Eq. 7.6). Such an equation, however, although technically correct, does not consider that, often, on real surfaces, different values for θ can be attained under dynamic conditions, i.e. upon motion of the liquid with respect to the solid or *vice versa*. The evaluation of dynamics of wetting in addition or alternative to statics, although studied and reported since the 1960s, has been often neglected in experimental research for decades and sometimes still skipped; notwithstanding, the dynamic situations in wetting are those effectively linked to applications. In particular, a so-called CA hysteresis exists, defined as the difference between the advancing and the receding CA, observed under dynamic conditions. The former is the angle when the liquid front is advancing over an unwetted portion of the surface, and it is the maximum angle measurable on a surface; the latter is the one observed while the liquid front is retracting from the previously wetted surface, i.e. the minimum observable angle. The measurement of hysteresis can be done either by forcing the liquid to increase/reduce the probing volume on a level surface or by evaluating the front and rear angles of a drop at rest on a tilted plane, and this will be better explained in Section 7.4.

Furmidge [35] first observed that such a difference between the front and the rear wetting state with respect to liquid (drop) motion generates a force opposing the weight of the drop and that is able, if the drop is small enough, to balance it. Thus, he modified the Young-Dupree equation to take into account the difference between the advancing, θ_a, and the receding angle, θ_r, as follows

$$W_{sl} = \gamma_l(\cos\theta_r - \cos\theta_a) \tag{7.17}$$

Thus, a drop will remain stuck on the surface as long as

$$\pi_b \gamma_l(\cos\theta_r - \cos\theta_a) \geq mg \sin \alpha \tag{7.18}$$

where α is the angle of inclination, m is the drop mass, g is the gravity acceleration and b is the radius of the solid liquid contact. From Eq. 7.18, it is evident that when tilting a plane, a drop placed on the top will start to slide when the angle α will be such that the gravity force, $mg \sin \alpha$, overtakes the adhesion one, first term of eq 7.18. In this view, the parameter directly correlated to the adhesive or sticky character is the CA hysteresis and not the mere single static angle.

However, what about the origin of such a hysteresis? As first observed by De Gennes [36], chemical, topographical, and morphological heterogeneities are the basis of the phenomenon. An ideal surface, perfectly smooth and made of only one type of atoms/moieties, possibly with no/poor interaction with the liquid, would not have any reason to show a difference between the advancing and the receding angle. This, for instance, is almost the case of a (freshly washed) flat polystyrene (PS) slab [15]. Instead, because heterogeneities exist on most real surfaces, the liquid-solid interfacial energy is characterized by several metastable states, each leading to different apparent CAs in a range defined by the advancing and receding CA.

Another view on hysteresis is the following: usually, a surface is composed of a variety of functional groups (the distribution can be more or less broad, depending on the surface) but homogeneously distributed in the space. Just considering the simple system where the liquid is water and air is the immersion fluid, according to their definition, θ_a probes a dry surface while θ_r probes a surface that has reached a stable state with the liquid. Hence, the advancing angle is more sensitive to lower-energy, apolar groups, whereas the receding angle is affected by high-energy, more polar ones: θ_a and θ_r must be different. For example, the receding angle on a plasma-treated polymer (when the treatment is addressed to hydrophilicity) is the only way to investigate high-energy-grafted functionalities. It should be considered that this is even more important when dealing with treated polymers that have been aged: as illustrated in Fig. 7.4, whereas the advancing angle can reveal hydrophobic recovery (WCA

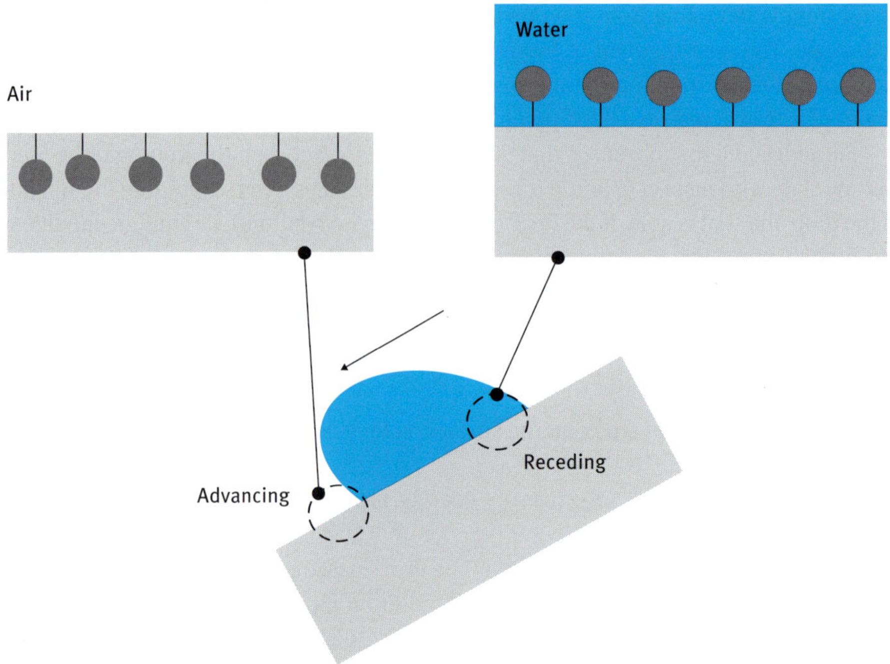

Fig. 7.4: Sketch of the different orientations of polar surface groups (grey circles) grafted on the surface of an unpolar polymer (e.g. by plasma treatment) in air, after a certain time from the treatment: advancing and receding angles are affected by the different configurations that can be achieved by the surface.

increases with time, as better explained in Section 7.5.3), the receding angle generally remains much lower because it retains the memory of the modification, suggesting that high-energy groups have been buried within the reorganized surface [37].

Meanwhile, there are also surfaces with little or no expectation of chemical heterogeneity that can nevertheless exhibit strong hysteresis; this might be possibly due to peculiar chain mobility upon wetting, leading to reversible interactions with the liquid drop. Typical is the case of commercial PTFE (PTFE), which, although being a conventional polymer with the highest advancing WCA, displays a quite high hysteresis (115°/90°) [38–39]. Taking into consideration Eq. 7.17, it is evident that the work of adhesion can be important for PTFE, and such particular behavior has led famous wetting scholars to state that "Teflon is hydrophilic": distinguishing between shear and tensile hydrophobicity, they found that Teflon is "shear" hydrophilic (but tensile hydrophobic) because the drop does not slide easily if the surface is tilted [40]. This is also found in the case of thin, smooth Teflon-like films deposited by conventional PECVD processes, which show higher hysteresis than homologous hydrocarbon surfaces [41].

Roughness has a strong influence on the static angle, and furthermore, it sensitively affects CA hysteresis, as will be better described in Section 7.5. This had been already shown by Johnson and Dettre [42], who reported (Fig. 7.5) a steep increase in both advancing and receding angles as a function of roughness, with the latter showing a non-monotonous trend. The increase in hysteresis on the first range in Fig. 7.5 is a consequence of a full wet contact between liquid and solid; thus, the presence of protrusions hinders the motion of the liquid (described as "pinning" of the drop); the following reduction of hysteresis is due to the establishment of non-wet states, where the drop is suspended on the top of the surface protrusions and sliding/rolling is favored. This behavior has been observed by several authors dealing with

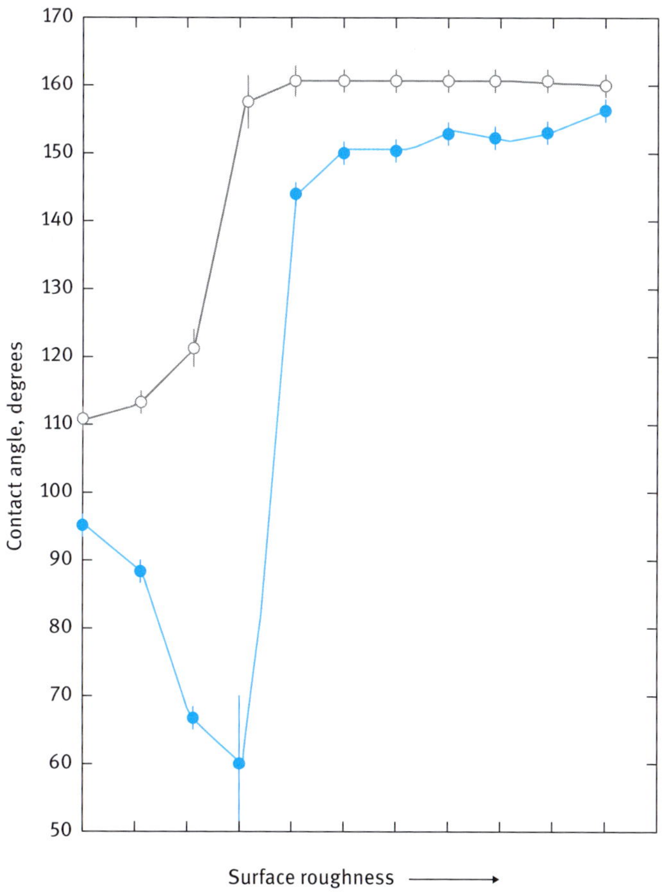

Fig. 7.5: CA of water on wax substrates as a function of the substrate roughness. Both the advancing angle (°) and the receding angle (•) are reported. Adapted with permission from Johnson RE and Dettre R, in Fowkes F, ed. *Contact angle, wettability, and adhesion*, Vol 43, 1964. Copyright 1964 American Chemical Society.

rough surfaces, in particular when tuning a topographical parameter (rms roughness, height, aspect ratio, density of the structures) in such a way that the non-wet regime can be established [15, 43].

From the above discussion, it is easily understood that the use of a simple static angle (often very close in value to the advancing angle) is often misleading, and it may hide an adhesive surface (indicated by a low receding angle), a situation that is unfavorable for hydrorepellent applications, for instance. That is why advancing and receding angles should always be measured.

The way CA hysteresis is represented and reported is particularly relevant when dealing with the concept of adhesion. As a consequence of Eq. 7.17, a mere difference between advancing and receding angles is not meaningful: a low work of adhesion is achieved when θ_a and θ_r are very close and their absolute values are high. Therefore, when numbers have to be compared, it is suggested to either calculate the difference between cosines ($\cos\theta_r - \cos\theta_a$) or simply report both absolute values, as done by most scholars [5, 10, 15, 16, 40, 41, 43].

7.4 Measurement methods for CA

Different methods have been reported for the measurement of CA, and for nonporous solids, two main approaches can be identified: direct geometric determination through an optical goniometer and force tensiometry. In goniometry, typically, the sessile drop of a test liquid on a solid substrate is observed and the tangent in the triple point determined. In addition, the observation of a captive bubble on a solid, both immersed in a liquid, can be included in this methodology. Tensiometry, meanwhile, is based on the measurement of the force of interaction between a solid and a liquid when in contact. In case the solid is porous, such as powders and the like, the wetting liquid can be absorbed by the material under investigation and another approach is used, based on capillarity and known as the Washburn method. This has been detailed in Section 7.2.1.3 and it will not be discussed here.

7.4.1 Direct measurement by optical goniometry

The most common technique of CA measurement is the direct measurement of the tangent angle at the three-phase contact point on a sessile drop placed on a solid and requires a quite simple apparatus. It consists basically of three components: a horizontal sample stage where the solid should be placed, a light source for the illumination of the drop sample region, and on the opposite side, an image capture system. In the old systems, the latter consisted of a simple optical observation through a goniometric graduated eyepiece, with a suitable magnification (Fig. 7.6). Afterward, to improve

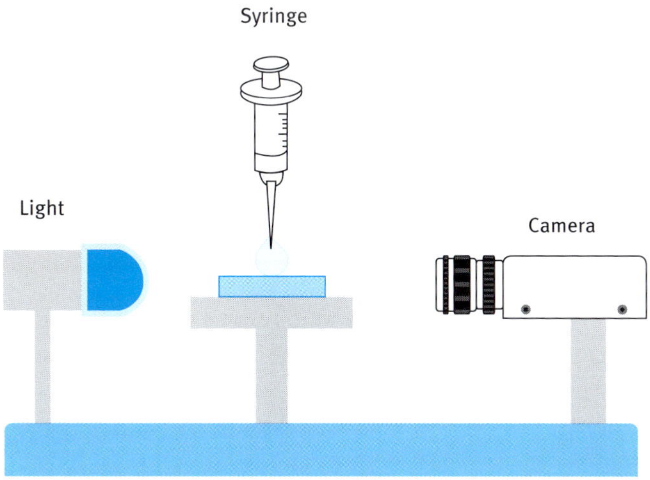

Fig. 7.6: Sketch of a CA goniometer.

reliability and accuracy of the analysis, the equipment has been implemented with a camera that allows the acquisition of either photographs or movies, which are useful in case the CA has a dynamic (CA hysteresis) or kinetic behavior (absorbing materials, such as textiles). Furthermore, the goniometer can be equipped with a manual or motor-driven graduated microsyringe for dispensing liquid drop, a suitable system for tilting the sample stage, and a containment chamber to keep the environment around the sample clean and eventually for conditioning the ambient to measure the CA at different temperatures or pressures. From the operative point of view, it should be considered that drops might be not symmetric, so it is advisable to measure the CA on both sides of the liquid drop profile. It is also recommended to measure the CA of drops in more points on the investigated material for a more accurate characterization of the wetting properties. It is quite recognized that this method cannot lead to accuracy better than ±2°.

In the static mode, a drop is placed with a fixed volume (typically 2–5 µl), and then the measurement is done supposing that the drop is not changing during the analysis. When the material is characterized by hysteresis, the advancing and receding CA evaluation can be obtained by measuring the CA by increasing and decreasing the volume of the drop during the analysis (keeping the needle of the syringe inside the liquid) or, alternatively, by placing a drop on an inclined plate with variable tilting angle, as illustrated in Fig. 7.7a–c. In the first case, the volume of the liquid drop placed on the sample is slowly increased; for small added volume, the measured CA increases but the triple-phase contact line remains stuck. For a certain drop size, the line front starts to move, but the CA does not increase more: that is the advancing CA (Fig. 7.7a). For the receding CA, the situation is reversed (Fig. 7.7b): the liquid is slowly withdrawn, with

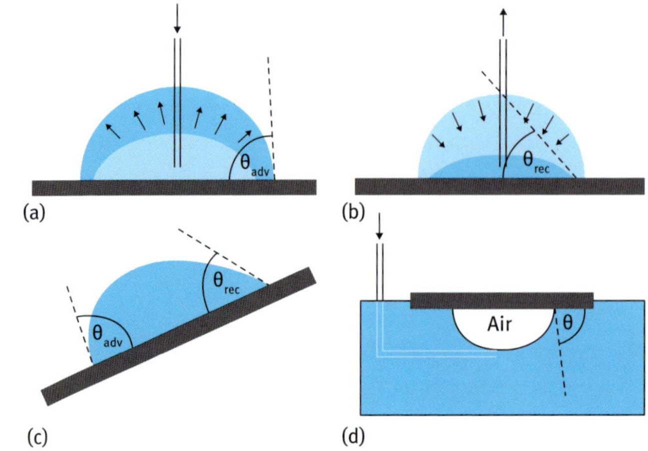

Fig. 7.7: Scheme of the determination of the advancing and receding angle by the sessile drop (a and b) and of the tilting method (c). Captive bubble method (d).

consequent reduction of the volume and the CA. When the contact line starts to recede, the acquired CA is the receding one. With a motor-driven syringe and a fast-speed camera, this acquisition is straightforward. In the "tilting plate method" (Fig. 7.7c), the liquid drop is placed on the plate, and the top and bottom CAs are measured during the slow tilting until the drop just begins to slide. At this step, the CAs measured at the lowest and highest points are the advancing and receding ones, respectively [44]. It should be mentioned here that some authors demonstrated that, in some cases, mostly on hydrophobic surfaces, the tilting plate method could lead to artifacts and the volume addition and withdrawal method should be preferred [45].

For a more accurate determination of the CA, an image of the drop on the solid surface is acquired; hence, a software for the elaboration of the drop shape image (normally provided with the goniometer) and for the contact line determination is important. Different methods have been developed to fit the contour of the whole drop or part of it, and basically, they can be divided in those fitting the curve with a geometric function and those approaches using the Young-Laplace equation, relating the pressure difference (across the liquid-air interface) to the shape of the surface drop [46].

Instead of placing a liquid sessile drop above the solid sample, an air bubble can be formed underneath the surface sample, which is immersed in the testing liquid (Fig. 7.7d): then the CA formed by the air bubble in the liquid can be directly measured. This method is referred to as the "captive bubble method." The addition of more air to the bubble through the needle, leads the surrounding liquid front to recede, and θ_r is obtained; on the contrary, air withdrawal from the bubble causes the liquid front to advance, and θ_a can be determined. The captive bubble method is similar to the sessile drop one, and it has the advantage of ensuring that the surface is in contact with a saturated atmosphere. Furthermore, it is much easier to monitor the temperature of the liquid in the captive bubble method than with sessile drops, which makes it possible to study the temperature dependence of CAs. However, the captive bubble method requires far more liquid than the sessile drop method.

7.4.2 Force tensiometry

Tensiometry indirectly measures CAs evaluating the forces involved when a sample of solid is brought into contact with a test liquid. A flat and thin sample is typically hung to a Wilhelmy balance, and the test liquid is raised to contact the solid. Upon the contact and during the immersion, the force measured by the balance changes, and it is recorded as a function of the depth. The detected force, F, depends on buoyancy and on wetting (once zeroed for the weight), and it can be defined as

$$F = \gamma_l p \cos\theta - gV\Delta\rho, \tag{7.19}$$

where γ_l is the surface tension of the test liquid, p is the perimeter of the contact line (supposed to be the same with the sample section), θ is the CA, V is the displaced liquid volume, $\Delta\rho$ is the difference in density between the liquid and air (or a second liquid), and g is the gravity acceleration. Considering Fig. 7.8, during an immersion-emersion cycle with a Wilhelmy balance measurement, the following events follows: (a) the sample approaches the liquid, and the force is zero; (b) the sample contacts the liquid surface, the liquid rises up along the surface if $\theta < 90°$, with a positive wetting force (the force is opposite for $\theta > 90°$); (c) the sample is further dipped, and the increase of buoyancy causes a decrease in the force detected on the balance; the force is measured for the advancing angle; (d) The sample is pulled out of the liquid after

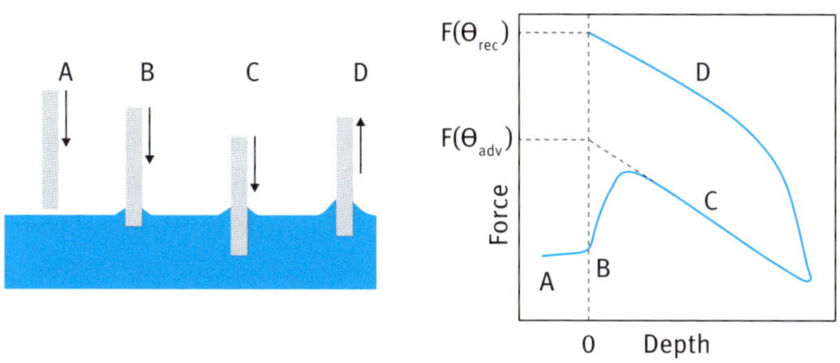

Fig. 7.8: Operating principle of Wilhelmy balance (left) and corresponding force/depth diagram (right). $F(\theta_{adv})$ and $F(\theta_{rec})$ force extrapolated at depth 0, giving the advancing and receding angles, respectively, when hysteresis is present.

having reached the desired depth; the force is measured for the receding angle. When hysteresis is present, the intercept of the advancing and receding curves, obtained with a regression fit, at depth = 0, gives the forces corresponding to the advancing and the receding angle. If there is no hysteresis behavior, the advancing and receding curves should overlap and the force extrapolated at depth 0 directly gives the CA.

7.4.3 Approach comparison

Optical goniometry is usually quite fast (few minutes for the direct acquisition of advancing and receding angle on one sample), and it can be used in many situations where force tensiometry cannot, provided that the sample has a relatively flat portion for testing. Testing can be done with a very small amount of liquid; hence, small samples are feasible, and a suitable chamber allows for measurements in severe conditions.

Meanwhile, some limited drawbacks can be highlighted. Identification of the tangent line (especially when an eye determination is operated) can affect the reproducibility of CA measurements, mainly at extreme CA. The conditions that produce advanced and receding angles are sometimes difficult to reproduce. It is challenging to measure CAs on single fibers, even if picoliter liquid dispenser are now available [47].

Considering force tensiometry, the main advantage is the direct achievement of an average value both in terms of force and surface on a sample wider than the ones usually tested with goniometry: this is particularly advantageous in industrial laboratories, where the typical size of the sample is several centimeters. Because a whole force/immersion curve is recorded, values can be averaged all over the surface of the sample or on a portion of it, depending on the region of the graph obtained. A further feature is the possibility of generating multiple wetting/dewetting cycles, gaining information on liquid-solid interaction (such as absorption or surface reorientation) [48]. Finally, analysis of fibers can be handled quite easily by force tensiometer.

There are two main drawbacks (more relevant than those of optical goniometry) in CA determination by tensiometry. First, enough liquid is necessary to immerse the solid sample (for common solvents this is quite straightforward). Second, the sample must satisfy certain constraints: the regular shape of a well-defined perimeter (see Eq. 7.19) and a size compatible with the tensiometer. Here, it should be stressed that, often, the surface morphology can be complex, and evaluating the true perimeter is not trivial: porous, rough samples and, in the worst case, membranes or textiles do not have regular geometry. On a rough hydrophilic surface, the liquid can permeate through the asperities and the probed perimeter is higher than the geometric one. Meanwhile, a super-hydrophobic rough surface has a reduced real contact with the liquid. Thus, in these cases, sessile drop method should be preferred.

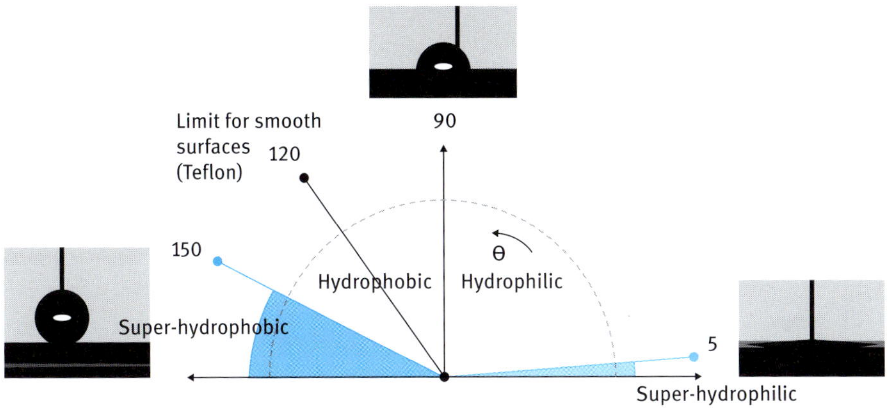

Fig. 7.9: Representation of the CA (θ) scale in the full range 0°–180°. The hydrophilic (0°–70°/90°), hydrophobic (70/90°–180°), super-hydrophilic (0°–5°), and super-hydrophobic (150°–180°) ranges are indicated with images of drops with θ = 2°, 90°, 170°.

7.5 Application of CA measurement

In Fig. 7.9, a representation of the CA scale in the full 0°–180° range is reported. Although based on conventional boundaries, it is useful to use a uniform language of the involved community. When no hysteresis is observed, surfaces with WCA much lower than 90° (somewhere lower than 60°) are considered hydrophilic. These are described as super-hydrophilic when the angle is lower than 5°, i.e. when a full spreading of the drop (formation of a liquid film) is observed. Above 90°, a surface is hydrophobic; it is considered super-hydrophobic when the angle exceeds 150°, i.e. when an almost spherical drop is formed. As it will be better discussed in Section 7.5.1, such a behavior can only arise by a combination of a low-surface-energy chemistry and a nano-microtexture because the angle limit for a flat surface is 120°.

7.5.1 From hydrophobic to water- and oil-repellent materials

In Tab. 7.2, the WCA for some common polymers has been reported. It is evident that, with the exception of polyvinyl alcohol (PVA) and to some extent PMMA, the main part have a significant hydrophobic behavior. In fact, such polymers are mainly composed of scarcely polar or polarizable groups as CH_x and above all CF_x or Si-CH_x units, leading to low surface energy. However, some fluorocarbon polymers and polydimethylsiloxane (PDMS) are the only conventional polymers exhibiting WCA higher than 110° [1]. A similar wetting behavior is also found in polymer films deposited using less conventional methods, such as low-temperature plasma processing. Plasma processes, addressed to provide surfaces with tailored functionalities, either with grafting or, more often,

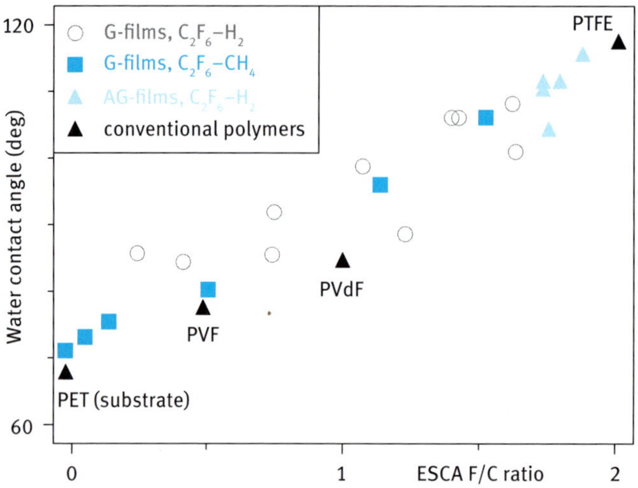

Fig. 7.10: Variation of the static WCA as a function of the fluorine over carbon surface atomic ratio for coatings that were plasma deposited in different conditions. G- and AG-films, deposited in glow and after-glow conditions. Some conventional polymers are shown for comparison. Reproduced with kind permission from Springer Science+Business Media B.V. Favia P, in: d'Agostino R, Favia P, Fracassi F, eds. *Plasma-processing of polymers. NATO ASI Ser E*, Kluwer Academic, Norwell (MA), 1997, 346. Copyright 1997 Springer [49].

with thin film deposition, are extensively applied to a wide range of materials. In this context, the plasma-enhanced chemical vapor deposition of Teflon-like coatings from fluorocarbon monomers (often with the addition of H_2) has been widely investigated in the 1980s to the 1990s. It was clear that plasma parameters able to tune the F/C ratio of the coatings are, in turn, active in tuning the surface hydrophobic character (i.e. the value of WCA), in the range 70°–120°, as shown in Fig. 7.10 [49].

More recently, specifically in the last 10 years, the interest in developing super-hydrophobic surfaces, with WCA higher than 150°, increased exponentially, as demonstrated by the huge number of published articles on this subject [1, 3, 5, 6, 8, 10, 15, 40, 41, 43]. However, merely playing with surface chemistry cannot lead to a WCA higher than 120°, and the technique of fabricating super-hydrophobic surfaces should be developed. Copying nature, "biomimetics", is often useful to surface engineers [6, 50]: the lotus leaf (microscopy images reported in Fig. 7.11 together with those of similar cases) as well as other examples of plants and insects cannot be wetted by water because its surface is made of nanobumps covered with a wax. Hence, nano/microtexturing is the key in achieving extreme wettability properties, and the understanding of the role of the roughness on CA is extremely important.

The description of the influence of roughness on WCA is mainly attributed to Wenzel [51], and separately, to Cassie and Baxter [52], with pioneering investigations pushed by the demand for waterproof textiles. The Wenzel model describes a liquid

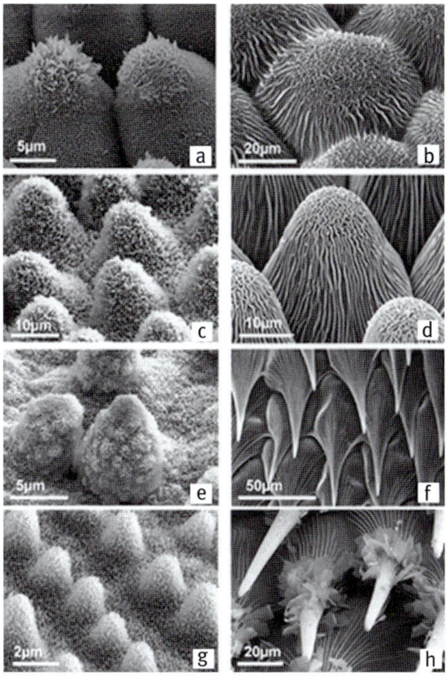

Fig. 7.11: SEM micrographs of nano/microstructures on plant surfaces. (a) *Colocasia esculenta*, (b) the seed surface of *Parodia alacriportana*, (c) *Euphorbia mysinites*, (d) the flower leaf of *Rosa montana*, (e) *Nelumbo nucifera* (lotus leaf), (f and h) *Sarracenia leucophylla*, and (g) *Oryza sativa*. Reproduced with permission from Koch K, Bhushan B, Barthlott W, *Soft Matter*, 4, 1943–63, 2008. Copyright 2008 Royal Society of Chemistry [50].

in contact with the whole solid surface, and the enhancement of hydrophobicity (but also hydrophilicity) is ascribed to the increased surface area of the *s-l* interface according to the following linear relationship:

$$\cos\theta_{\mathrm{w}} = r\cos\theta_{\mathrm{flat}}, \tag{7.20}$$

where θ_{w} is the observed angle on the rough surface, θ_{flat} is the equilibrium CA on the corresponding flat surface, and r, with $r > 1$, is the ratio of the actual surface area to the projected area of the surface. This model can account for liquid CA higher than 120°, but not for super-hydrophobic surfaces on which the drop can freely move: in fact, in such a description, because the surface contact area is increased, the rough asperities will hinder the displacement of the liquid (Fig. 7.12).

The Cassie-Baxter (CB) model, in general, describes the liquid CA of surfaces composed of two different phases according to

$$\cos\theta_{\mathrm{CB}} = fa\cos\theta_a + (1 - fa)\cos\theta_b, \tag{7.21}$$

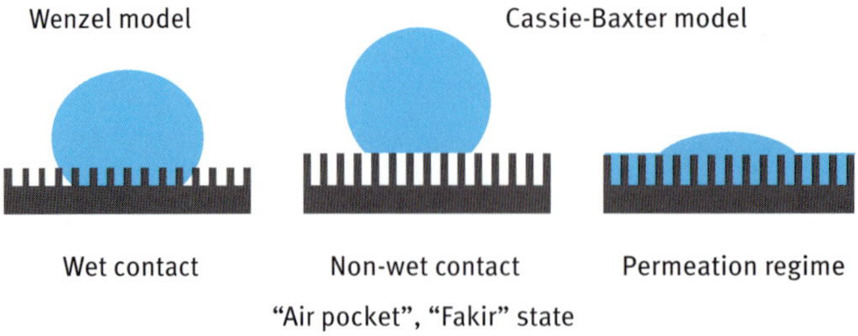

Fig. 7.12: Liquid drop on a rough hydrophobic surface according to the Wenzel model (left) and CB (center). Imbibition/permeation CB state on rough hydrophilic surface (right).

where f_a is the fraction of phase a, and θ_a and θ_b are the CA on the respective pure phases, and it can be extended to the case where the surface is composed of n phases. For a very rough and hydrophobic surface, it is possible that the liquid cannot pass through the asperities, but the drops remain pinned onto the tops of the hills, thus probing a composite surface made of solid, s, and air, and Eq. 7.21 can be written as follows:

$$\cos\theta_{CB} = fs\cos\theta_s + (1 - fs)\cos\theta_{\text{air}}. \tag{7.22}$$

Because $\cos\theta_{air}$ is –1, Eq. 7.23 is obtained:

$$\cos\theta_{CB} = fs\cos\theta_s - (1 - fs). \tag{7.23}$$

Equation 7.23, thus, gives the CA of a liquid placed on a rough surface in the so-called fakir way, or "air pocket" one, as depicted in Fig. 7.12.[1]

It is important to highlight that whatever the r value in the Wenzel equation, the CA increases only if the value on the corresponding flat surface is higher than 90°. Meanwhile, because the factor (1 – fs) is present in the CB equation, the measured θ_{CB} angle can be higher than θ_s, even for materials with CA lower than 90° (not very hydrophilic to completely wet the asperities – air should remain trapped between the liquid and the valley; otherwise, the composite CB model cannot stand). As an example, Di Mundo *et al.* [41] showed that a flat hydrocarbon coating deposited by plasma, with a WCA around 80°, leads to advancing WCA as high as 145° when deposited on a rough PS sample. More recently, the same authors demonstrated that a

1 In case the interface between the liquid and the solid phase (the top of the asperities) is rough, the term $fs\cos\theta_s$ in Eq. 7.23, should be multiplied by a suitable Wenzel roughness factor, r. In this chapter, it is assumed that the top of the asperities is smooth enough to consider $r = 1$. Similarly, in Eq. 7.21, the fraction of each phase should be multiplied by the rough factor of each phase.

plasma-deposited silicone-like coating with a WCA of 70° on a flat surface achieved a value of 110° when deposited on nanotextured polycarbonate (PC) [53]. Hence, water-repellent surfaces can be obtained, playing at the same time with chemistry and texturing to have surfaces with hydrophobic asperities where air is trapped between, as demonstrated by spectroscopic techniques [54]. Further, as a consequence of the non-wet contact, such surfaces are characterized by a slippery behavior of water drops, thus by a low WCA hysteresis, besides high WCA. In fact, considering a drop sliding on a tilting plate, as reported in Fig. 7.7c, in an air pocket regime, the energy barrier of the maxima between the metastable states describing such phenomenon decreases because most of the drop lies on air: the energy gained when the front of the contact line moves is very close to the energy necessary to leave the wetted region on the rear part. Therefore, to identify a super-hydrophobic water-repellent behavior, the measurement of a high static WCA is not enough: the roughness and chemistry could be such that a Wenzel state, a wet contact, can be obtained, preventing the drop motion, and characterized by a high hysteresis. A good characterization of a water slippery surface should be based on the evaluation of the advancing and receding WCA: super-hydrophobic water-repellent surfaces show hysteresis lower than 15° [55]. In general, features with high aspect ratio and short spacing are needed to avoid water collapsing into the grooves; therefore, at this regard, the use of the mere rms roughness as criterion to drive water repellency could be misleading.

Both bottom-up and top-down approaches have been proposed for the preparation of nano- and micro-super-hydrophobic surfaces. Photoinitiated copolymerization of butyl methacrylate and ethylene dimethacrylate in the presence of cyclohexanol and 1-decanol, allows the deposition of a nanoporous coating as shown in Fig. 7.13. Levkin and colleagues [56] showed that, during the deposition, the inert

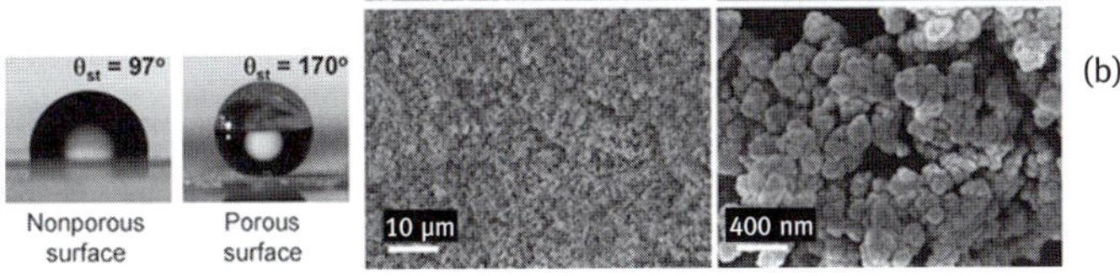

Fig. 7.13: Shape of water droplets formed on porous and nonporous polymer layers and SEM images of the super-hydrophobic porous polymers: (a) poly(butyl methacrylate-*co*-ethylene dimethacrylate) and (b) poly(styrene-*co*-divinylbenzene). Reproduced with permission from Levkin PA, Svec F, Fréchet JMJ, *Adv Funct Mater*, 19, 1993–8, 2009. Copyright 2009 Wiley-VCH Verlag GmbH & Co. KGaA, Weinheim [56].

solvent acts like a porogen leaving such rough surface and leading to WCA as high as 172° with negligible hysteresis. Many articles deal with patterning *via* lithographic techniques followed by deposition (PECVD, or self-assembling or the like) of a low-energy polymeric film [8, 57]. Chang-Hwan Choi and Chang-Jin Kim examined a wide range of nanostructures on silicon by interferometric lithography and subsequent etching as reported in Fig. 7.14 a–c. The deposition of a thin Teflon-like coating made the samples with the higher-aspect-ratio structures super-hydrophobic [57].

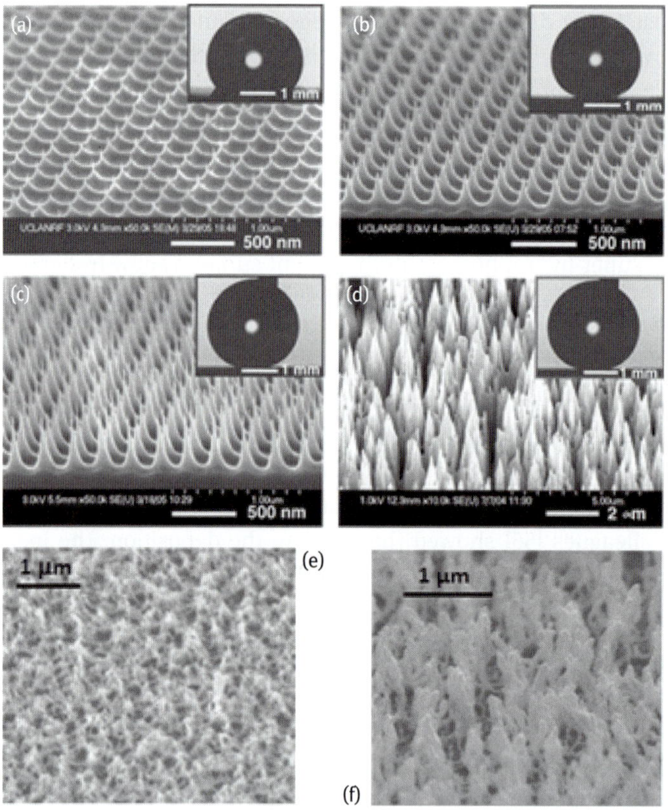

Fig. 7.14: SEM images of nanotextured samples for application to super-hydrophobic surfaces. (a–c) Regular nanopost structures created on Si by the interference lithography followed by DRIE. (d) Random nanopost structures created by the black silicon method. (b–d) Nanostructures with height higher than 200 nm show extreme super-hydrophobicity; (a) short structures do not. Reproduced with permission from Choi C-H, Kim C-J, *Nanotechnology*, 17, 5326–33, 2006. Copyright 2006 IOP Publishing [57]. (e and f) Super-hydrophobic plasma nanotextured PS and PC, with the latter coated with a plasma-deposited silicone-like thin film. PS: Adapted with permission from Di Mundo R, Palumbo F, d'Agostino R, *Langmuir*, 4, 5044–51, 2008. Copyright 2008, American Chemical Society. [15]. PC: Adapted with permission from Palumbo F, Di Mundo R, Cappelluti D, d'Agostino R, *Polymers*, 8, 118–26, 2011. Copyright 2011 Wiley-VCH Verlag GmbH & Co. KGaA, Weinheim [10].

A widely diffuse single-step top-down approach is etching a material under high-ion-bombardment conditions (reactive ion etching), producing stochastic-shaped nanostructures, sometimes in the form of pillars (Fig. 7.14d–f). CF_4-plasmas have been tested for etching/roughening polybutadiene [58], SF_6/O_2-plasmas for PDMS [59], CF_4/O_2-plasmas for PE [60], and for PS [15, 41]. In general, the grafting of functional groups occurs along the nonhomogeneous etching, thus, if the density of the fluorinated groups is optimized, the suitable hydrophobic chemistry can be combined with the nanotexture.

Oleophobic surfaces need a separate description because the surface tension of oil/organic liquids is lower than that of water. Therefore, to get a oleophobic material at a solid-air-oil interface, the surface energy should be very low, and in turn, the correct choice of chemistry and texturing is even more difficult. Jung and Bhushan [61] produced hierarchical structures, composed of micrometric patterned pillars covered with nanometric vertical platelets of *n*-hexatriacontane, and a final coating of *n*-perfluoroeicosane ($C_{20}F_{42}$) by thermal evaporation. An hexadecane CA as high as 115° could be achieved by a well-designed structures morphology. Tuteja *et al.* [62, 63] demonstrated that micrometric reentrant structures, as those shown in Fig. 7.15, coated with fluorodecyl POSS and alike, can trap air, thus leading to octane advancing CA of 163° (145° receding) and also to the design of "omniphobic" surfaces repellent to a wide range of liquids such as acetic acid, toluene, and hexadecane. A valuable and recent review on oleophobic surfaces and their applications can be found in [64].

7.5.2 Hydrophilic to super-hydrophilic materials

It has been already stressed that polymers, as typical cases of soft matter, are normally characterized by a low energy both in their bulk and in their surface, thus generally expressing high WCA. If this property represents an advantage for most of polymer applications, because it turns into good inertness, it appears as a limitation when

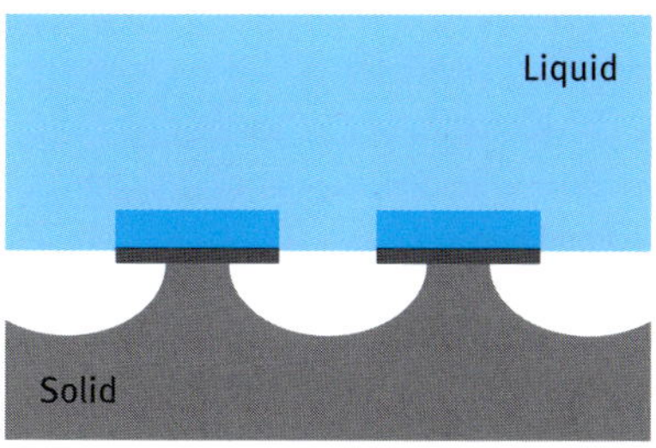

Fig. 7.15: Reentrant, hoodoo-like microstructures necessary for oleophobic behavior of surfaces when properly coated with low-surface-energy film.

gluing, painting, dyeing, washing, or functionalizing (for typical device purpose) their surfaces is a relevant process.

It is well known that surfaces bearing functionalities with asymmetric charge distribution (-OH, -NH_2, -CO_2H, -C=O) can express high surface energy. Thus, several surface modification techniques or methods have been addressed to the addition of such groups to polymer surfaces. Plasma processes (often corona discharge treatments, very common in industry), thanks to the low temperature and easily scalable nature of the process, have been thoroughly investigated, and actually used, with O_2, NH_3, N_2/H_2, and H_2O feed gases for imparting hydrophilicity to PE, PET, PTFE, and others [65–67].

Research interest toward super-hydrophilic surfaces is particularly recent (i.e. the last 5–6 years), mostly for the emerging applications of microfluidic and antifog surfaces. The latter can be obtained with the coalescence of the condensed microdroplets up to the formation of a liquid film when the WCA of the surface is particularly low. Figure 7.16 shows photographs of condensed water vapor on a transparent plastic with the different hydrophilic and super-hydrophilic charactertistics of the surface [68].

Unlike the super-hydrophobic surfaces, a WCA lower than 5° can be also achieved on flat surfaces if a high chemical affinity with water is established (e.g. strong acid-base interactions or strong permanent charge separations). In this regard, it is worth mentioning the natural case (from plant kingdom) of super-hydrophilic leaves, which exist both with a relatively smooth surface (those permanently wet) and with micro-bump-rich (aquaplaning) or micro-porous (water-absorbing) surface [69]. However, to produce a super-hydrophilic material surface, the strategy of combining a moderate hydrophilic chemistry with textured morphology is preferred. This is probably pushed by the necessity of reducing aging effects of the hydrophilic chemistry (see Section 7.5.3). In this frame, plasma nanotexturing by O_2 etching, which generates

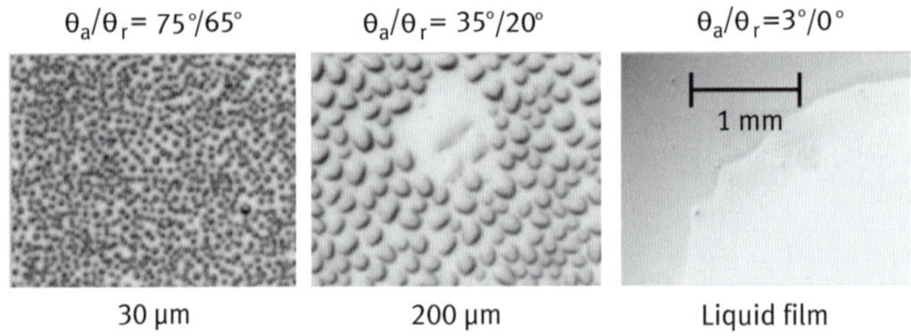

Fig. 7.16: Photographs of condensed water drops on the surface of a transparent plastic (PC) exposed to water vapor. The material has been treated by a patented plasma process [68]. As the WCA decreases (reported over each picture), the mean width of the condensed water drops increases, up to the point of coalescence and formation of a liquid film (antifog condition) as in the brighter region of the image on the right. Under each image, the average diameter of drops is reported.

nano-pillars with a high O/C surface ratio, has been utilized onto polymers such as polyethyl methacrylate (PMMA) [70] or photoresist epoxy Novolak SU-8 [71]. This strategy configures the so-called permeation wetting regime, which has been depicted as a hydrophilic CB regime because the liquid permeates the textured solid and the drop faces a composite surface consisting of both solid and liquid [72]. In fact, as shown in Fig. 7.12 (right side), the liquid penetrates underneath and between the asperities, and the drop probes a surface made of water and solid. Since the CA of water with itself is equal to 0°, an equation similar to Eq. 7.23 can be derived:

$$\cos\theta_{CB} = fs\cos\theta_s + (1 - fs). \tag{7.24}$$

Analyzing Eq. 7.24, it can be observed that surfaces that follows this regime are extremely wettable even if the WCA on the corresponding flat surface, θ_s, is not particularly low.

7.5.3 Hydrophobic recovery of hydrophilic surfaces

Highly hydrophilic surfaces (WCA lower than 30°–40°) are, in air, intrinsically unstable and are thus subject to "hydrophobic recovery," which is the tendency to reduce the energy by decreasing the solid/air interfacial energy. This phenomenon occurs through different "aging" mechanisms, which are more or less important depending on the surface characteristics. Among these, the most frequent are

- reorientation of surface functional groups to have the low-energy ones facing the atmosphere and the high-energy ones buried in the bulk of the polymer;
- absorption on the hydrophilic surface of hydrocarbon molecules present in the atmosphere, as a consequence of their low surface energy.

The former mechanism is particularly relevant in polymers because they are characterized by high-chain/bond mobility and are, in fact, demonstrated to be favored by temperature increase [65].

Typically, plasma-treated polymers with hydrophilic modifications show a steep increase (even 30°–40°) of the WCA in the first (3–5) days after the treatment, then a slighter increase up to a plateau reached in 1–2 weeks. However, although increased, the final value of the CA generally is still lower than that for the untreated polymer. A typical example is shown in Fig. 7.17. Several efforts in optimizing working conditions to limit aging effects have been made by academic and industrial researchers [65, 73].

In case of super-hydrophilic micro/nanostructured modifications, such as those reported in Section 7.5.2 [70, 71], different aging kinetics have been observed: the WCA stabilizes in 2–3 months at values between 30° and 60°. That is the reason why, also in

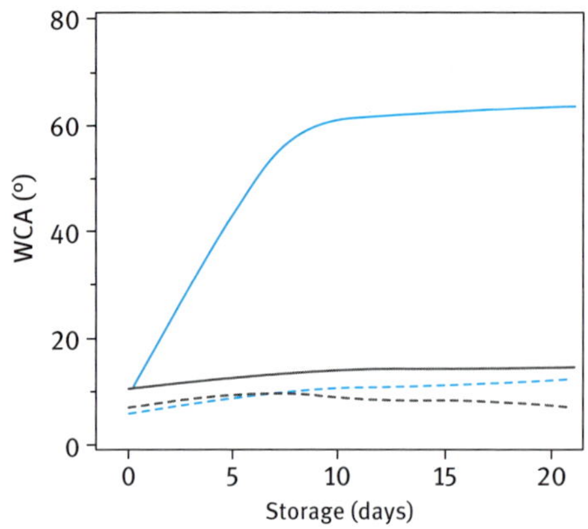

Fig. 7.17: WCA time evolution at room temperature for a PC sample after oxygen plasma treatment. Blue line, sample stored in air; black line, stored in water. Both advancing, continuous line, and receding angle, dashed line, are reported.

the plasma processing science community, new strategies are under study to get stable super-hydrophilic polymer surfaces. One of these approaches is the deposition of a highly hydrophilic inorganic coating, such as a well-defined SiO_x film, onto a tailored nanotextured surface [10].

It is worth highlighting that the hydrophobic recovery is not an aging phenomenon intrinsic of the material but is related to the interaction between the material and its environment. As a matter of fact, when the surface of a polymer is made more hydrophilic and is stored in water, the WCA shows a negligible variation compared with the as-treated sample, as can be observed in Fig. 7.17. Another important consideration should be done regarding the hysteresis of the WCA. As already discussed in Section 7.3, the advancing and receding angles show a different behavior when the surface is stored (as usually) in air: whereas the advancing angle shows the typical increasing trend, the receding one increases only slightly. Such an effect, rarely reported in literature due to the long unawareness about receding measurements (though already observed [37]), suggests that high-energy groups remain within the reorganized surface and are "recalled" by the wetting action.

Some authors [73–75] tried to develop a model for rationalizing the time evolution of the hydrophobic recovery, which is particularly useful in describing the behavior of plasma-treated polymer surfaces, that could be applied also to other kinds of processing leading to hydrophilic surfaces. Once polar groups are grafted on the surface of a polymer, because the surface becomes composed of polar and apolar moieties, the equilibrium CA, θ_E, can be written, according to Eq. 7.23, as

$$\cos\theta_E = fp\cos\theta_p + (1 - fp)\cos\theta_{np}, \tag{7.25}$$

where f_p is the fraction of the surface area covered with polar groups and θ_p and θ_{np} are the WCA relative to the completely polar surface and the apolar one, respectively. Soon after the functionalization process, *fp* will start to decrease, as a consequence of the aging process, leading to the increase of the WCA. In general, an exponential decay can be supposed for the polar fraction

$$f_p(t) = f_p^e + f_p^o e^{-t/\tau}, \tag{7.26}$$

where f_p^e is the equilibrium polar fraction, the sufficiently stable part of the polar surface that will not change during the aging process, for example, because it is sufficiently cross-linked or blocked by steric hindrance. f_p^o, instead, is the initial fraction of surface involved in the aging, and τ is the time constant of polar groups decay. Equation 7.25, taking into account Eq. 7.26, can be arranged as follows:

$$\cos\theta_E\,(t) = (f_p^e + f_p^t e^{-t/\tau})\cos\theta_p + (1 - f_p^e + f_p^o e^{-t/\tau})\cos\theta_{np}. \tag{7.27}$$

This equation can be simplified and solved with the Marquardt-Levenberg algorithm to find suitable values for f_p^e, f_p^o, and τ, considering that $\cos\theta_p$ is 1, supposing the polar fraction completely hydrophilic, whereas $\cos\theta_{np}$ is that on the untreated polymer. A complete description of the solving procedure of Eq. 7.27 can be found in [75]. It is important to highlight that to limit hydrophobic recovery, a surface treatment leading to high f_p^e should be optimized – basically, the initial fraction of stable grafted polar groups; in fact, the final WCA will depend on how many polar functionalities persist on the surface. Concerning the aging rate, determined by the inverse of τ, two different approaches can be followed: fast rearrangement kinetics, i.e. complete aging in 1–2 days, could allow a better handling of the samples for subsequent processing. Meanwhile, very slow kinetics allow the sample be considered sufficiently stable for the period of investigation. Favia *et al.* [73] thoroughly investigated plasma treatments of PTFE: they showed that H_2-plasma treatments followed by NH_3-plasma ones lead to stable wettable PTFE surfaces with limited aging and long shelf life. Stable WCA as low as 40° with a f_p^e reaching a value of 82% could be achieved. This was explained with the cross-linking capacity of the H_2-plasma leading to a decrease of the mobility of the surface polymer chains, and as a consequence the amino groups grafted in the second step can remain stable in their position.

7.5.4 CA on porous surfaces

For porous materials such as powders, the Washburn method can be used for evaluating the wettability, measuring the rise of the liquid in a capillary tube. For materials, such as membranes or adsorbents, the direct measurement of the CA from the sessile drop method can be very difficult, not only for the aforementioned hysteresis issues but also for the absorption and spreading phenomena. Consider what happens on a paper filter once a drop of liquid is placed: the drop contact diameter increases, spreading over the surface; meanwhile, because the paper is porous, water will be absorbed, filling the pores and penetrating underneath (Fig. 7.18). Hence, in a period that depends on the nature of the liquid and the paper, the characteristics of the probing drop (CA, volume diameter, height) will change. If either the spreading or the absorption rate is too fast compared with the operator observation, the measurement of the CA is questionable. However, observation of the drop characteristics variation, better accomplished with a goniometer equipped with a high speed camera, can be very useful even when fast spreading or absorption kinetics are observed. In fact, recording the drop image evolution, besides wettability evaluation, can give information on the absorbency and spreading behavior of such materials often developed for application exploiting such properties: adsorbents, filters, writing papers, and so on. The combination of absorption and spreading is analogous to what is observed on nanotextured hydrophilic materials in the permeation regime described in Section 7.5.2. The main difference is that, in porous materials the observed phenomenon is not limited to the top surface of the solid (e.g. just consider the case of a water drop on a membrane, permeating through the pores of the solid).

Because nature very rarely helps investigators, it should also be considered that the evaporation of the liquid contributes to drop volume changes. Thus, it is important

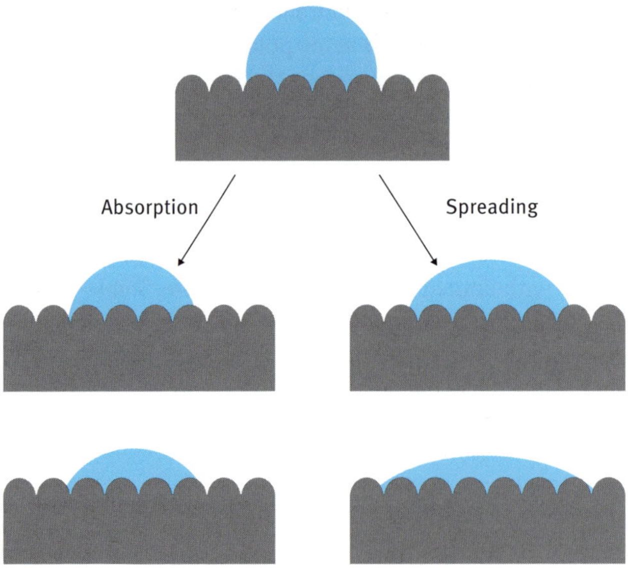

Fig. 7.18: Drop of liquid in contact with a porous surface. Depending on the nature of the solid, absorption (decrease of volume and constant contact area) or spreading (constant volume and increase of the contact area) can be observed.

in such measurements to inhibit the evaporation, using low-vapor pressure liquids or saturating the ambient around the drop. As an alternative, some authors suggest to take into account the effect of evaporation placing a drop of the same liquid in the same conditions onto a non-absorbing material (e.g. PTFE for water). The liquid evaporation rate on such a reference sample can be considered for the mathematical correction of the volume variation on the surface under investigation. In general, both spreading and absorption can develop on a porous surface [76, 77]. However, the former does not affect the volume, but it widens the contact area, and absorption is commonly slower because the liquid molecules should diffuse beneath the substrate surface.

Farris et al. [76] deeply investigated spreading/absorbance behavior over different biopolymers, polysaccharides (chitosan, pullulan, and pectine) and pig skin-gelatin-coated PET. They observed water drop evolution on the time scale of 60 s and found that, for some materials, the increase in liquid contact area (spreading) was more evident than the decrease of volume (absorption). Through trigonometric correlation between the drop shape and the CA, they could separate the spreading contribution from the absorption one and concluded that the considered biopolymers, except chitosan, whose surface presented pores, show negligible absorption characteristics.

Dastjerdi *et al.* [77] optimized the synthesis of a composite amino-functionalized polysiloxane coating with amphiphilic features. It is interesting to highlight that, with a contribution due to the hierarchical roughness, the water drop evolution of coatings deposited onto TiO_2-coated PET textiles is very peculiar. Although appearing slippery (advancing WCA as high as 150° with a hysteresis lower than 3°) in a matter of few minutes, the surface becomes absorbent because of the polymer chains' mobility, which pushes the hydrophilic amino groups toward water.

7.5.5 Acid-base characterization of polymeric surfaces

Wetting measurements can be used to get information about the acid-base behavior of surfaces. According to the van Oss theory (Section 7.2.1.2.2), the shape assumed by a probe liquid drop onto a solid depends also on the acid-base interactions at the interface, and with a similar formalism, it is possible to write

$$W_{sl} = \gamma_l (1 + \cos\theta) = W_{sl}^{LW} + W_{sl}^{AB} \tag{7.28}$$

It is straightforward that once W_{sl}^{LW} is calculated, in agreement to what described in Section 7.2, W_{sl}^{AB} can be obtained from a measurement of CA. In fact, as shown by Huttinger *et al.* [78], for a more correct determination of acid-base work of adhesion, a CA "sampling" in the whole pH-surface "titration" scale is necessary. For this purpose, a classical CA technique (sessile drop or Wilhelmy method) is utilized, where acid and base aqueous solutions with pH values ranging from 1 to 14 are used as probe liquid, instead of pure water. Here, the work of adhesion is defined as a function of the pH. In particular, as reported in Fig. 7.19, four cases can be distinguished: (a) corresponds to a nonpolar surface for which acid-base interactions, W_{sl}^{AB}, do not apply; (b) and (c) show a surface with acid and base groups, respectively; (d) is expected to exhibit amphoteric character. More often, in later published works, CA variation alone, i.e. without calculating the derived quantity W_{sl}, has been reported as a function of the pH [79–81].

From these data, an estimation of the pK_a (or pK_b) of the surface is sometimes inferred by simply considering the pH value at the sigmoid inflection point [80] or by more complex thermodynamic treatment of the CA data [82].

The approach of using CA to characterize acid-base behavior is extensively utilized for the analysis of grafted (surface-treated) polymer sheets. Arefi-Khonsari *et al.* [83, 84] investigated the possibility of tuning, by plasma treatments, the work of

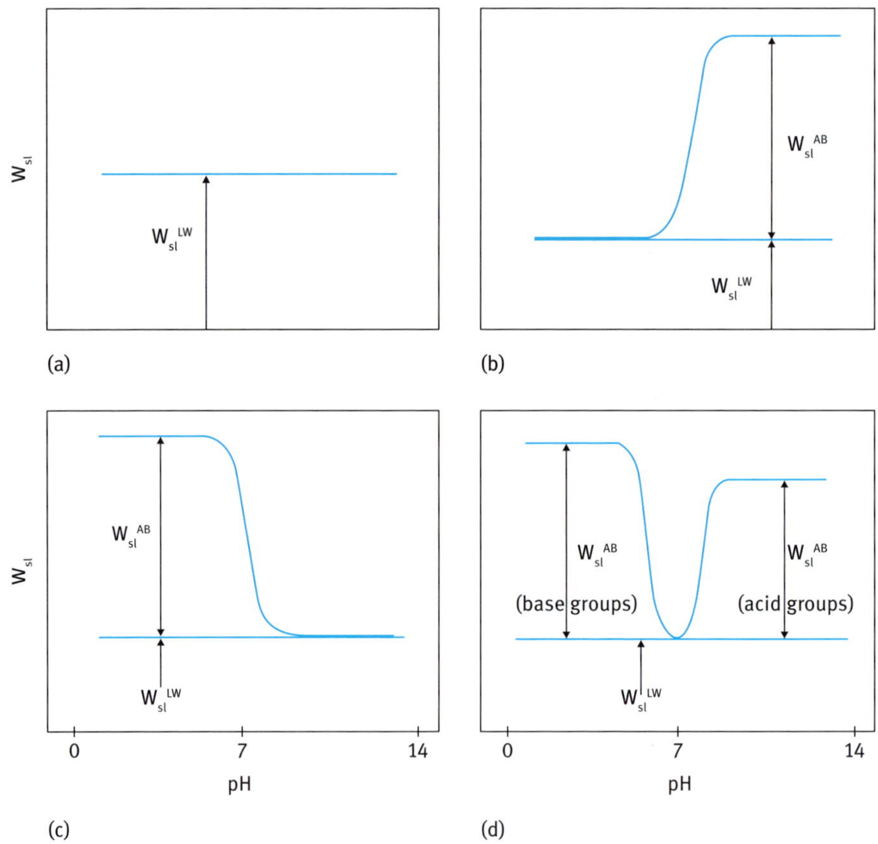

Fig. 7.19: Work of adhesion as a function of the pH of the testing aqueous solution for a (a) nonpolar, (b) acid, (c) basic, and (d) amphoteric material.

adhesion of poly(propylene). Whitesides *et al.* [85] used CA titrations, among other diagnostics, to investigate the behavior of acid and base functional groups (carboxylic acids, amines, and others) grafted by wet methods at the surface of low-density PE, and they determined the ionization extent of the grafted groups as well as the relationship between WCA values and density of surface groups.

It is worth highlighting that this kind of measurements does not leave the surface unchanged because it involves irreversible reactions (which are mainly ionization reactions) between the probe liquid and the test surface. That is why during measurements, a so-called reactive spreading can be observed, and when reported, the receding angles are always very low (near zero) [80]. Another consideration is that the CA-based titration method is not utilizable for granules, powders, and porous materials such as absorbers, whose absorption properties to gas/vapors or liquids are strongly dependent on the capability of forming acid-base bonds with their molecules. In such cases, other titration methods are used. For example, for activated carbon, carbon black (as well as graphene and carbon nanotubes), the Bohem titration [86] is often reported, and it is based on the use of acid or base water solutions that the granules are soaked up and back-titrated at the end of the reaction.

References

[1] Garbassi F, Morra M, Occheillo E. Polymer surfaces from physics to technology. New York: Wiley; 1994.

[2] Määttänen A, Ihalainen P, Bollström R, Toivakka M, Peltonen J. Wetting and print quality study of an inkjet-printed poly(3-hexylthiophene) on pigment coated papers. Colloids Surf A 2010;367:76–84.

[3] Su B, Li M, Lu Q, Toward understanding whether superhydrophobic surfaces can really decrease fluidic friction drag. Langmuir 2010;26:6048–52.

[4] Becker H, Gärtner C. Polymer microfabrication technologies for microfluidic systems. Anal Bioanal Chem 2008;390:89–111.

[5] Mortazavi M, Nosonovsky M. Adhesion, wetting and superhydrophobicity of polymeric surfaces. In: Zeng H. ed. Polymer adhesion, friction, and lubrication. 1st edn. Hoboken: Wiley; 2013;177–226.

[6] Barthlott W, Schimmel T, Wiersch S, et al. The Salvinia paradox: superhydrophobic surfaces with hydrophilic pins for air retention under water. Adv Mater 2010;22:2325–8.

[7] Strobel M, Lyons C S, Mittal KL, Plasma surface modification of polymers: relevance to adhesion. Utrecht (The Netherlands): VSP; 1994.

[8] Corbella C, Portal S, Rubio-Roy M, et al. Surface structuring of diamond-like carbon films by colloidal lithography with silica sub-micron particles. Diamond Relat Mater 2010;19:1124–30.

[9] Young T. An essay on the cohesion of fluids. Philos Trans R Soc London; 1805;95:65–87.

[10] Palumbo F, Di Mundo R, Cappelluti D, d'Agostino R. Superhydrophobic and superhydrophilic polycarbonate by tailoring chemistry and nano-texture with plasma processing plasma process. Polymers 2011;8:118–26.

[11] Bacharouche J, Haidara H, Kunemann P, Vallat MF, Roucoules V. Singularities in hydrophobic recovery of plasma treated polydimethylsiloxane surfaces under non-contaminant atmosphere. Sensors Actuators A 2013;197:25–9.

[12] Igor Krupaa, Volkan Cecenb, Abderrahim Boudennec, Jan Prokešd, Igor Nováke. The mechanical and adhesive properties of electrically and thermally conductive polymeric composites based on high density polyethylene filled with nickel powder. Mater Des 2013;51:620.
[13] Stamboulides C, Englezos V, Hatzikiriakos SG. The ice friction of polymeric substrates. Tribol Int 2012;55:59–67.
[14] Süzer S, Argun A, Vatansever O, Aral O. XPS and water contact angle measurements on aged and corona-treated PP. J Appl Polym Sci 1999;74:1846–50.
[15] Di Mundo R, Palumbo F, d'Agostino R. Nanotexturing of polystyrene surface in fluorocarbon plasmas: from sticky to slippery superhydrophobicity. Langmuir 2008;24:5044–51.
[16] Coulson SR, Woodward I, Badyal JPS. Super-repellent composite fluoropolymer surfaces. J Phys Chem B 2000;104:8836–40.
[17] Kaynak A, Mehmood T, Dai XJ, Magniez K, Kouzani A. Study of radio frequency plasma treatment of PVDF film using Ar, O2 and (Ar + O2) gases for improved polypyrrole adhesion. Materials 2013;6:3482–93.
[18] Li X, Chen G, Ma X, et al. Preparation of a super-hydrophobic poly(vinyl chloride) surface via solvent-non solvent coating. Polymer 2006;47:506–9.
[19] Wakeham W, Assael M, Marmur A, et al. Material properties: measurement and data. In: Springer handbook of experimental fluid mechanics. Springer-Verlag Berlin Heideberg; 2007:85—178.
[20] Zisman WA. Relation of the Equilibrium Contact Angle to Liquid and Solid Constitution. ACS Adv Chem Ser 1964;43:1–50.
[21] Girifalco LA, Good RJ. A Theory for the Estimation of Surface and Interfacial Energies. I. Derivation and Application to Interfacial Tension. J. Phys. Chem 1957;61:904–909.
[22] Kwok DY, Neumann AW. Contact angle measurement and contact angle interpretation. Adv Colloid Interface Sci 1999;81:167–249.
[23] Fowkes FM. Dispersion force contributions to surface and interfacial tensions, contact angles, and heats of immersion. In: Gould RF, ed. Contact angle, wettability, and adhesion. ACS Adv Chem Ser. Washington (DC): American Chemical Society; 1964;43:99–111.
[24] Owens DK, Wendt RC. Estimation of the surface free energy of polymers. J Appl Polym Sci 1969;13:1741–7.
[25] Wu S. Calculation of interfacial tension in polymer systems. J Polym Sci C 1971;34:19 –30.
[26] Li D, Neumann AW. Contact angles on hydrophobic solid surfaces and their interpretation. J Colloid Interface Sci 1992:148:190–200.
[27] Kwok DY, Neumann AW. Contact angle measurement and contact angle interpretation. Adv Colloid Interface Sci 1999;81:167–249.
[28] Żenkiewicz M. Methods for the calculation of surface free energy of solids. J Achieve Mater Manuf Eng 2007;24:137–45.
[29] Kitazaki Y, Hate T. Surface-chemical criteria for optimum adhesion II. The variability of critical surface tension (γc) and its choice. J Adhes 1972;4:123–132.
[30] van Oss CJ, Good RJ, Chaudhury MK. The role of van der Waals forces and hydrogen bonds in "hydrophobic interactions" between biopolymers and low energy surfaces. J Colloid Interface Sci 1986;111:378–90.
[31] van Oss CJ, Chaudhury MK, Good RJ. Interfacial Lifshitz-van der Waals and polar interactions in macroscopic systems. Chem Rev 1988;88:927–41.
[32] Good RJ. Contact angle, wetting, and adhesion: a critical review. J Adhes Sci Technol 1992:6:1269 –302.
[33] Schultz J, Tsutsumi K, Donnet J-B. Surface properties of high-energy solids: I. Determination of the dispersive component of the surface free energy of mica and its energy of adhesion to water and n-alkanes. J Colloid Interface Sci 1977;59:272–7.

[34] Marmur A. Solid-surface characterization by wetting. Annu Rev Mater Res 2009;39:473–89.
[35] Furmidge CGL. Studies at phase interfaces. I. The sliding of liquid drops on solid surfaces and a theory for spray retention. J Colloid Sci 1962:17:309–24.
[36] De Gennes PG. Wetting: statics and dynamics. Revs Mod Phys 1985;57:827–63.
[37] Morra M, Occhiello E, Garbassi F. Contact angle hysteresis on oxygen plasma treated polypropylene surfaces. J Colloid Interface Sci 1989;132:504–6.
[38] Gotoh K, Nakata Y, Tagawa M, Tagawa M. Wettability of ultraviolet excimer-exposed PE, PI and PTFE films determined by the contact angle measurements. Colloids Surf A 2003;224:165–173.
[39] Dwight DW, Riggs WM. Fluoropolymer surface studies. J Colloid Interface Sci 1974;47:650.
[40] Gao L, McCarthy TJ. Teflon is hydrophilic. comments on definitions of hydrophobic, shear versus tensile hydrophobicity, and wettability characterization. Langmuir 2008;24:9183–8.
[41] DiMundo R, Palumbo F, d'Agostino R. Influence of chemistry on wetting dynamics of nanotextured hydrophobic surfaces. Langmuir 2010;26:5196–201.
[42] Johnson RE, Dettre R. Contact Angle Hysteresis. In: Fowkes F, ed. Contact angle, wettability, and adhesion. Vol 43. Washington (DC): ACS; 1964:112–145.
[43] Vourdas N, Tserepi A, Gogolides E. Nanotextured super-hydrophobic transparent poly(methyl methacrylate) surfaces using high-density plasma processing. Nanotechnology 2007;18:125304.
[44] Extrand CW, Kumagai Y. Contact angles and hysteresis on soft surfaces. J Colloid Interface Sci 1996;184:191–200.
[45] Pierce E, Carmona FJ, Amirfazli A. Understanding of sliding and contact angle results in tilted plate experiments. Colloids Surf A 2008;323:73–82.
[46] Yuan YH, Lee TR. Contact angle and wetting properties. In: Surface science techniques. Vol 51. Springer-Verlag Berlin Heideberg; 2013:3–34.
[47] Taylor M, Urquhart AJ, Zelzer M, Davies MC, Alexander MR. Picoliter water contact angle measurement on polymers. Langmuir 2007;23:6875–8.
[48] Moghaddam MS, Wålinder M, Claesson PM, Swerin A. Multicycle Wilhelmy plate method for wetting properties, swelling and liquid sorption of wood. Langmuir 2013;29:12145–53.
[49] Favia P. In: d'Agostino R, Favia P, Fracassi F, eds. Plasma-processing of polymers. NATO ASI Ser E. Norwell (MA): Kluwer Academic; 1997:346.
[50] Koch K, Bhushan B, Barthlott W. Diversity of structure, morphology and wetting of plant surfaces. Soft Matter 2008:4:1943–63.
[51] Wenzel TN. Resitance of solid surfaces to wetting by water. Indust Eng Chem 1936;28:988–94.
[52] Cassie ABD, Baxter S. Wettability of porous surfaces. Trans Farad Soc 1944;40:546–51.
[53] Palumbo F, Di Mundo R, d'Agostino R. Plasma nanotexturing combined with thin film deposition: an enabling technology for stable superhydrophilic materials. In: Proceedings of the 20th International Symposium on Plasma Chemistry; 2011.
[54] Checco A, Hofmann T, Di Masi E, Black CT, Ocko BM. Morphology of air nanobubbles trapped at hydrophobic nanopatterned surfaces. Nanoletters 2010;10:1354–8.
[55] McHale G, Shirtcliffe NJ, Newton MI.Contact-angle hysteresis on super-hydrophobic surfaces. Langmuir 2004;20:10146–9.
[56] Levkin PA, Svec F, Fréchet JMJ. Porous polymer coatings: a versatile approach to superhydrophobic surfaces. Adv Funct Mater 2009;19:1993–8.
[57] Choi C-H, Kim C-J. Fabrication of a dense array of tall nanostructures over a large sample area with sidewall profile and tip sharpness control. Nanotechnology 2006;17:5326–33.
[58] Woodward I, Schofield WCE, Roucoules V, Badyal JPS. Super-hydrophobic surfaces produced by plasma fluorination of polybutadiene films. Langmuir 2003;19:3432–8.

[59] Tsougeni K, Tserepi A, Boulousis G, Constantoudis V, Gogolides E. Control of nanotexture and wetting properties of polydimethylsiloxane from very hydrophobic to super-hydrophobic by plasma processing. Plasma Processes Polym 2007;4:398–405.
[60] Fresnais J, Chapel JP, Poncin-Epaillard F. Synthesis of transparent superhydrophobic polyethylene surfaces. Surf Coat Technol 2006;200:5296–5305.
[61] Jung YC, Bhushan B. Wetting behavior of water and oil droplets in three-phase interfaces for hydrophobicity/philicity and oleophobicity/philicity. Langmuir 2009;25:14165–73.
[62] Tuteja A, Choi A, Ma M, et al. Designing superoleophobic surfaces. Science 2007;318:1628–1.
[63] Pan S, Kota AK, Mabry JM, Tuteja A. Superomniphobic surfaces for effective chemical shielding. J Am Chem Soc 2013;135:578–81.
[64] Kesong L, Ye T, Lei J. Bio-inspired superoleophobic and smart materials: design, fabrication, and application. Progr Mater Sci 2013;58:503–64.
[65] Inagaki N, Narushim K, Ejima S, et al. Hydrophobic recovery of plasma-modified film surfaces of ethylene-co-tetrafluoroethylene co-polymer. J Adhes Sci Technol 2003;17:1457–75.
[66] Favia P, Stendardo MV, d'Agostino R. Selective grafting of amine groups on polyethylene by means of NH3-H2 RF glow discharges. Plasmas Polym 1996;1:91–112.
[67] Kim J, Chaudhury MK, Owen MJ. Hydrophobicity loss and recovery of silicone HV insulation. IEEE Trans Dielectr Electr Insul 1999;6:695–702.
[68] WO/2012/020295. Optical elements having long-lasting hydrophilic and anti-fog properties and method for their preparation; 2012.
[69] Koch K, Barthlott W. Superhydrophobic and superhydrophilic plant surfaces: an inspiration for biomimetic materials. Philos Trans R Soc A 2009;367:1487–509.
[70] Tsougeni K, Vourdas N, Tserepi A, Gogolides E. Mechanisms of oxygen plasma nanotexturing of organic polymer surfaces: from stable super hydrophilic to super hydrophobic surfaces. Langmuir 2009;25:11748–59.
[71] Walther F, Davydovskaya P, Zurcher S, et al. Stability of the hydrophilic behavior of oxygen plasma activated SU-8. J Micromech Microeng 2007;11:524–31.
[72] Querè D. Non-sticking drops. Rep Prog Phys 2005;68:2495–532.
[73] Favia P, Milella A, Iacobelli L, d'Agostino R. Plasma pretreatments and treatments on polytetrafluoroethylene for reducing the hydrophobic recovery. In: d'Agostino R, Favia P, Oehr C, Wertheimer MR, eds. Plasma processes and polymers. Wiley-VCH Verlag; 2005:271–280.
[74] Gengenbach TR, Xie X, Chatelier RC, Griesser H. Evolution of the surface composition and topography of perfluorinated polymers following ammonia-plasma treatment. J Adhes Sci Technol 1994;8:305–28.
[75] Chatelier RC, Xie X, Gengenbach TR, Griesser HJ. Quantitative analysis of polymer surface restructuring. Langmuir 1995;11:2576–84.
[76] Farris S, Introzzi L, Biagioni P, Holz T, Schiraldi A, Piergiovanni L. Wetting of biopolymer coatings: contact angle kinetics and image analysis investigation. Langmuir 2011;27:7563–74.
[77] Dastjerdia R, Montazerb M, Stegmaierc T, Moghadamd MB. A smart dynamic self-induced orientable multiple size nano-roughness with amphiphilic feature as a stain-repellent hydrophilic surface. Colloids Surf B 2012;91:280–90.
[78] Huttinger KJ, Hohmann-Wien S, Sciferling M. A modified sessile drop technique for the determination of the interactions at carbon-liquid interfaces. Carbon 1991;29:449–55.
[79] Favia P, De Vietro N, Di Mundo R, Fracassi F, d'Agostino R. Tuning the acid/base surface character of carbonaceous materials by means of cold plasma treatments. Plasma Process Polym 2006;3:66–74.
[80] Lee TR, Carey RI, Biebuyck HA, Whitesides GM. The wetting of monolayer films exposing ionizable acids and bases. Langmuir 1994;10:741–9.

[81] Holmes-Farley SR, Bain CD, Whitesides GM. Wetting of functionalized polyethylene film having ionizable organic acids and bases at the polymer-water interface: relations between functional group polarity, extent of ionization, and contact angle with water. Langmuir 1988;4:921–31.
[82] Badre C, Mayaffre A, Letellier P, Turmine M. Modification of the wettability of a polymeric substrate by pH Effect. Determination of the surface acid dissociation constant by contact angle measurements. Langmuir 2006;22:8424–30.
[83] Khonsari FA, Tatoulian M, Shahidzadeh N, et al. In: d'Agostino R, Favia P, Fracassi F, eds. Plasma-processing of polymers. NATO ASI E. Norwell (MA): Kluwer Academic, 1997:165.
[84] Shahidzadeh N, Chehimi MM, Arefi-Khonsari F, Amouroux J, Delamar M. Evaluation of acid-base properties of ammonia plasma-treated polypropylene by means of XPS. Plasmas Polym 1996;1:27–45.
[85] Whitesides GM, Biebuyck HA, Folkers JP, Prime KL. Acid-base interactions in wetting. J Adhes Sci Technol 1991;5:57.
[86] Boehm HP. Chemical Identification of Surface Groups. In: Eley D, Pines H, Weisz PB, eds. Advances in catalysis. New York (NY): Academic Press; 1966:179–194.

Eva Bittrich and Klaus-Jochen Eichhorn

8 Advances of spectroscopic ellipsometry in the analysis of thin polymer films-polymer interfaces

8.1 Introduction

Spectroscopic ellipsometry (SE), a non-invasive technique, is well placed for the investigation of organic films and interfaces, including polymers. In an ellipsometric measurement, the change in the polarization state of light upon reflection from a surface is detected. Spectroscopic investigations from the ultraviolet (UV) to the infrared (IR) range are possible, which provide a profound set of data for the analysis of polymer film dielectric functions. Different conventional ellipsometry setups, such as standard polarizer-compensator-sample-analyzer (PCSA) configuration, but also advanced setups such as total internal reflection ellipsometry (TIRE), *in situ* cells, and in-line monitoring are applied. Ellipsometric investigation of polymer films can provide various information on the film properties: on the one hand, the dielectric function (optical constants) can be modeled for as complex architectures as polymers in nanostructured surfaces or for semiconducting polymers used in organic solar cells and light-emitting diodes. On the other hand, parameters such as the nanoparticle (NP) content in composite films or the amount of protein molecules adsorbing at polymer surfaces can be retrieved from modeled ellipsometric data.

Polymer films are prepared in a huge variety of special film structures such as polymer brushes [1], multilayers [2–4], cross-linked polymer films [5], blends [6], polymer-NP composites [7], or other nanostructured films [8, 9]. Besides the determination of refractive index n and film thickness d for dry polymer films, other basic problems in the characterization of such films were addressed in past decades by ellipsometry: among them are surface roughness [10, 11], glass transition behavior [12], swelling in aqueous and non-aqueous solvents [13, 14], and adsorption processes at polymer surfaces [15, 16]. After a short introduction on ellipsometry, we will discuss recent advances in the investigation of optical, structural, and thermodynamic properties by spectroscopic ellipsometry (SE) in the whole UV-VIS-IR spectral range.

8.1.1 Basics of ellipsometry

Light, in its physical nature, is described by the electromagnetic theory for transverse plane waves. Because the induced magnetization can be neglected in most ellipsometric experiments, the electric field $\vec{E}$ is the most important quantity for the description of the light wave. We refer to the literature for a comprehensive mathematical description [17]. One aspect important for ellipsometry is the polarization state of light. The light wave can be completely unpolarized (e.g. sunlight), partially polarized, or totally polarized, characterized by the degree of polarization. In an unpolarized wave, the orientation of the electric field vector $\vec{E}$ changes randomly with time, where, in a polarized wave, the orientation of $\vec{E}$ is determined for every time step. The wave can have linear, circular, or, in general, elliptical polarization [18, 19].

When the light is reflected at a planar surface, the wave vector $\vec{K}$ of the incident beam and the surface normal define the plane of incidence (see Fig. 8.1).

With respect to this plane, $\vec{E}$ can be expressed by two components: E_p parallel and E_s perpendicular to the plane of incidence. Both components have an amplitude $A_{s,p}$ and a phase $\delta_{s,p}$. The polarization state of the wave is described by the amplitude ratio A_p/A_s and the phase difference $\delta_p - \delta_s$. Amplitude ratio and phase difference change in the standard ellipsometric reflection experiment. Thus, two ellipsometric angles Δ and Ψ are defined as

$$\Delta = (\delta^r_p - \delta^r_s) - (\delta^i_p - \delta^i_s) \tag{8.1}$$

$$\tan \psi = \frac{A^r_p / A^r_s}{A^t_p / A^t_s}, \tag{8.2}$$

where the index r indicates the reflected beam and the index i the incident beam.

Upon interaction with the surface material, the wave changes velocity and can be absorbed by the material. This is expressed by the complex index of refraction N or the complex dielectric function ε:

$$N(\lambda) = n(\lambda) + ik(\lambda), \tag{8.3}$$

$$\varepsilon(E) = \varepsilon_1(E) + i\varepsilon_2(E) = (n + ik)^2, \tag{8.4}$$

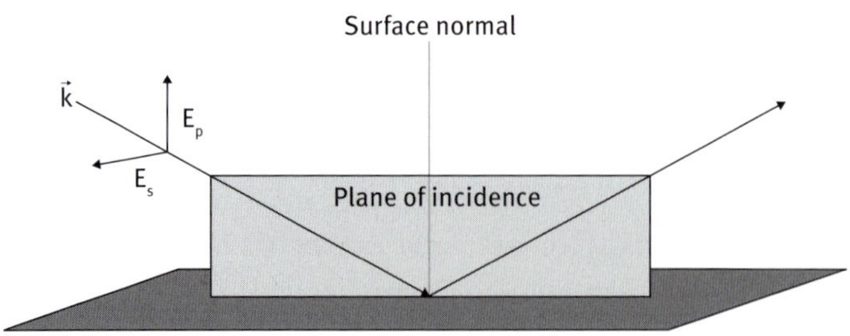

Fig. 8.1: Sketch of the incident and reflected beam and the surface normal spanning the plane of incidence.

where n is the index of refraction, k is the extinction coefficient, and ε_1 and ε_2 are the real and imaginary parts of the dielectric function, respectively. Usually, the materials scientist uses the dielectric function ε dependent on the photon energy E, where the polymer chemist or physicist is often interested in the complex refractive index N *vs* the wavelength λ. Conversion between these quantities is as follows:

$$\varepsilon_1 = n^2 - k^2,\ \varepsilon_2 = 2nk \tag{8.5}$$

$$E = \frac{hc}{\lambda}, \tag{8.6}$$

where h is the Planck constant and c is the velocity of light [20]. Furthermore, the extinction coefficient k can be expressed in terms of the absorption coefficient α well known in optical spectroscopy as

$$k = \frac{\lambda}{4\pi}\alpha. \tag{8.7}$$

The reflection of the light wave at a thin film on a substrate is described by the total complex reflection coefficients R_s for the *s*- and R_p for the *p*-component of the electrical field, which depend on the complex index of refraction of the media (ambient, film, and substrate), on the film thickness, on incident and refractive angles, and on the wavelength λ of light (see [18] for details). The total reflection coefficients are connected to the ellipsometric angles by the "fundamental equation of ellipsometry"

$$\tan \psi e^{i\Delta} = \rho = \frac{R_p}{R_s}. \tag{8.8}$$

Note that both expressions $\tan\psi e^{i\Delta}$ and ρ are complex numbers. In practice, an optical model for the substrate layer system is created, and theoretical values for $\tan\Psi$ and Δ are calculated. These values are compared with the measured quantities, and the model is iteratively refined. Because of the spectroscopic measurement at various wavelengths, a comprehensive set of data is available.

For ellipsometry, often, the incident angle or the range of incident angles in a variable angle (VASE) experiment is chosen near to the Brewster or polarizing angle. At this angle of incidence, the reflectance $\Re_p = |R_p|^2 = 0$ and only the *s*-polarized wave is reflected from the surface. Additionally, the angle between the reflected and the transmitted beams is 90°.

Because ellipsometry uses changes in the polarization state of the light wave for the characterization of a surface or a film, some basic components are, in principle, sufficient to make an ellipsometer: a light source, a polarizing element (e.g. polarizer P, compensator C) to fix the polarization state of the light before the sample, and a device to measure changes in the polarization after reflection on the sample (e.g. analyzer A, detector). A typical PCSA ellipsometer configuration is sketched in Fig. 8.2.

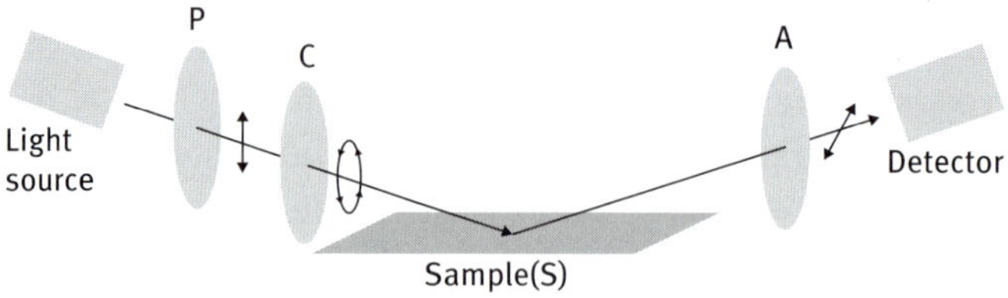

Fig. 8.2: Sketch of a typical ellipsometer setup with light source, polarizer (P), compensator (C), sample (S), analyzer (A), and detector.

The compensator is used to additionally modify the state of polarization before the sample, for example, to generate circular polarization. Configurations also exist with rotating elements (analyzer or compensator) or with phase modulation elements [17]. For IR-SE, an ellipsometer setup is combined with a Fourier transform spectrometer (see [21] for details). Highly sophisticated techniques using Stokes parameters and Jones and Mueller matrix algorithms to fully describe the polarization are shown in [22].

8.2 New ellipsometric methods, techniques, and aspects

8.2.1 Optical dispersion

A set of standard optical dispersion relations for (n,k) or $(\varepsilon_1,\varepsilon_2)$ used for the description of polymer films is listed in Tab. 8.1.

The Cauchy dispersion

$$n(\lambda) = A + \frac{B}{\lambda^2} + \frac{C}{\lambda^4} + \ldots, \; k = 0 \tag{8.9}$$

commonly used for transparent polymer films [16]. For some polymers, an absorption tail in the near-UV range is observed, where the extinction coefficient $k(\lambda)$ increases exponentially. In this case, as the simplest model, the Cauchy-Urbach can be tried [18, 23] (Fig. 8.3).

Polymers with $k \neq 0$ in the near-UV or VIS range can also be modeled very effectively by a B-spline model [24, 25]. This is a purely mathematical description of the optical dispersion *via* spline curves (piecewise-defined smooth polynomial functions). For materials with $k \neq 0$, Kramers-Kronig (KK) consistency is important to ensure the physical validity of the model, where the KK relations connect the imaginary part and the real part of the optical function [19]. The B-spline model can be made KK consistent. However, for Cauchy and Cauchy-Urbach models, KK consistency is not considered. Alternatively to the Cauchy model, the KK-consistent Sellmeyer model can be used [26].

Tab. 8.1: Standard dispersion relations for polymer films

Model	Description	Example
Cauchy	$n(\lambda)$ for transparent films $k(\lambda)=0$	Fluorohydrocarbon polymer film [16]
Cauchy-Urbach	$n(\lambda)$ described by Cauchy model $k(\lambda) \neq 0$: model of absorption tail for energies below the band gap energy of the material	Poly(tetrafluoroethylene) thin film [23] (Fig. 8.3)
Oscillator	$k(\lambda) \neq 0$, e.g. semiconducting polymers absorbing in the VIS range; description of molecular vibrations in IR range Defined oscillator lineshape in $\varepsilon_2(\lambda)$ (Gaussian, Tauc-Lorentz, etc.) and KK-consistent calculation of $\varepsilon_1(\lambda)$	Poly(3-butylthiophene) film [27] (Fig. 8.4)
B-spline	$\varepsilon_1(\lambda)$ and $\varepsilon_2(\lambda)$ modeled by mathematical spline curves Parameters: number of node points/node spacing Can be calculated KK consistent	Any polymer film

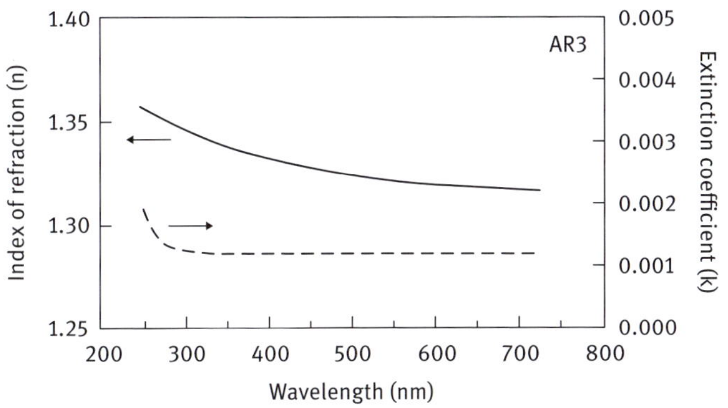

Fig. 8.3: Wavelength-dependent refractive index and extinction coefficient of a poly(tetrafluoroethylene) (PTFE) thin film prepared by hot filament chemical vapor deposition (HFCVD) using hexafluoropropylene oxide (HFPO) as precursor gas. The optical dispersion is describe by a Cauchy-Urbach model accounting for an absorption tail ($k > 0$) in the near UV. Reprinted with permission from Lau KKS, Caufield JA, Gleason KK, *J Vac Sci Technol A*, 18, 2404–11, 2000. Copyright 2000 American Vacuum Society [23].

Different oscillator models exist to model absorption bands of polymers over the whole spectral range [26, 28], among them Gauss, Lorentz, or Tauc-Lorentz oscillators. In contrast to the mathematical description with a B-spline, such an analytical expression of the dielectric function *via* a set of fit parameters can give further information on the electronic/chemical properties of the material. Figure 8.4 gives an example of the optical dispersion of a semiconducting polymer with an electronic transition in

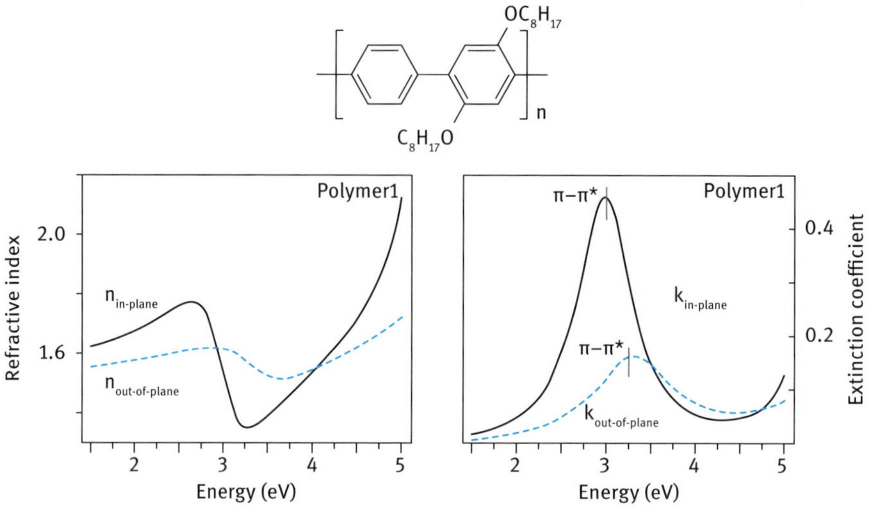

Fig. 8.4: Structure (top) of poly(arylenephenylene) and uniaxial anisotropic refractive index (left) and extinction coefficient (right) of a spin-coated thin film. SE spectra were fitted to two sets of a double Lorentzian oscillator. Adapted with permission from Losurdo M, Giangregorio MM, Capezzuto P, *et al.*, *Macromolecules*, 36, 4492–7, 2003. Copyright 2003, American Chemical Society [27].

the VIS range [27]. The film is uniaxial anisotropic with an absorption band above the fundamental absorption edge due to a π-π* electronic transition. The fundamental band gap energy is the energy necessary to excite an electron from the highest occupied molecular orbital to the lowest unoccupied molecular orbital of the polymer material. In principle, k is equal to zero below E_g and has positive values above E_g. However, for thin films, often, a nonzero absorption below the fundamental band gap edge is observed, for example, due to localized defect states [27]. Another approach to describe the dispersion of absorbing polymers is the standard critical point (SCP) model [29]. Campoy-Quiles *et al.* [30] found the SCP model superior to characterize the dielectric function of a conjugated polymer [poly (9,9-dioctylfluorene)] film.

To describe the effective dielectric functions of heterogeneous media, effective medium approximations (EMAs) were developed [31]. The individual dielectric functions of components a and b contribute to an effective ε according to the underlying model assumptions. Besides a linear combination, more advanced approaches were developed leading to the Maxwell-Garnett, the Bruggeman, and the Lorentz-Lorenz expression (Tab. 8.2).

Tab. 8.2: Two-component EMA with model assumptions

Theory	Assumptions
Lorentz-Lorenz	Mixing of a and b on atomic scale in host medium vacuum
Maxwell-Garnett	b is the host medium and $f_b \gg f_a$
Bruggeman	$f_b \approx f_a$

The EMA according to Bruggeman can be described by

$$0 = f_a \frac{n_a^2 - n^2}{n_a^2 + 2n^2} + f_b \frac{n_b^2 - n^2}{n_b^2 + 2n^2} \qquad (8.10)$$

where f_a and f_b are the volume fractions of two components a (e.g. solvent) and b (e.g. polymer) and $n_{a,b}$ are the corresponding refractive indices, with n as the total effective refractive index of the heterogeneous medium. Typically, f_a and f_b are the fitting parameters and $n(\lambda)_{a,b}$ is known for the components. More than two-component EMAs are also possible. However, correlations between the volume fractions of the individual components are strongly increasing with the number of components.

As certain polymer chains can be found in an oriented state in the cast film, anisotropy of polymer films is a common feature [32]. The films can have uniaxial or biaxial anisotropy. A uniaxial anisotropic polymer film has its optical axis parallel to the surface normal and has different optical constants in the plane of the film (ordinary: n_o, k_o) and perpendicular to it (extraordinary: n_e, k_e). Uniaxial anisotropy can be modeled from a standard SE experiment, where $\Delta n = |n_o - n_e|$ describes the birefringence of the film. To characterize biaxial polymer films, with different optical constants along all three axis of space, "generalized ellipsometry" (GE) has to be done [33]. For GE, either the Jones matrix or Mueller matrix formalism is used. The first one provides the mathematical description for the reflection at a layered surface in the case of nondepolarizing samples, where the latter is needed if depolarization occurs. The dielectric function is replaced by the full nine-element anisotropic dielectric function tensor (see [34] for details).

8.2.2 In situ setups

8.2.2.1 Liquid cells for measurements in solution

Liquid cells exist in the form of batch and flow cells [35]. Batch cells with a fixed angle of incidence are used to study swelling of polymer films and also the adsorption of proteins to these films [16, 36]. Flow cells are, in principle, used for the same investigations. Common problems of liquid cells are concentration gradients in the batch geometry, which can have a great effect on kinetics and equilibrium value of adsorption processes, and turbulent flow as well as shearing forces in flow cells, depending on the flow rate. To address the latter, a especially-designed flow cell using stagnation point flow conditions was reported, where the solution and interface conditions are well defined [37]. Thus, these cells are best suitable to measure the swelling and adsorption kinetics. The drawback of flow cells is the relatively high need for material (solution and adsorbent). For more features of *in situ* cells that can have an influence on the measurement, we refer to the literature [35].

For *in situ* measurement in the IR range, a different cell design than in the VIS range is needed [38]. An example of such a cell is sketched in Fig. 8.5. Because the

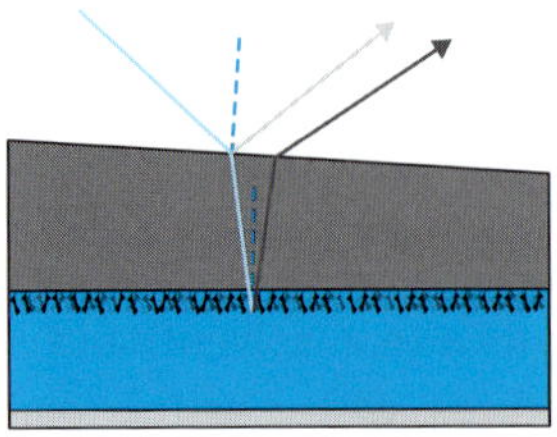

Fig. 8.5: Sketch of an IR *in situ* cell. Reprinted with permission from Aulich D, Hoy O, Luzinov I, *et al.*, *Langmuir*, 26, 12926–32, 2010. Copyright 2010, American Chemical Society [38].

penetration depth of IR radiation in water is very limited (in the micrometer range), the investigation of the solid-liquid interface is done *via* the IR-transparent silicon substrate. Usually, a silicon wedge is used to separate front and backside reflection.

8.2.2.2 Coupling ellipsometry with quartz crystal microbalance

Quartz crystal microbalance with dissipation monitoring (QCMD) is an acoustic measurement technique [39, 40]. The substrate consists of a piezoelectric quartz crystal, where a mechanical oscillation of the crystal lattice is induced by an applied alternating current. Adding or removing a surface layer on this crystal results in changes in the overtone frequencies Δf of the oscillating mass, and in case of a viscous layer, dissipation ΔD occurs. These parameters can be measured for a set of overtones and modeled to obtain absorbed mass and/or elastic data of the layer. Special considerations in the modeling of elastic, hydrated layers are necessary. A common approach for such viscous layers (e.g. swollen polymer layers) is a continuum viscoelastic model, based on theoretical work of Voinova and colleagues [41–43].

The combination of SE and QCMD provides an extended set of optical and mechanical data of the film. In addition, surface masses/adsorbed amounts calculated individually from both techniques can be correlated [44]. The latter was used to retrieve stepwise changes in the amount of hydration upon the buildup of polyelectrolyte multilayers [45]. The comparison of ellipsometric optical mass m_{opt} (i.e. the mass of the polymer deposited) and the mass m_{QCMD} calculated from QCMD measurements is displayed in Fig. 8.6a. In m_{QCMD}, the solvent in the film is included. From the comparison of these quantities, the percentage of hydration can be obtained (Fig. 8.6b).

For thick polymer layers with a known bulk dispersion, the solvent content in the film can be retrieved from both techniques, i.e. from the ellipsometric thickness and the viscoelastic thickness, respectively. However, systematic differences can occur, for example, if the density of the film is not homogeneous but varies along the surface normal or if the film-solution interface is diffuse [46]. In principle, a correlation of the optical (n,k) to the mechanical properties (viscosity η, shear modulus μ) of a polymer film is possible but not yet reported.

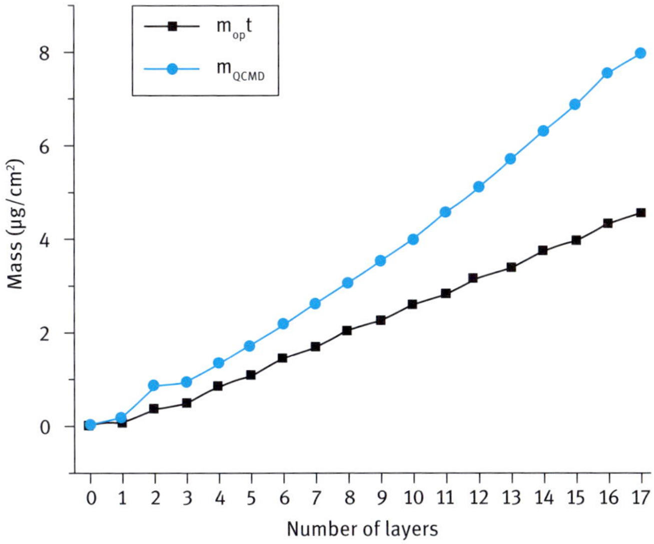

(a)

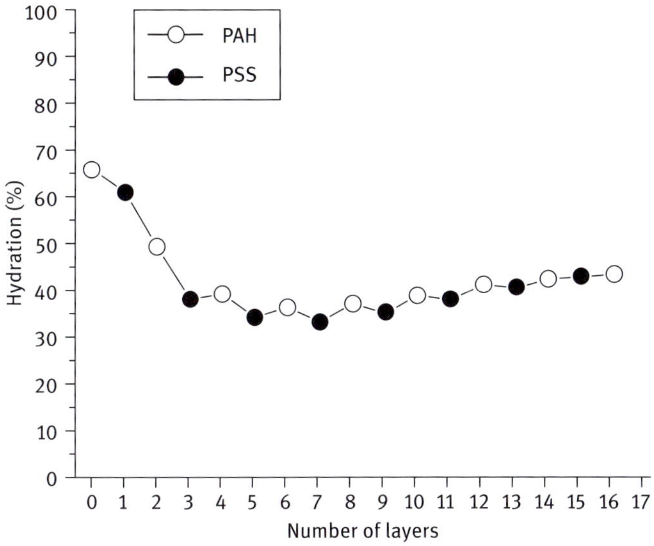

(b)

Fig. 8.6: (a) Comparison of m_{opt} (■) and m_{QCMD} (•) for poly(allylamine hydrochloride)/poly(sodium 4-styrenesulfonate) multilayers. The polymer surface was washed in water between deposition steps. (b) Hydration (in percentage) of the multilayer for each step. Reprinted with permission from Iturri Ramos JJ, Stahl S, Richter RP, Moya SE, *Macromolecules*, 43, 9063–70, 2010. Copyright 2010, American Chemical Society [45].

8.2.2.3 Total internal reflection ellipsometry (TIRE)

Because of its enhanced sensitivity to changes in the ellipsometric parameters Δ and tanψ, TIRE is often used to investigate the adsorption of proteins to various surfaces [47] and to utilize patterned surfaces with biomolecules as biosensors [48].

A common setup of a TIRE experiment is sketched in Fig. 8.7. In addition to the ellipsometry setup, a prism and a especially designed flow cell are needed (parts 6 and 8 in Fig. 8.7). With these devices, the adsorption of biomolecules could be performed directly on the glass prism-solution interface [48]. However, in most cases, a thin metal (e.g. gold) layer is coated to the glass prism or to a glass slide (part 7) to utilize the surface plasmon resonance (SPR) technique for enhanced sensitivity [49]. Thus, TIRE commonly refers to ellipsomety in internal reflection mode with SPR. An additional metal-coated glass slide is favorable because adsorption of molecules is kept away from the prism and single use of the glass slide is possible. The cleaning of the prism or the metal layer on the prism is not trivial. To reduce interface reflections, the refractive indices of the prism and the slide should be the same, and the glass slide is mounted with a matching index oil. Various prism forms are used for the internal reflection including triangular, trapezoidal, and semicylindrical shapes.

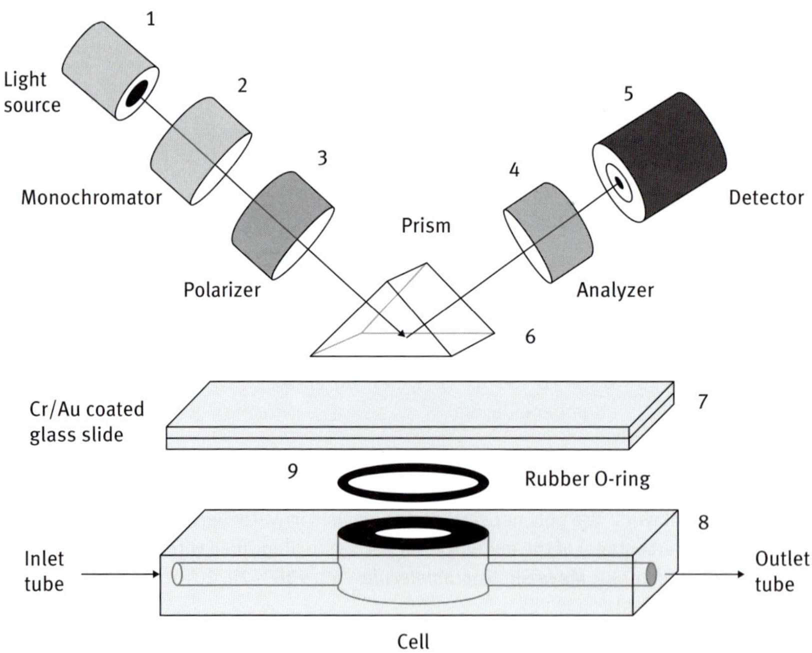

Fig. 8.7: Setup for a TIRE experiment with triangular prism. Reprinted with permission from Basova T, Tsargorodskaya A, Nabok A, *et al.*, *Mater Sci Eng C*, 29, 814–8, 2009. Copyright 2009 Elsevier [50].

For a given set of angle of incidence, metal layer thickness, and wavelength, the *p*-polarized component of the electric field incident through the glass prism on the metal film couples to a surface plasmon wave, a collective oscillation of the conducting electrons. This excitation can be described in terms of a quasi-particle, the surface plasmon polariton (SPP) [51]. The field of the SPP is maximal at the metal-liquid interface and decreases into both media. This leads to a minimum in the reflectance curve, often called the SPR dip. Adsorbing a thin protein film to the metal layer leads to a shift in the wavelength of the SPR dip, as displayed in Fig. 8.8, for the example of ferritin adsorption to a gold layer [52]. Changes in Δ of more than 90° are observed compared with few degrees in standard external SE experiments [53].

For each wavelength, a different set of system parameters (e.g. angle of incidence, metal layer thickness) leads to the resonance situation. Thus, TIRE is commonly used in single-wavelength mode, but a scan of wavelength is, in principle, possible. Due to its potential for biosensing, it is suitable to combine imaging ellipsometry with TIRE [54].

8.2.3 In-line monitoring

In-line monitoring with ellipsometry is promising due to the non-invasiveness of the technique and the fast recording of measurement data. Here, in-line stands for a real-time *in situ* measurement integrated in a manufacturing process. Such real-time *in situ* ellipsometry is, on the one hand, reported for the monitoring of reaction processes in a vacuum chamber, e.g. plasma treatment of polycarbonate substrates for surface functionalization [55]; on the other hand, SE can be adapted to the monitoring

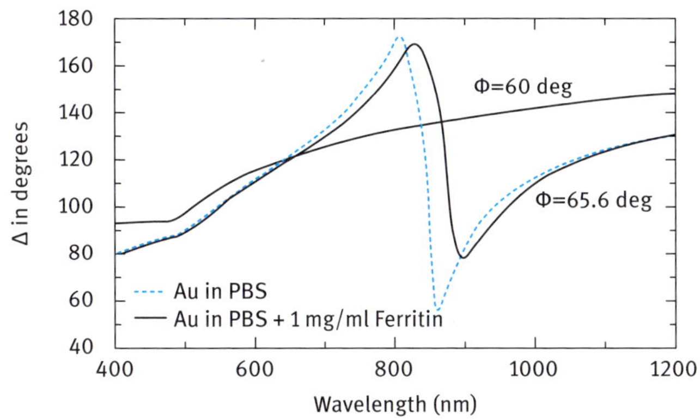

Fig. 8.8: Comparison of Δ for an Au layer surface in PBS buffer with (solid line) and without (dashed line) adsorbed ferritin at 65.6° (SPR) and 60° angle of incidence. Reprinted with permission from Poksinski M, Arwin H, *Thin Solid Films*, 455, 716–21, 2004. Copyright 2004 Elsevier [52].

of roll-to-roll (R2R) fabrication processes done on flexible materials, for example, PET foils [56, 57].

8.2.3.1 Monitoring processes in a vacuum chamber

Challenges for using an ellipsometer together with a vacuum chamber lie in the minimization of vibrations of the optical elements, in the quality of optical windows from the ellipsometer to the vacuum chamber, and, in general, in the attachment of the ellipsometric parts to the vacuum setup in the correct optical geometry. Suitable optical windows consist of BaF_2 [56], and firm attachment of the optical elements of the ellipsometer to the vacuum chamber by suitable optical ports ensures minimal vibration.

8.2.3.2 R2R fabrication processes

R2R processes can be used in the fabrication of organic electronic devices [organic photovoltaics (OPVs) and organic light-emitting diodes (OLEDs)] on flexible foils. Here, in-line ellipsometry can contribute to the quality control. A simplified fabrication setup with integrated ellipsometer is displayed in Fig. 8.9.

The foil represents a moving sample, and thus, very fast acquisition times are necessary. Logothetidis *et al.* [57] used an array detector with 32 fiber optics and achieved a 100-ms integration time for one multi-wavelength spectrum. To ensure stable measurement conditions, the ellipsometer is located between two directional rolls,

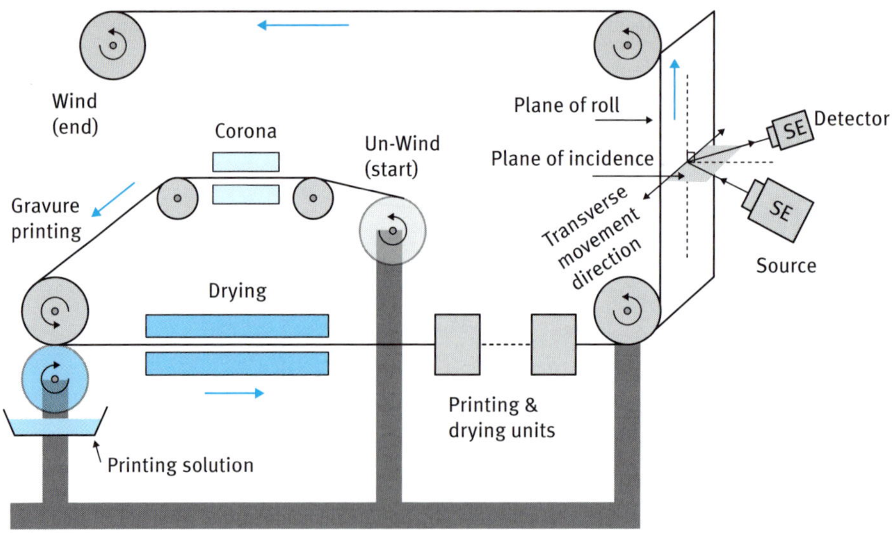

Fig. 8.9: Scheme of a laboratory scale R2R fabrication system with in-line ellipsometry unit. Reprinted with permission from [57]. Copyright 2013 Elsevier.

where the foil is measured on a vertical path. With this setup, the substrate is ensured to be flat and without wrinkles. The latter would lead to a defocusing of the reflected light beam. Another challenge is represented by vibrations due to the moving rolls of the in-line fabrication setup. The vibration damping for the setup displayed in Fig. 8.9 was done by putting the whole R2R system on an anti-vibration stage [57].

8.2.4 Micropatterned films

Micropatterned polymer/biopolymer films are, for example, appealing for localized cell adhesion [58, 59], tissue engineering [60], or, in general, for biosensor applications [61]. Here, the ellipsometric characterization of the heterogeneity of the film structure (thickness and optical constants) in the VIS range [59, 61, 62] as well as chemical heterogeneity (IR range) is interesting [63]. The heterogeneity of micropatterned films can be accessed by imaging/mapping ellipsometry.

8.2.4.1 VIS imaging ellipsometry

A typical setup of a VIS imaging ellipsometer includes objective and a spatially resolving CCD detector (Fig. 8.10) [64, 65].

With the objective, the irradiated area of the sample is projected onto the detector with spatially resolved optical information. Thickness and/or refractive index maps are recorded with a resolution as high as 1–2 μm [65]. Imaging is often combined with null ellipsometry, where selected areas of the sample map fulfill nulling conditions of polarizer, compensator, and analyzer. Thus, the light signal is extinguished at the detector (dark image areas) [66]. The example displayed in Fig. 8.10 works with a laser

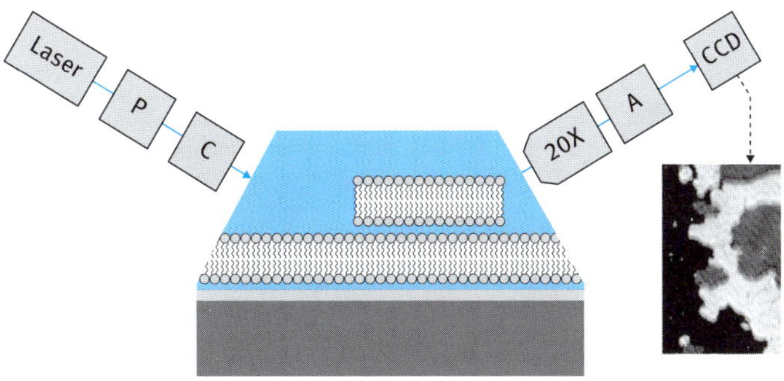

Fig. 8.10: Configuration of a single-wavelength imaging null-ellipsometer with polarizer (P), compensator (C), objective (20×), analyzer (A), and CCD camera. Additionally, an example map of a lipid bilayer is displayed. Reprinted with permission from Bruun Nielsen MM, Cohen Simonsen A, *Langmuir*, 29, 1525–32, 2013. Copyright 2013, American Chemical Society [65].

source in the single-wavelength mode. However, spectroscopic measurements are the state of the art [67].

Very recently, the combination of imaging ellipsometry and grazing incidence small angle X-ray scattering at the synchrotron beamline BW4 (Hamburg, Germany) was reported for the investigation of microstructured conducting polymer films [68, 69].

8.2.4.2 Microfocus-mapping IR ellipsometry

For the mapping of structured surfaces in the IR range, the combination of SE with light sources based on synchrotron radiation was developed [70]. These light sources have up to three times higher brilliance than IR light sources in the laboratory and enable spatial resolutions down to 10 times of the wavelength (e.g. down to 100 μm at BESSYII, Berlin, Germany) [71]. Thus, experiments with both high sensitivity and high lateral resolution can be done, which is not possible with laboratory instruments. The brilliance of a light source is defined by the number of photons impinging on a unit sample area per time unit.

In Fig. 8.11, an example of the application of microfocus-mapping IR-SE for the study of lateral homogeneity of an organic surface layer is displayed [72]. In another example, the chemical "fingerprint" (characteristic vibrational bands) of a linker and a peptide molecule were used to resolve the lateral position of the surface components, giving insight into the photoelectrochemical reaction process of the linker molecule with the surface [73].

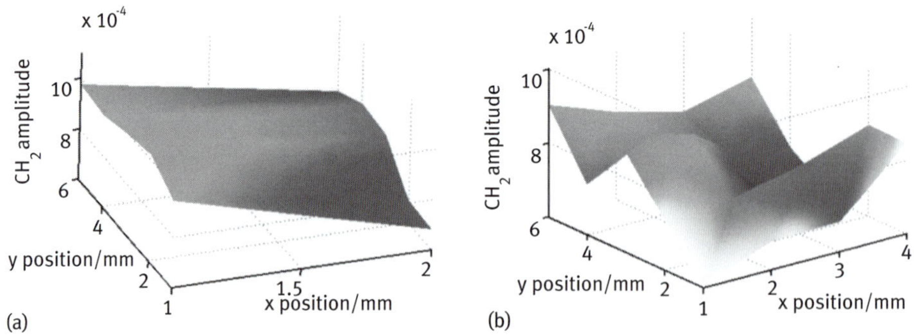

Fig. 8.11: 2D maps of hexadecanethiol (HDT) on Au (a) and GaAs (b). Displayed is the CH_2 amplitude at 2922 cm^{-1} from the tanΨ spectra. Lateral step width is 1 mm. Reprinted with permission from Rosu DM, Jones JC, Hsu JWP, *et al.*, *Langmuir*, 25, 919–23, 2009. Copyright 2009, American Chemical Society [72].

8.3 Selected architectures of polymer films, blends, and composites

8.3.1 Polymer blends and cross-linked polymer films

Because IR-SE gives access to the structural and chemical properties of interfaces/polymer films, it is well placed to characterize polymer blends, and the cross-linking process in polymer films.

For polymer blends, the miscibility of the single components dependent on temperature was investigated by monitoring the formation of a mixing layer in a bilayer system of miscible polymers [74, 75]. Due to interdiffusion at the polymer-polymer interface, the blend was formed. Hinrichs *et al.* [75] studied the poly(*n*-butyl methacrylate) (PnBMA)/poly(vinyl chloride) (PVC) blend after annealing of a PnBMA/PVC double layer. In Fig. 8.12, the measured and simulated $\tan\Psi$ and Δ spectra with the successful three-layer model are displayed. The optical constants n and k for the single phases were determined beforehand by combining IR and VIS SE. For the mixed phase, n and k were simulated by weighing the individual optical constants with the thickness of the components PnBMA (100 nm) and PVC (106.6 nm) in the original bilayer.

Cross-linking in polymer films was investigated by IR-SE for catalytic processes [76] and possible cross-linking by plasma treatment [77–79]. Simpson *et al.* [76] investigated the catalytic cross-linking of vinyl-terminated poly(dimethylsiloxane) (PDMS) with a trimethylsilyl-terminated poly(hydrogen methylsiloxane) cross-linker at the presence of platinum as catalyst. A $\cos(\Delta)$ spectrum of a 14.6-µm-thick film at the beginning of cross-linking at 80°C is displayed in Fig. 8.13a. Vibration bands are (1) the SiH stretching mode at 2150 cm^{-1}, (2) the asymmetric and (3) symmetric CH_3

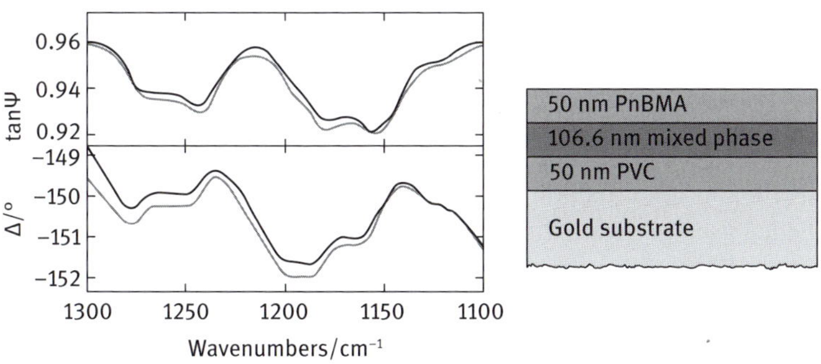

Fig. 8.12: Measured $\tan\Psi$ and Δ spectra (black) and simulated spectra (gray) of a three layer model of PnBMA and PVC in a double-layer conformation with interdiffusion blend layer after annealing at 110°C at 10 min. Reprinted with permission from Hinrichs K, Gensch M, Nikonenko N, Pionteck J, Eichhorn K-J, *Macromol Symp*, 230, 26–32, 2005. Copyright 2005 Wiley-VCH Verlag GmbH & Co. [75].

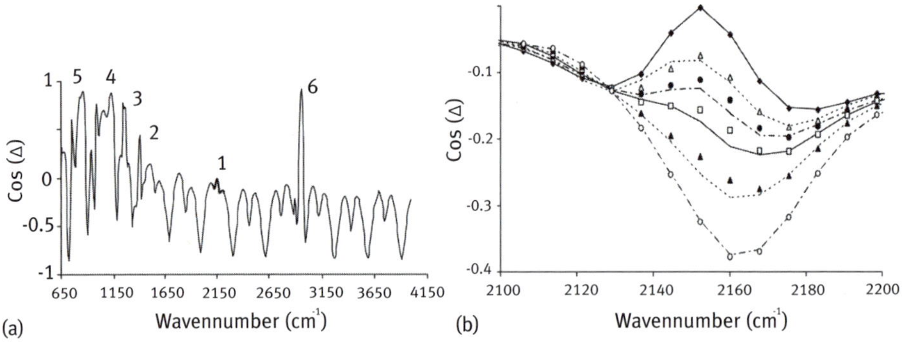

Fig. 8.13: Cos(Δ) spectrum of a 14.6-μm-thick PDMS coating at the beginning of cross-linking at 80°C (a), and the decrease of the SiH stretching vibration band at 2150 cm^{-1} during cross-linking (b). Measurements were taken at 0 (dark squares), 1.2 (light triangles), 3.7 (dark circles), 4.9 (light squares), 13.7 (dark triangles), and 182 min (light circles). The lines are the best fit to a Bruggeman EMA, where the PDMS is considered to be a blend of a fully cured component (without SiH groups) and an unreacted component (with SiH groups). Reprinted with permission from Simpson TRE, Parbhoo B, Keddie JL, *Polymer*, 44, 4829–38, 2003. Copyright 2003 Elsevier [76].

deformations, (4) the Si-O-Si stretching mode, (5) CH_3 rocking mode and Si-C stretching mode, and (6) the asymmetric CH_3 stretching mode. To analyze the completion of the cross-linking reaction, they evaluated the decrease in the concentration of unreacted SiH groups in the cross-linking agent by monitoring the decrease of the characteristic SiH stretching vibration band at 2150 cm^{-1} (Fig. 8.13b). Coatings between 1 and 27 μm in thickness were investigated, and Simpson *et al.* found the reaction rate to be dependent on the thickness of the coating.

Plasma treatment of polymer surfaces is commonly done to improve the adhesion properties on polymers. With IR-SE, chain scissions in an overlayer could be detected that could lead to cross-linking [78, 79]. The overlayer thickness depends on the treatment time and on applied voltage for the plasma treatment and can increase to several hundred nanometers [79]. Logothetidis and colleagues [78] combined IR-SE with VIS-far UV SE and correlated changes in the chemical bonding in the overlayer of polymer membranes with modifications in characteristic features of electronic transitions.

Next to the advanced investigation of cross-linking with IR-SE, a combination of VIS ellipsometry with Fourier transform IR (FTIR) measurements is common [80–82]. Ellipsometry is hence used to derive thickness and optical constants of cross-linked polymer films, where FTIR spectroscopy is used to monitor chemical processes *via* characteristic vibration bands in the cross-linking procedure.

8.3.2 T_g in thin polymer films of different architectures: Confinement effects

Thin polymer films can have different properties than in the bulk. One representative parameter is the glass transition temperature T_g, which is of technological relevance for amorphous polymers [83]. The glass transition of a polymer is not a real phase transition but rather a solidification process. With decreasing temperature, the relaxation time of the polymer molecules increases until it is of the order of the experimental time and the polymer material appears solid.

As the film thickness decreases, the surface-to-volume ratio increases. Thus, polymers become geometrically confined and interface interactions become more important for the overall behavior of the film. For the influence of confinement on T_g controversial results exist, and some researchers even doubt any influence for films thicker than 10 nm but claim the existence of phenomena that mimic confinement effects.

In older investigations, the T_g of polystyrene (PS) was found to decrease with reduced film thickness (<100 nm) [12]. However, in recent publications, no such shift of T_g was reported [84, 85]. T_g can be determined by ellipsometry in analyzing the kinks of the ellipsometric angles $\Delta(T)$ and $\tan\Psi(T)$, of the film thickness $d(T)$, or of the refractive index $n(T)$, or in fitting the data and analyzing the maximum of the second derivative of the fit. In Fig. 8.14, temperature-dependent thicknesses for PS films for two different molecular weights and five film thicknesses each as well as the second derivatives of the fits are displayed [85].

Tress *et al.* [85] found no significant influence of molecular weight and film thickness on the value of T_g. Small deviations from the bulk T_g were within error ranges. They compared the determination of T_g by SE measurements at thin PS films with broadband dielectric spectroscopy results and found coincidence in the absence of

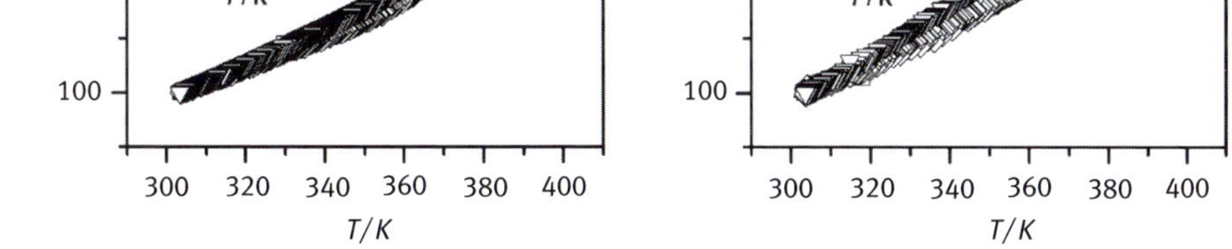

Fig. 8.14: Temperature dependence of the thickness d of PS films determined by SE for a molecular weight of (a) 27,500 g/mol [thicknesses: 234 nm (■), 121 nm (Δ), 86 nm (★), 54 nm (♦), 26 nm (∇)] and (b) 58,900 g/mol [thicknesses: 360 nm (■), 138 nm (Δ), 59 nm (★), 26 nm (♦), 12 nm (∇)]. Thicknesses are normalized at 300 K. Insets show the second derivative of the interpolated thickness data (ninth-order polynomial fitting). Reprinted with permission from Tress M, Erber M, Mapesa EU, *et al.*, *Macromolecules*, 43, 9937–44, 2010. Copyright 2010, American Chemical Society [85].

significant alterations in T_g for decreasing film thickness. The controversial results from the literature are discussed to result from preparation and treatment artifacts such as chemical degradation [86], water sorption [86], plasticizing due to remaining solvent [87], thermal history of the sample [88], and agglomeration of solvent at the interface between polymer and substrate [89]. Thus, confinement effects are not expected to influence the T_g in PS films thicker than 10 nm [89].

Erber *et al.* [90, 91] additionally investigated the influence of polymer architecture, i.e. polymer backbone and functional groups, on the T_g of polyester thin films. They found no significant influence of the topology (linear, star-like, hyperbranched polymer) on the T_g for PS thin films [90], but significant shifts in the T_g occurred when changing the interaction between polymer and substrate due to different functional groups at aromatic-aliphatic hyperbranched polyesters (Fig. 8.15) [89, 91]. Introducing attractive interactions led to an increase of the T_g up to 14 K, where repulsive interactions let to a depression as low as 9 K.

8.3.3 Polymer-NP composites

With the addition of NPs to a polymer material, its mechanical, optical, or electrical properties can be changed [92]. For example, self-healing materials could be created by controlling particle size and particle coating. Also, noble metal NPs are utilized in plasmonics using the localized SPR effect [93], where the polymer matrix can serve as a template for the NP growth [94]. Special routes to the formation of polymer-NP composites are the introduction of NPs into polymer brushes [95, 96] and the segregation of NPs at phase boundaries of structured block-copolymer films [92].

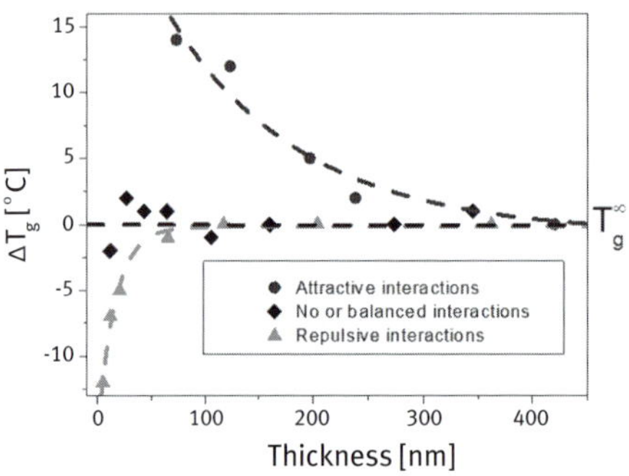

Fig. 8.15: Effect of interfacial interactions between functionalized aromatic-aliphatic hyperbranched polyesters and the substrate (Si/SiO_x) on T_g. Shifts of T_g(ΔT_g) dependent on the film thickness are displayed. Functional groups were -OH (squares), -OBz (circles), and -OSi (stars). Adapted from Erber M, Tress M, Eichhorn, K-J, In: Ellipsometry of Functional Organic Surfaces and Films, Germany, Springer, 2014 [89].

Standard SE can be used to describe the polymer-NP composites. The volume fraction/fill factor of NPs in the polymer matrix is commonly derived by the Maxwell-Garnett EMA, with the restrictions that particle sizes should be well below the wavelength of light (for the VIS range <20 nm) and that no interactions between particles in the host material occur (low fill factors) [94, 96–98]. Under these prerequisites, the bulk dielectric function of both materials (host matrix and NPs) can be used for the EMA [96, 98]. For a non-absorbing polymer matrix, usually a two component Cauchy dispersion is used, determined for the corresponding thick polymer film without NPs. For the dielectric function of the particles, the use of averaged literature values for iron oxide bulk is reported [96], where for noble metal particles, a modified Drude model was used [98].

Oates and Christalle [94, 98] utilized polymer matrixes as growing templates for silver NPs. They monitored NP formation with time, where the increase of the plasmon absorption peak in ε_2 reflects the increase in particle size at selected time spots (Fig. 8.16a) [98]. After NP growth, the polymer matrix was evaporated to achieve an array of freestanding particles at the substrate. A considerable shift in the plasmon resonance can be observed upon removing the polymer (Fig. 8.16b).

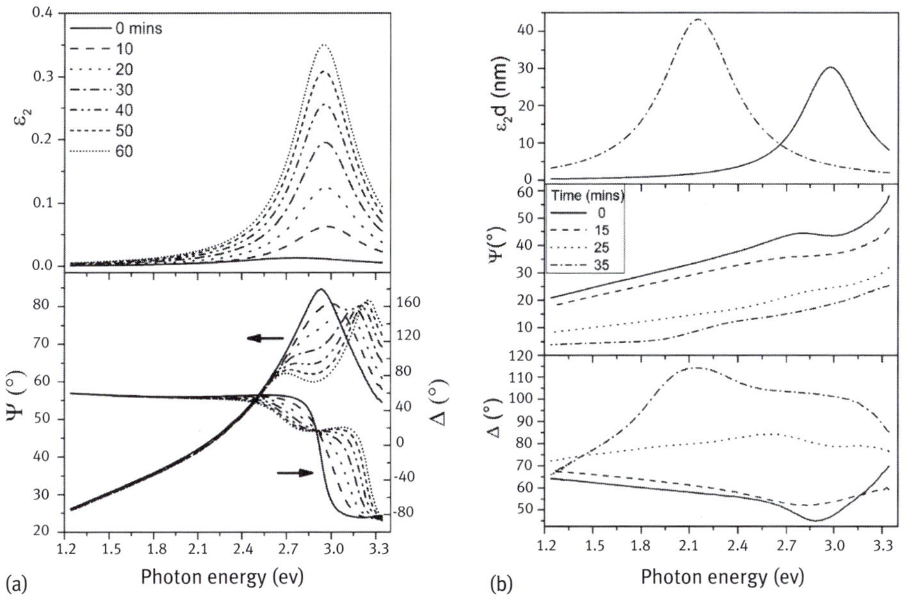

Fig. 8.16: (a) Imaginary part ε_2 of the dielectric function (top) of Ag particles in poly(vinyl alcohol) (PVA) films upon NP growth with time at 120°C. Corresponding ellipsometric data Ψ and Δ (bottom). (b) Shift of the plasmon absorbance upon evaporation of the polymer matrix. Shown is ε_2 (top) scaled with the film thickness *d* for 0 min (solid line) and 35 min (dashed line) of the evaporation process, Ψ (middle), and Δ (bottom) for 0, 15, 25, and 35 min. Reprinted with permission from Oates TWH, Christalle E, *J Phys Chem C*, 111, 182–7, 2007. Copyright 2007 Elsevier [98].

Freestanding arrays of noble NPs are highly interesting for biosensors and are used in surface-enhanced Raman scattering [99]. Due to the enhanced local electrical field at the particle surfaces, the adsorption of Raman-active biomolecules to the NPs can be detected very sensitively.

8.3.4 Polymers in nanostructured surfaces

A novel approach in polymer film preparation is the incorporation of polymers into glancing angle electron beam deposition (GLAD)-structured surfaces [100, 101]. In GLAD, a particle flux is inserted on a substrate at oblique angles, and adatom diffusion is limited [102]. Because of the self-shadowing effect, slanted columns can be grown, leading to "porous" nanostructured surfaces. By rotation of the substrate stage in discrete steps or continuously at a defined speed, more sophisticated structures such as chevrons, staircases, or helices can be sculptured. Figure 8.17 shows cross-sectional scanning electron microscopy (SEM) images of a titanium chevron structure as prepared (a) and after coating with a polymer (b).

Next to their unique mechanical properties [103], their optical properties, especially their tunable birefringence is explored [104, 105]. By the infiltration of these structures with polymer (e.g. semiconducting) as well as functionalization with polymer brushes, nanohybrid interfaces can be created, greatly enhancing their potential for applications such as nano-electronics or sensorics [100, 101].

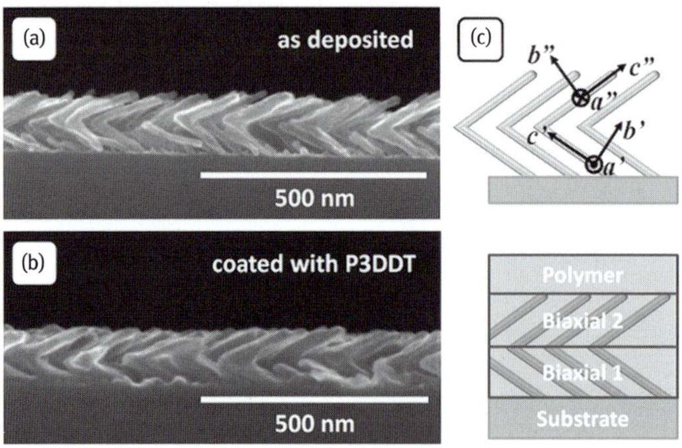

Fig. 8.17: SEM images of cross sections of TI chevron structures as prepared (a) and after infiltration with poly(3-dodecylthiophene) (P3DDT) (b). (c) Scheme of the chevron structure with sample coordinate system along major polarizability axes (top) and ellipsometric layer model (bottom). Reprinted with permission from Schmidt D, Müller C, Hofmann T, *et al.*, *Thin Solid Films*, 519, 2645–9, 2011. Copyright 2011 Elsevier [101].

For the optical characterization of these biaxial anisotropic layered surfaces (Fig. 8.17c), standard SE has to be replaced by GE [105]. Schmidt *et al.* [101] recorded the Mueller matrix elements $M_{kl}(\lambda)$ normalized to M_{11} with a dual rotating compensator ellipsometer for multiple angles of incidence and sample rotation. The chevron-like Ti film was modeled with two anisotropic layers on the silicon substrate. Parameters of the model are the thickness of each biaxial layer and the three major elements ε_a, ε_b, and ε_c of the dielectric tensor in the sample coordinate system. The orientation of the nanostructures is described by a set of three Euler angles that are the elements of the rotation matrix from Cartesian coordinates to the sample coordinate system of the individual layers. After infiltration of the GLAD structures with polymer, an additional polymer top layer was introduced in the model. The geometry of the chevron structures is not affected by the polymer infiltration. A complete set of optical constants (n_i,k_i) of the Ti chevron structures before and after hybridization with polymer could be obtained for all axes a, b, and c in the sample coordinate system. SEM investigations are in very good agreement with the structural information derived from the ellipsometric modeling.

Recently, Kasputis *et al.* [100] reported on novel *in situ* GE experiments. They characterized for the first time the swelling of poly(acrylic acid) brushes grafted inside and on top of slanted silicon GLAD structures. By varying the pH of the ambient buffer solution between pH 7.3 (swelling of the PAA brush) and pH 3.7 (deswelling), they monitored changes in the Mueller matrix elements. Mueller matrix measurements were recorded for wavelengths between 400 and 900 nm at three angles of incidence (65°, 70°, and 75°) and three substrate orientations (0°, 45°, and 180°). With an anisotropic Bruggeman EMA (AB-EMA), the volume fraction of brush inside the slanted columns could be evaluated (Fig. 8.18, black squares). A top layer above the AB-EMA layer is used in the model to consider the polymer chains grafted on top and swelling out of the slanted columns (Fig. 8.18, white squares). As they coupled their *in situ* GE with QCM-D measurements, periodic changes in the surface mass were obtained simultaneously and confirmed the swelling/deswelling of the PAA brushes.

8.4 Polymer layers absorbing in the VIS spectral range

8.4.1 Chemical modification with dye molecules

Including dye molecules, absorbing in the VIS range, into polymer films provides new analytical functionality [106] as well as altered surface properties of these films [107–109]. Rauch *et al.* [109] modified temperature-sensitive poly(*N*-isopropylacrylamide) (PNIPAAm) brushes with the fluorescent dye rhodamine B and studied *in situ* the changes in film thickness, effective optical properties (n,k), and water content upon deswelling of the PNIPAAm brushes by temperature. They observed a tuning of the effective extinction coefficient of the rhodamine-PNIPAAm system dependent on

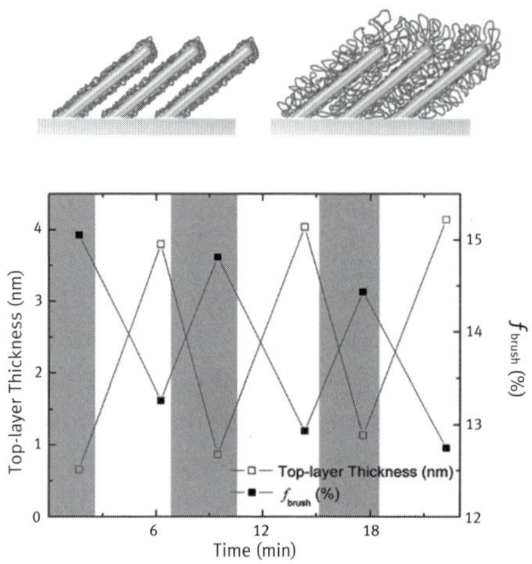

Fig. 8.18: Scheme of the slanted columnar thin films modified by swellable poly (acrylic acid) (PAA) brushes (top) and top-layer thickness as well as the vol% of the brushes for swelling/deswelling cycles (bottom). Parameters of the swollen brush are indicated with light, of the deswollen brush with gray background. Reprinted with permission from Kasputis T, Koenig M, Schmidt D, *et al.*, *J Phys Chem C*, 117, 13971–80, 2013. Copyright 2013, American Chemical Society [100].

the water content of the swollen polymer brush (Fig. 8.19b). This effect was concluded to be solely due to the change in water content upon deswelling of the PNIPAAm brush. This is an example of how *n* and *k* obtained by SE depend on the film conditions and how changes in the environment/constitution of the film affect its optical properties, although the dye concentration is constant. The ellipsometric parameters Δ and Ψ of the PNIPAAm-rhodamine film were modeled by a box layer using two Gaussian oscillators.

An increase of sensitivity in Δ and Ψ could be found in the absorbing region of the dye in the modified PNIPAAm brushes (Fig. 8.19a). A similar effect was used to increase the sensitivity of SE in the study of protein adsorption [106]. Nejadnik and Garcia reacted absorbed protein (bovine serum albumin) with Coomassie brilliant blue G, which enhanced the sensitivity of subsequent measured Δ and Ψ values by about 2.5 in the absorbing region of the dye compared with the transparent protein film. They used a Lorentz oscillator model for their analysis.

8.4.2 Semiconducting polymers and blends for OPV and OLED

OPV and OLED are two of the most promising applications of organic electronics [110]. The use of semiconducting polymers for the preparation of such typically multilayered structures is highly desirable due to the low cost, easy processing, high optical

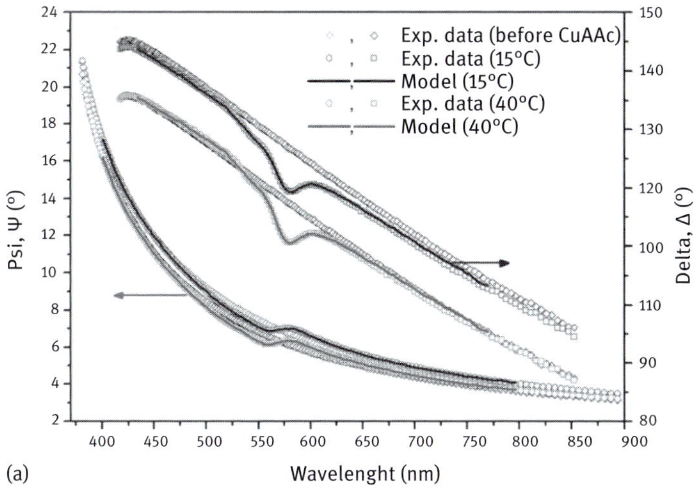

(a)

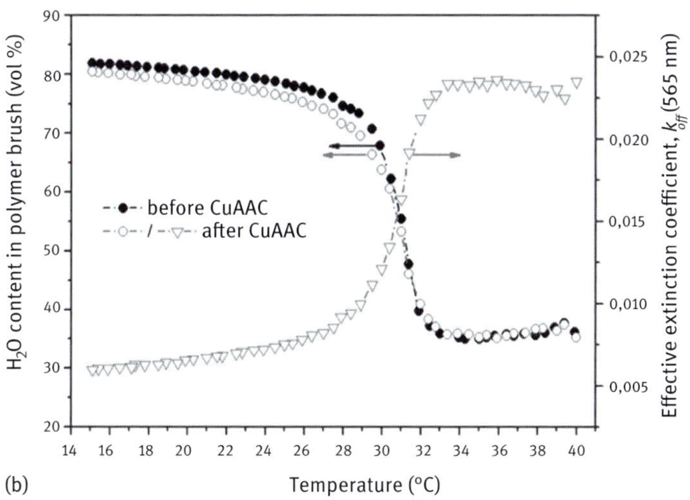

(b)

Fig. 8.19: (a) *In situ* ellipsometric parameters Δ and Ψ dependent on wavelength for the rhodamine B-PNIPAAm system swollen at 15°C and 40°C and before modification with dye. (b) Water content of the swollen PNIPAAm film before and after functionalization with rhodamine B (copper-catalyzed azide-alkyne cycloaddition, or CuAAc) and effective extinction coefficient of the rhodamine B-PNIPAAm system dependent on temperature Rauch S, Eichhorn K-J, Oertel U, Stamm M, Kuckling D, Uhlmann P, *Soft Matter*, 8, 10260–70, 2012. Reproduced by permission of the Royal Society of Chemistry (http://pubs.rsc.org/en/content/articlelanding/2012/sm/c2sm26571k) [109].

transparency, and tunability of mechanical properties of polymers. Intense research has been dedicated to this field in the recent decade, and a multitude of new publications on SE characterization of semiconducting polymers exists. Important examples of the field are SE measurements on polymer-fullerene blends suitable as photoactive nanolayer in OPV [111–121], on polymers for transparent electrodes such as poly(3,4-ethylenedioxythiophene):poly(styrenesulfonate) (PEDOT:PSS) [122–127], and on polyfluorene gain media for OLED [128–130].

Out of the huge number of published SE studies on semiconducting polymers, we want to focus in this section on polymer:fullerene blends for OPV and highlight fundamental advances in the investigation of the morphology of the blend films.

The photoactive nanolayer is the core of the OPV, and a donor-acceptor blend is used to absorb photons and generate electric charges most successfully [131]. The blend consists of an electron donor such as poly(3-hexylthiophene) (P3HT) and an electron acceptor such as methanofullerene derivatives (PCBM) [111, 112, 114–121]. The optical and electronic properties of the blend depend on its morphology. The regioregular P3HT forms microcrystalline structures by self-organization, and the formation of a phase-separated morphology with PCBM domains in the crystalline P3HT is desired [132]. For optimal generation of charges, the interfacial area between PCBM and P3HT should be maximized. More fullerene (n-type semiconductor) should be placed near the cathode as well as more polymer (p-type semiconductor) near the anode for optimal charge collection by the electrodes [110]. Thus, detailed characterization of the morphology is needed for optimization of the blend structure.

Because semiconducting polymers typically have electronic transitions in the VIS range, SE is well suited to study the correlation between optical and electronic properties. Optical dispersions for a P3HT:PCBM blend and the pristine layers are displayed in Fig. 8.20 [121]. Madsen *et al.* calculated the refractive index and extinction coefficient of the blend layer from the pure phase dispersion using a Bruggeman EMA. To apply the EMA, the dielectric functions of the pristine films have to be known. SE data of pure P3HT and PCBM films were modeled with a generalized oscillator model using up to three Tauc-Lorentz oscillators. Karagiannidis *et al.* [120] modeled the effective dielectric function of the blend layer directly using five Tauc-Lorentz oscillators. Engmann *et al.* [114, 116] modeled the optical anisotropy of the P3HT phase and used optical dispersions of anisotropic P3HT (high- and low-ordered phases) and of isotropic PCBM in their EMAs. Besides the generalized oscillator approach, Campoy-Quiles and colleagues [30, 129] suggested the SCP model based on critical points of zero to be best suited for the description of the optical dispersion of conjugated polymers such as polyfluorene.

Next to the modeling of the dielectric function of the blend, SE provides the means for non-invasive analysis of the blend morphology. Campoy-Quiles *et al.* [112] reported for the first time a gradient in composition with higher P3HT concentration at the air-polymer interface and higher PCBM concentration at the blend-electrode interface and investigated aging of the morphology. They modeled the vertical composition profile of the blend using two Bruggeman EMA layers. Aging of the blend morphology, which highly affects the efficiency of the photo-generation of charges,

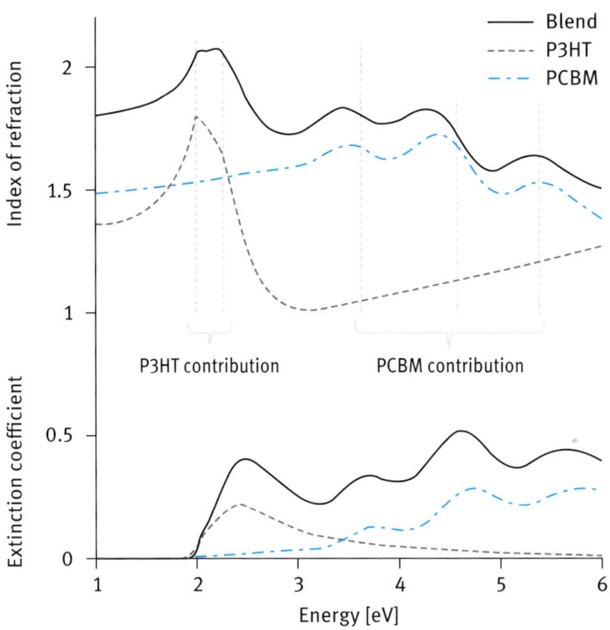

Fig. 8.20: Optical dispersions of pristine P3HT (dotted line), pristine PCBM (dashed/dotted line), and the blend layer (solid line). The opical dispersion for the blend was calcuated by an EMA model from the pure phase dispersions. Reprinted with permission from Madsen MV, Sylvester-Hvid KO, Dastmalchi B, *et al.*, *J Phys Chem C*, 115, 10817–22, 2011. Copyright 2011, American Chemical Society [121].

was investigated by annealing the samples. An increase in the degree of crystallinity of P3HT as well as a diffusion of PCBM crystals to the surface of the blend layer was reported.

8.5 Swelling and adsorption processes: Proteins and stimuli-responsive polymers

8.5.1 Swelling of stimuli-responsive polymer layers

Stimuli-responsive polymers such as polyelectrolytes or polymers with a lower critical solution temperature (LCST) can be used to prepare smart functionalized surfaces [133]. Polyelectrolytes are characterized by ionizable groups at each monomer unit, where polymers with LCST deswell above a critical temperature [134, 135]. The unique swelling behavior of these layers can be utilized for various applications such as smart optical systems, drug delivery, tissue engineering, and sensorics [136].

Stimuli-responsive polymer films can be prepared as hydrogels [137], polyelectrolyte multilayers [138], or polymer brushes [134, 135, 139].

By the analysis of *in situ* VIS-SE measurements, swollen layer thickness, refractive index, and solvent content can be derived [36, 140, 141]. For the modeling of swollen polymer brushes with wet thicknesses below *ca* 100 nm, the use of a simple box model with sharp interfaces, as well as more advanced graded EMA description [140], and a density profile approach [141] were reported. For thicker brushes, a complementary error function $\varphi(z)$ defined as

$$\varphi(z) = \frac{1}{2}\left(1 - erf\left(\frac{z-d}{w}\right)\right) \tag{8.11}$$

was used to model the brush segment density profile (Fig. 8.21), where φ is the polymer volume fraction, z is the coordinate normal to the surface, d is the point of inflection, w is a measure of the smoothness of the brush edge, and $\varphi(z)$ approaches the box model for small w [142].

Kooij *et al.* [141] reported on the temperature-sensitive deswelling of PNIPAAm brushes with dry thicknesses between 8 and 250 nm. They used a box model with Cauchy dispersion for the description of n of the collapsed brush above the LCST. Below the LCST, a two-layer model with an outer gradient profile in n (exponential decrease in the Cauchy parameter A) was necessary to adequately fit the ellipsometric

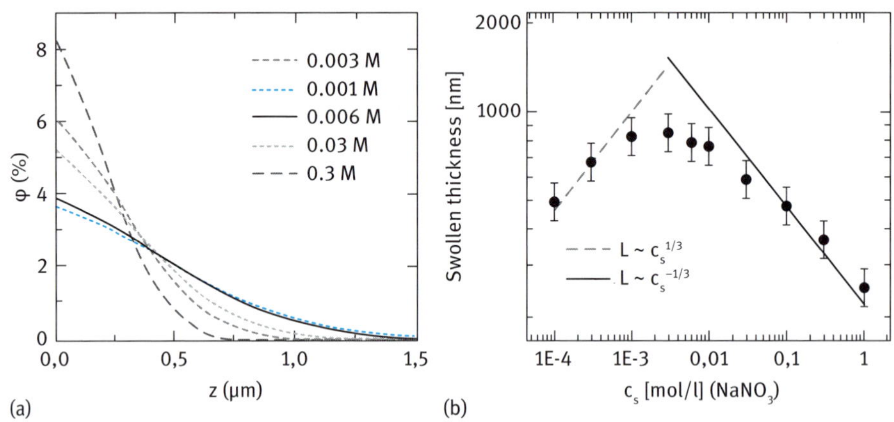

Fig. 8.21: Segment density profiles of a weak polyacid brush (PMAA) swollen at pH 4 in aqueous solution dependent on salt concentration (a). The profiles were obtained from model calculations on ellipsometric spectra using the complementary error function. The swollen thickness at individual salt concentration is determined as twice the first momentum of the segment density profiles (b). Weak polyacid brushes have a maximum swollen thickness at intermediate salt concentrations. The dashed and solid lines in (b) represent the expected scaling behavior. Reprinted with permission from Biesalski M, Johannsmann D, Rühe J, *J Chem Phys*, 117, 4988–15, 2002. Copyright 2002 AIP Publishing LLC [142].

data (Fig. 8.22a). Figure 8.22b shows the different spectral sensitivity of the ellipsometric parameters with temperature, and in Fig. 8.22c, the temperature dependence of the dense brush layer near the substrate (filled symbols) and the more diluted layer

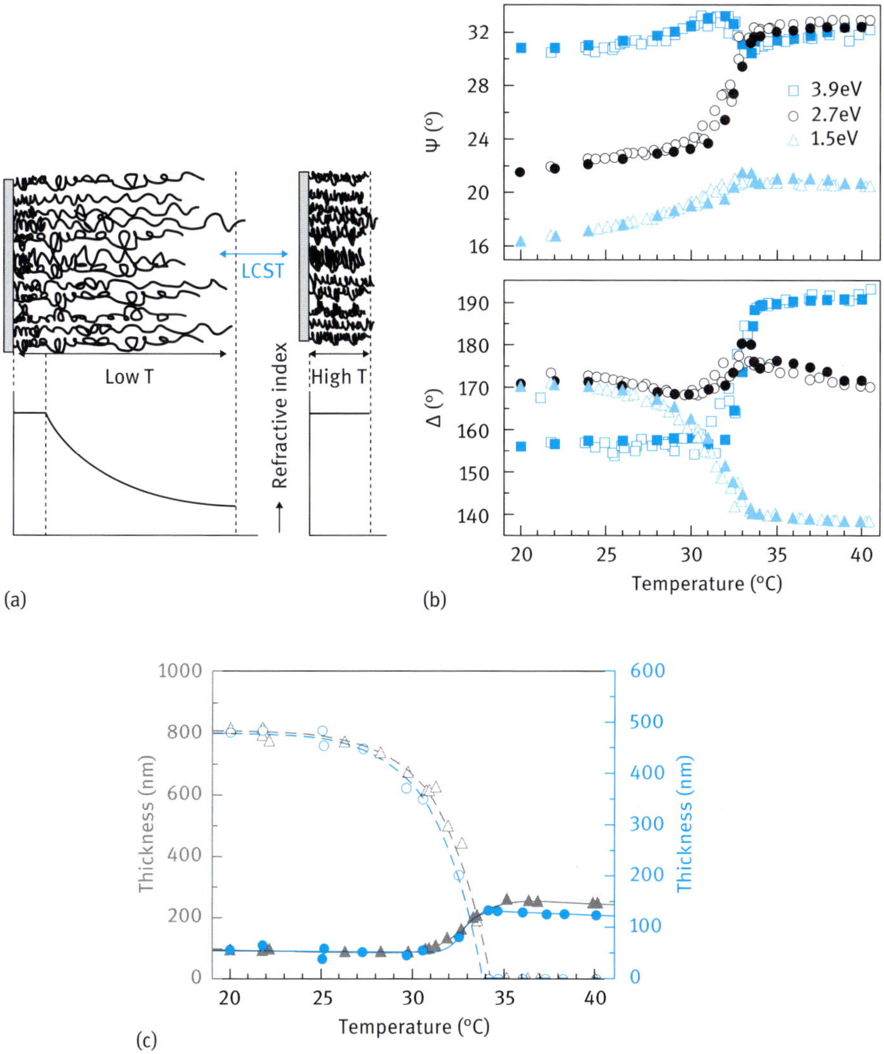

Fig. 8.22: (a) Scheme of the deswelling of PNIPAAm with the LCST and assumed model profile of the refractive index. (b) Measured (open symbols) and simulated (filled symbols) Δ and Ψ dependent on temperature for three different photon energies for a brush with a grafting density of approximately 0.27 chains/nm^2. (c) Temperature-dependent swollen brush thickness as modeled from the ellipsometric spectra. Results for brushes with two different grafting densities are displayed: triangles/left/gray: 0.69 chains/nm^2, and circles/right/blue: 0.27 chains/nm^2. Adapted with permission from Kooij ES, Sui X, Hempenius MA, Zandvliet HJW, Vancso GJ, *J Phys Chem B*, 116, 9261–8, 2012. Copyright 2012, American Chemical Society [141].

near the brush-solution interface (open symbols). Thus, a decreasing segment density of the brush with higher distances from the substrate is modeled.

The modeling of ellipsometric data for hydrogel layers can be done just as well. Junk *et al.* [143] reported on the use of optical gradient profiles for micrometer-thick PNIPAAm hydrogels, and Toomey *et al.* [144] used a three-parameter error function to model the interface roughness of a 1.2-μm-thick hydrogel in water. For very thin hydrogels in the nanometer-range successful modeling was achieved with a box model [145].

With *in situ* IR-SE changes in vibration bands due to swelling or dissociation, processes can be correlated to thickness and solvent content [146–148]. Furchner *et al.* [148] investigated the swelling of PNIPAAm brushes below and above the LCST in the IR range. They used a multilayer box model and described the effective dielectric functions of the PNIPAAm brush layer by a Bruggeman EMA in a three-step fitting procedure. First, the starting values for the swollen brush thickness and the water content were roughly estimated to match the characteristic vibrational bands [amide I and amide II bands of PNIPAAm and the water stretching band $\nu(H_2O)$] of measured and simulated $\tan\Psi$ spectra. Then, in the dielectric function of PNIPAAm, the oscillator positions of the amide bands had to be shifted, reflecting the interaction of the polymer molecules with water. In the last step of modeling, the thickness of the swollen brush and the water fraction were refined by minimizing the amplitude difference of the characteristic vibrational bands between measured and simulated spectra. Varying the water content in the model proved the high sensitivity of the $\nu(H_2O)$ band on the amount of water in the brush layer because distinct changes in the band amplitude can be observed for 2% deviation of the water content (Fig. 8.23).

8.5.2 Protein adsorption at soft polymer surfaces

Proteins are biopolymers consisting of polypeptide chains with a defined primary, secondary, and ternary structures. They are characterized by molecular weight, shape, state of hydration, and their electrochemical behavior. The investigation of protein adsorption to solid surfaces by ellipsometry is extensively reviewed in the literature [35, 149–153]. Recently, IR-SE was used to obtain the dielectric function of human serum albumin (HSA)/anti-HSA multilayers on silicon substrate in the IR range as well as to study structural changes upon heating above 100°C [153]. Because its enhanced sensitivity to adsorption processes, TIRE is increasingly used to characterize protein adsorption to solid surfaces as well as to create biosensors based on the detection of adsorbed protein molecules [49, 54] (see Section 8.2.2).

Investigation of protein adsorption at soft polymer surfaces by SE provides new challenges due to the swollen state of the polymer layer. For the modeling of the adsorbed amount of protein (Γ, in mg/m^2) at soft polymer surfaces, several approaches were reported recently [154–156]. The calculations of Γ, presented here, are all based

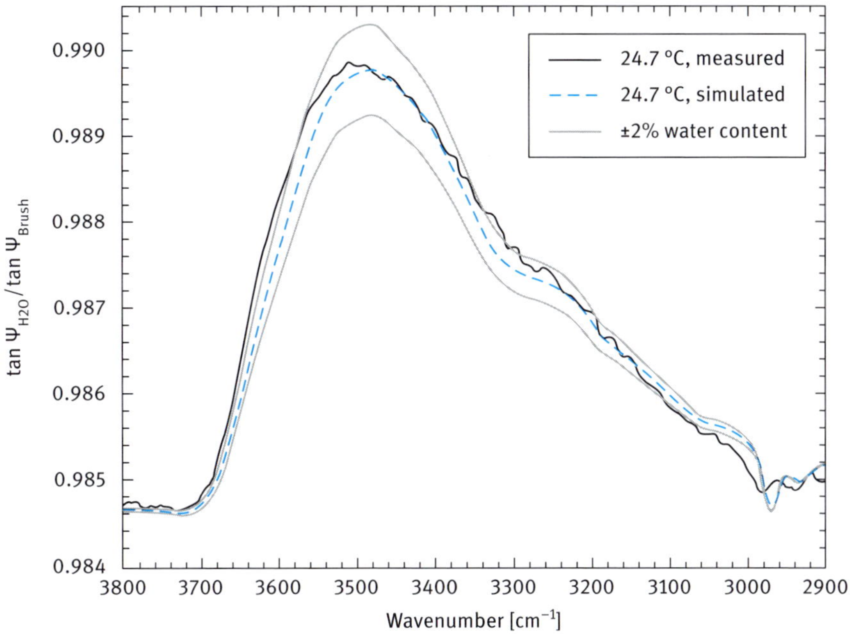

Fig. 8.23: Measured and simulated tanΨ spectrum of the water OH stretching band $\nu(H_2O)$ referenced to tanΨ_{brush} at 24.7°C below the LCST of PNIPAAm. Gray lines reflect the sensitivity of $\nu(H_2O)$ toward changes of the water content in the swollen PNIPAAm brush about 2%. Reprinted with permission from Furchner A, Bittrich E, Uhlmann P, Eichhorn K-J, Hinrichs K, *Thin Solid Films*, 541, 41–5, 2013. Copyright 2013 Elsevier [148].

on the model of de Feijter, which was developed for the adsorption of organic molecules at water-air interfaces [157]. It was later shown that this model can be adopted to adsorption processes of proteins at solid surfaces [150, 158]. At soft surfaces, different types of protein adsorption (primary, secondary, and ternary) can occur, and the de Feijter approach has to be modified (Tab. 8.3).

Formulas based on the de Feijter equation consider the increase of molecule concentration at the adsorbing surface expressed by an increase in the optical density. These equations can be used for adsorption processes in aqueous solution. However, a prerequisite is that dn/dc of the adsorbing biomacromolecule stays constant over a wide range of concentrations c, which was proven for proteins [157].

The application of modified de Feijter formulas is still subject to research, where good agreement could be found for $\Gamma_{I,III}$ between ellipsometry and radio assays [155]. For this approach, the overall layer thickness stays constant in the adsorption experiment and only the refractive index of the film is changing due to penetration of molecules into the film. For the calculation of Γ_{comb}, which considers an increase in the film thickness upon adsorption, there is still no reference experiment. However, we found good agreement with other more complex calculations [156]. For both modified

Tab. 8.3: List of models based on the de Feijter approach used to calculate the adsorbed amount Γ for different protein adsorption processes on soft polymer surfaces

Adsorption process	Formula	Assumptions
Secondary adsorption  Black: substrate Light gray: polymer film Dark gray: protein molecules	$\Gamma_{II} = d\frac{n - n_{amb}}{dn/dc}$ d: layer thickness of protein n: refractive index of protein layer n_{amb}: refractive index of liquid ambient dn/dc: refractive index increment of protein	dn/dc constant for wide range of protein concentration Low swelling of polymer film Cauchy dispersion $n(\lambda)$ for protein layer as best-fit result or fixed at constant value from literature, e.g. $n = 1.375$ [149] or $n = 1.575$ [159]
Primary adsorption at substrate or ternary adsorption in film 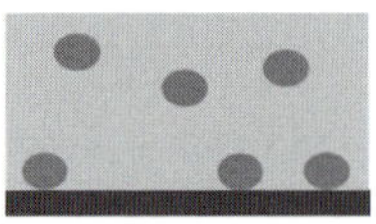	$\Gamma_{I,III} = d_{pol}\frac{n_{ads} - n_{pol}}{dn/dc}$ d_{pol}: d of swollen polymer layer before adsorption n_{ads}: n after adsorption n_{pol}: n of swollen polymer layer	Thickness of swollen polymer layer in adsorption process constant One-layer model with fixed thickness: d_{pol} = const Cauchy dispersion for $n_{ads}(\lambda)$ and $n_{pol}(\lambda)$
Secondary, primary, and/or ternary adsorption 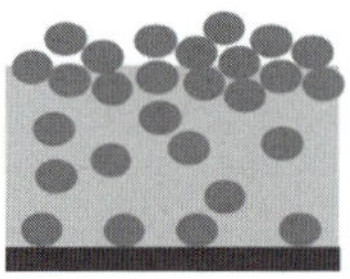	$\Gamma_{comb} = \Gamma_{I,III} + (d_{ads} - d_{pol})\frac{n_{ads} - n_{amb}}{dn/dc}$ d_{ads}: thickness of composite polymer-protein layer after adsorption	One-layer model with changing thickness upon adsorption Cauchy dispersion for $n_{ads}(\lambda)$ and $n_{amb}(\lambda)$

de Feijter approaches, a combined polymer-protein composite layer is modeled from the ellipsometric data.

Xue *et al.* [155] used the expression for $\Gamma_{I,III}$ to simulate changes in Δ and Ψ for given adsorbed amounts of protein between 0 and 3.5 mg/m^2 at PNIPAAm brushes and at the underlying self-assembled monolayer (SAM). These brushes were found to be highly resistive against the adsorption of the protein HSA for certain molecular weight and grafting densities of PNIPAAm [160]. Protein resistance is desirable for the formation of inert surfaces or the design of surfaces that provide specific interactions due to tailored functional groups. Xue *et al.* also reported on SE measurements of the adsorption of lysozyme on PNIPAAm brushes with different grafting densities and molecular weight, and changes in the ellipsometric parameters at $\lambda = 546$ nm upon adsorption are presented in Fig. 8.24. At the given wavelength, the sensitivity of Δ and Ψ toward protein adsorption was highest. Included in Fig. 8.24 are the simulated shifts for 1 mg/m^2 protein adsorption at the brushes and for 0.1 mg/m^2 protein

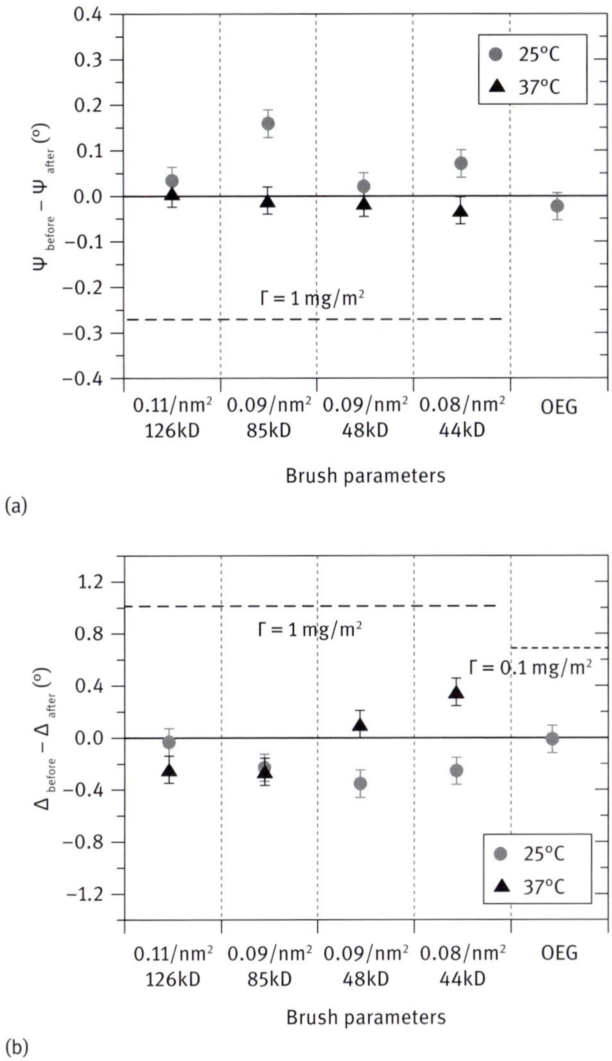

Fig. 8.23: Changes in the ellipsometric parameters Ψ (a) and Δ (b) at $\lambda = 546$ nm upon adsorption of lysozyme to PNIPAAm brushes with different molecular weight and grafting density as well as to the underlying SAM layer for 25°C (below the LCST of PNIPAAm) and for 37°C (above the LCST of PNIPAAm). Simulated shift of Δ and Ψ for 1 mg/m^2 adsorbed protein at the brush (dashed line) and for 0.1 mg/m^2 at the SAM layer (dotted line) are included. Reprinted with permission from Xue C, Yonet-Tanyeri N, Brouette N, Sferrazza M, Braun PV, Leckband DE, *Langmuir*, 27, 8810–8, 2011. Copyright 2011, American Chemical Society [155].

adsorption at the SAM underlayer. They concluded that protein adsorption at the PNIPAAm brushes was below 1 mg/m^2 at all conditions. Simulated Δ and Ψ shifts of the 0.5 mg/m^2 protein adsorption at the brushes were barely distinguishable.

Thus, the sensitivity of the SE measurements was limited to protein amounts higher than the detection limit. For the SAM underlayer, the detection limit was as good as 0.1 mg/m^2, although this simulated adsorbed amount produced no shift in Ψ, only a shift in Δ. To overcome this sensitivity limit investigations of protein resistance at soft polymer surfaces would also greatly profit from TIRE experiments.

References

[1] Advincula RC, Brittain WJ, Caster KC, Rühe J. Polymer brushes. Weinheim (Germany): Wiley-VCH Verlag; 2004.

[2] Lvov I. Electrostatic layer-by-layer assembly of proteins and polyions. In: Lvov Y, Möhwald H, eds. Protein architecture. New York (NY): Marcel Dekker; 2001:125–67.

[3] Boudou T, Crouzier T, Ren K, Blin G, Picart C. Multiple functionalities of polyelectrolyte multilayer films: new biomedical applications. Adv Mater 2010;22:441–67.

[4] Schaaf P, Voegel J-C, Jierry L, Boulmedais F. Spray-assisted polyelectrolyte multilayer buildup: from step-by-step to single-step polyelectrolyte film constructions. Adv Mater 2012;24:1001–16.

[5] Simpson TRE, Parbhoo B, Keddie JL. The dependence of the rate of crosslinking in poly(dimethyl siloxane) on the thickness of coatings. Polymer 2003;44:4829–38.

[6] Robeson LM. Polymer blends. Munich (Germany): Carl Hanser Verlag; 2007.

[7] Hanemann T, Szabó DV. Polymer-nanoparticle composites: from synthesis to modern applications. Materials 2010;3:3468–517.

[8] Horechyy A, Nandan B, Zafeiropoulos NE, et al. A step-wise approach for dual nanoparticle patterning via block copolymer self-assembly. Adv Funct Mater 2013;23:483–90.

[9] Schmidt D, Müller C, Hofmann T, et al. Optical properties of hybrid titanium chevron sculptured thin films coated with a semiconducting polymer. Thin Solid Films 2011;519:2645–9.

[10] Koh J, Lu YW, Wronski CR, et al. Correlation of real time spectroellipsometry and atomic force measurements of surface roughness on amorphous semiconductor. Appl Phys Lett 1996;69:1297–9.

[11] Keddie JL, Meredith P, Jones RAL, Donald AM. Kinetics of film formation in acrylic latices studied with multiple-angle-of-incidence ellipsometry and environmental SEM. Macromolecules 1995;28:2673–82.

[12] Keddie JL, Jones RAL, Cory RA. Size-dependent depression of the glass transition temperature in thin films. Europhys Lett 1994;27:59–64.

[13] Ruths J, Essler F, Decher G, Riegler H. Polyelectrolytes I: polyanion/polycation multilayers at the air/monolayer/water interface as elements for quantitative polymer adsorption studies and preparation of hetero-superlattices on solid surfaces. Langmuir 2000;16:8871–8.

[14] Habicht J, Schmidt M, Rühe J, Johannsmann D. Swelling of thick polymer brushes investigated with ellipsometry. Langmuir 1999;15:2460–5.

[15] Gölander C-G, Kiss E. Protein adsorption on functionalized and ESCA-characterized polymer films studied by ellipsometry. J Colloid Interface Sci 1988;121:240–53.

[16] Werner C, Eichhorn K-J, Grundke K, Simon F, Grählert W, Jacobasch H-J. Insights on structural variations of protein adsorption layers on hydrophobic fluorohydrocarbon polymers gained by spectroscopic ellipsometry (part I). Colloid Surf A 1999;156:3–17.

[17] Tompkins HG, Irene EA. Handbook of ellipsometry. Norwich (NY): William Andrew; 2005.

[18] Tompkins HG, McGahan, WA. Spectroscopic ellipsometry and reflectometry. New York (NY): Wiley; 1999.

[19] Humlíček J. Polarized light and ellipsometry. In: Tompkins HG, Irene EA, eds. Handbook of ellipsometry. Norwich (NY): William Andrew; 2005,3–91.
[20] Kurt A, Demirelli K. A Study on the optical properties of three-armed polystyrene and poly(styrene-b-isobutyl methacrylate). Polym Eng Sci 2010;50:268–77.
[21] Röseler A. Spectroscopic infrared ellipsometry. In: Tompkins HG, Irene EA, eds. Handbook of ellipsometry. Norwich (NY): William Andrew; 2005, 763–798.
[22] Garcia-Caurel E, Ossikovski R, Foldyna M, et al. Advanced Mueller ellipsometry instrumentation and data analysis. In: Losurdo M, Hingerl K, eds. Ellipsometry at the nanoscale. Berlin (Germany): Springer; 2013, 31–143.
[23] Lau KKS, Caufield JA, Gleason KK. Variable angle spectroscopic ellipsometry of fluorocarbon films from hot filament chemical vapor deposition. J Vac Sci Technol A 2000;18:2404–11.
[24] Johs B, Hale JS. Dielectric function representation by B-splines. Phys Stat Sol A 2008;205:715–9.
[25] Wang T, Pearson AJ, Lidzey DG, Jones RAL. Evolution of structure, optoelectronic properties, and device performance of polythiophene: fullerene solar cells during thermal annealing. Adv Func Mater 2011;21:1383–90.
[26] Collins RW, Ferlauto AS. Optical physics of materials. In: Tompkins HG, Irene EA, eds. Handbook of ellipsometry. Norwich (NY): William Andrew; 2005, 93–235.
[27] Losurdo M, Giangregorio MM, Capezzuto P, et al. Study of anisotropic optical properties of poly(arylenephenylene) thin films: dependence on polymer backbone. Macromolecules 2003;36:4492–7.
[28] Chapter 3: Optical properties of materials and layered structures. In: Tompkins HG, McGahan WA, eds. Spectroscopic ellipsometry and reflectometry. New York (NY): Wiley; 1999, 23–34.
[29] Campoy-Quiles M, Heliotis G, Xia R, et al. Ellipsometric characterization of the optical constants of polyfluorene gain media. Adv Funct Mater 2005;15:925–33.
[30] Campoy-Quiles M, Nelson J, Bradley, DDC, Etchegoin PG, Dimensionality of electronic excitations in organic semiconductors: a dielectric function approach. Rev B 2007;76:235206.
[31] Aspnes DE. Optical properties of thin films. Thin Solid Films 1982;89:249–62.
[32] Campoy-Quiles M, Nelson J, Etchegoin PG, et al. On the determination of anisotropy in polymer thin films: a comparative study of optical techniques. Phys Stat Sol C 2008;5:1270–3.
[33] Benecke C, Seiberle H, Schadt M. Determination of director distributions in liquid crystal polymer-films by means of generalized anisotropic ellipsometry. Jpn J Appl Phys 2000;39:525–31.
[34] Schubert M. Theory and application of generalized ellipsometry. In: Tompkins HG, Irene EA, eds. Handbook of ellipsometry. Norwich (NY): William Andrew; 2005, 637–717.
[35] Arwin H. Ellipsometry in life sciences. In: Tompkins HG, Irene EA, eds. Handbook of ellipsometry. Norwich (NY): William Andrew; 2005, 799–850.
[36] Bittrich E, Burkert S, Müller M, Eichhorn K-J, Stamm M, Uhlmann P. Temperature-sensitive swelling of poly(N-isopropylacrylamide) brushes with low molecular weight and grafting density. Langmuir 2012;28:3439–48.
[37] Mora MF, Nejadnik MR, Baylon-Cardiel JL, Giacomelli CE, Garcia CD. Determination of a setup correction function to obtain adsorption kinetic data at stagnation point flow conditions. J Colloid Interface Sci 2010;346:208–15.
[38] Aulich D, Hoy O, Luzinov I, et al. In situ studies on the switching behavior of ultrathin poly(acrylic acid) polyelectrolyte brushes in different aqueous environments. Langmuir 2010;26:12926–32.
[39] Buttry DA, Ward MD. Measurement of interfacial processes at electrode surfaces with the electrochemical quartz crystal microbalance. Chem Rev 1992;92:1355–79.
[40] Höök F, Kasemo B, Nylander T, Fant C, Scott K, Elwing H. Variations in coupled water, viscoelastic properties, and film thickness of a Mefp-1 protein film during adsorption and cross-linking: a quartz crystal microbalance with dissipation monitoring, ellipsometry, and surface plasmon resonance study. Anal Chem 2001;73:5796–804.

[41] Voinova MV, Jonson M, Kasemo B. Dynamics of viscous amphiphilic films supported by elastic solid substrates. J Phys Condens Matter 1997;9:7799–808.
[42] Voinova MV, Jonson M, Kasemo B. "Missing mass" effect in biosensor's QCM applications. Biosens Bioelectr 2002;17:835–41.
[43] Eisele, NB, Andersson FI, Frey S, Richter RP. Viscoelasticity of thin biomolecular films: a case study on nucleoporin phenylalanine-glycine repeats grafted to a histidine-tag capturing QCM-D sensor. Biomacromolecules 2012;13:2322–32.
[44] Richter RP, Rodenhausen KB, Eisele NB, Schubert M. Coupling spectroscopic ellipsometry and quartz crystal microbalance to study organic films at the solid-liquid interface. In: Hinrichs K, Eichhorn K-J, eds. Ellipsometry of functional organic surfaces and films. Berlin (Germany): Springer; 2014.
[45] Iturri Ramos JJ, Stahl S, Richter RP, Moya SE. Water content and buildup of poly(diallyldimethylammonium chloride)/poly(sodium 4-styrenesulfonate) and poly(allylamine hydrochloride)/poly(sodium 4-styrenesulfonate) polyelectrolyte multilayers studied by an in situ combination of a quartz crystal microbalance with dissipation monitoring and spectroscopic ellipsometry. Macromolecules 2010;43:9063–70.
[46] Domack A, Prucker O, Rühe J, Johannsmann D. Swelling of a polymer brush with a quartz crystal resonator. Phys Rev E 1997;56:680–9.
[47] Poksinski M, Arwin H. Total internal reflection ellipsometry: ultrahigh sensitivity for protein adsorption on metal surfaces. Opt Lett 2007;32:1308–10.
[48] Liu L, Chen Y, Meng Y, Chen S, Jin G. Improvement for sensitivity of biosensor with total internal reflection imaging ellipsometry (TIRIE). Thin Solid Films 2011;519:2758–62.
[49] Poksinski M, Arwin H. Total internal reflection ellipsometry: ultrahigh sensitivity for protein adsorption on metal surfaces. Opt Lett 2007;32:1308–10.
[50] Basova T, Tsargorodskaya A, Nabok A, et al. Investigation of gas-sensing properties of copper phthalocyanine films. Mater Sci Eng C 2009;29:814–8.
[51] Arwin H. TIRE and SPR-enhanced SE for adsorption processes. In: Hinrichs K, Eichhorn K-J, eds. Ellipsometry of functional organic surfaces and films. Berlin (Germany): Springer;
[52] Poksinski M, Arwin H. Protein monolayers monitored by internal reflection ellipsometry. Thin Solid Films 2004;455:716–21.
[53] Mårtensson J, Arwin H, Nygren H, Lundström I. Adsorption and optical properties of ferritin layers on gold studied with spectroscopic ellipsometry. J Colloid Interface Sci 1995;174:79–85.
[54] Liu L, Chen Y, Meng Y, Chen S, Jin G. Improvement for sensitivity of biosensor with total internal reflection imaging ellipsometry (TIRIE). Thin Solid Films 2011;519:2758–62.
[55] Laskarakis A, Kassavetis S, Gravalidis C, Logothetidis S. In situ and real-time optical investigation of nitrogen plasma treatment of polycarbonate. Nucl Instrum Methods Phys Res Sec B 2010;268–460.
[56] Laskarakis A, Logothetidis S. In-line Quality Control of organic thin film fabrication on rigid and flexible substrates. In: Hinrichs K, Eichhorn K-J, eds. Ellipsometry of functional organic surfaces and films. Berlin (Germany): Springer; 2014.
[57] Logothetidis S, Georgiou D, Laskarakis A, Koidis C, Kalfagiannis N. In-line spectroscopic ellipsometry for the monitoring of the optical properties and quality of roll-to-roll printed nanolayers for organic photovoltaics. Sol Ener Mater Sol Cells 2013;112:144–56.
[58] Telford AM, Meagher L, Glattauer V, Gengenbach TR, Easton CD, Neto C. Micropatterning of polymer brushes: grafting from dewetting polymer films for biological applications. Biomacromolecules 2012;13:2989–98.
[59] Chien H-W, Chang T-Y, Tsai W-B. Spatial control of cellular adhesion using photo-crosslinked micropatterned polyelectrolyte multilayer films. Biomaterials 2009;30:2209–18.
[60] Huang NF, Thakar RG, Wong M, Kim D, Lee RJ, Li S. Tissue engineering of muscle on micropatterned polymer films. Conf Proc IEEE Med Biol Soc 2004;7:4966–9.

[61] Niu Y, Jin G. Surface modification methods to improve behavior of biosensor based on imaging ellipsometry. Appl Surf Sci 2013;281:84–8.
[62] Schmaljohann D, Nitschke M, Schulze R, Eing A, Werner C, Eichhorn K-J. In situ study of the thermoresponsive behavior of micropatterned hydrogel films by imaging ellipsometry. Langmuir 2005;21:2317–22.
[63] Hinrichs K, Gensch M, Esser N, et al. Analysis of biosensors by chemically specific optical techniques. Chemiluminescence-imaging and infrared spectroscopic mapping ellipsometry. Anal Bioanal Chem 2007;387:1823–9.
[64] Chang M-J, Pang C-R, Liu J, et al. High spatial resolution label-free detection of antigen-antibody binding on patterned surface by imaging ellipsometry. J Colloid Interface Sci 2011;360:826–33.
[65] Bruun Nielsen MM, Cohen Simonsen A. Imaging ellipsometry of spin-coated membranes: mapping of multilamellar films, hydrated membranes, and fluid domains. Langmuir 2013;29:1525–32.
[66] Jin G, Jansson R, Arwin H. Imaging ellipsometry revisited: Developments for visualization of thin transparent layers on silicon substrates. Rev Sci Instrum 1996;67:2930–6.
[67] Matković A, Beltaos A, Milićević M, Ralević U, Vasić B. Spectroscopic imaging ellipsometry and Fano resonance modeling of grapheme. J Appl Phys 2012;112:123423.
[68] Meier R, Chiang H-Y, Ruderer MA, et al. In situ film characterization of thermally treated microstructured conducting polymer films. J Polym Sci B 2012;50:631–41.
[69] Körstgens V, Meier R, Ruderer MA, et al. Note: Grazing incidence small and wide angle x-ray scattering combined with imaging ellipsometry. Rev Sci Instrum 2012;83:076107.
[70] Gensch M. Brilliant infrared light sources for micro-ellipsometric studies of organic thin films. In: Hinrichs K, Eichhorn K-J, eds. Ellipsometry of functional organic surfaces and films. Berlin (Germany): Springer; 2014.
[71] Gensch M, Korte EH, Esser N, Schade U, Hinrichs K. Microfocus-infrared synchrotron ellipsometer for mapping of ultra thin films. Infrared Phys Technol 2006;49:74–7.
[72] Rosu DM, Jones JC, Hsu JWP, et al. Molecular orientation on octanedithiol and hexadecanethiol monolayers on GaAs and Au measured by infrared spectroscopic ellipsometry. Langmuir 2009;25:919–23.
[73] Hinrichs K, Gensch M, Esser N, et al. Analysis of biosensors by chemically specific optical techniques. Chemiluminescence-imaging and infrared spectroscopic mapping ellipsometry. Anal Bioanal Chem 2007;387:1823–9.
[74] Duckworth P, Richardson H, Carelli C, Keddie JL. Infrared ellipsometry of interdiffusion in thin films of miscible polymers. Surf Interface Anal 2004;36:33–41.
[75] Hinrichs K, Gensch M, Nikonenko N, Pionteck J, Eichhorn K-J. Spectroscopic ellipsometry for characterization of thin films of polymer blends. Macromol Symp 2005;230:26–32.
[76] Simpson TRE, Parbhoo B, Keddie JL. The dependence of the rate of crosslinking in poly(dimethyl siloxane) on the thickness of coatings. Polymer 2003;44:4829–38.
[77] Vallon S, Drévillon B, Poncin-Epaillard F. In situ spectroellipsometry study of the crosslinking of polypropylene by an argon plasma. Appl Surf Sci 1997;108:177–85.
[78] Laskarakis A, Gravalidis C, Logothetidis S. FTIR and Vis-FUV real time spectroscopic ellipsometry studies of polymer surface modifications during ion beam bombardment. Nucl Instrum Methods Phys Res B 2004;216:131–6.
[79] Laskarakis A, Kassavetis C, Gavalidis C, Logothetidis S. In situ and real-time optical investigation of nitrogen plasma treatment of polycarbonate. Nucl Instrum Methods Phys Res B 2010;268:460–5.
[80] Zhou H, Toney MF, Bent SF. Cross-linked ultrathin polyurea films via molecular layer deposition. Macromolecules 2013;46:5638–43.
[81] Lilge I, Schönherr H. Covalently cross-linked poly(acrylamide) brushes on gold with tunable mechanical properties via surface-initiated atom transfer radical polymerization. Eur Polym J 2013;49:1943–51.

[82] Griesser T, Wolfberger A, Daschiel U, et al. Cross-linking of ROMP derived polymers using the twophoton induced thiol-ene reaction: towards the fabrication of 3D-polymer microstructures. Polym Chem 2013;4:1708–14.

[83] Baeurle SA, Hotta A, Gusev AA. On the glassy state of multiphase and pure polymer materials. Polymer 2006;47:6243–53.

[84] Mapesa EU, Erber M, Tress M, et al. Glassy dynamics in nanometer thin layers of polystyrene. Eur Phys J 2010;189:173–80.

[85] Tress M, Erber M, Mapesa EU, et al. Glassy dynamics and glass transition in nanometric thin layers of polystyrene. Macromolecules 2010;43:9937–44.

[86] Serghei A, Kremer F. Unexpected preparative effects on the properties of thin polymer films. Progr Colloid Polym Sci 2006;132:33–40.

[87] Perlich J, Körstgens V, Metwalli E, Schulz L, Georgii R, Müller-Buschbaum P. Solvent content in thin spin-coated polystyrene homopolymer films. Macromolecules 2009;42:337–44.

[88] Serghei A, Kremer F. Metastable states of glassy dynamics, possibly mimicking confinement-effects in thin polymer films. Macromol Chem Phys 2008;209:810–7.

[89] Erber M, Tress M, Eichhorn K-J. Glass transition of polymers with different architectures in the confinement of nanoscopic films. In: Hinrichs K, Eichhorn K-J, eds. Ellipsometry of functional organic surfaces and films. Berlin (Germany): Springer; 2014.

[90] Erber M, Georgi U, Müller J, Eichhorn K-J, Voit B. Polystyrene with different topologies: study of the glass transition temperature in confined geometry of thin films. Eur Polym J 2010;46:2240–6.

[91] Erber M, Khalyavina A, Eichhorn K-J, Voit BI. Variations in the glass transition temperature of polyester with special architectures confined in thin films. Polymer 2010;51:129–35.

[92] Balazs AC, Emrick T, Russell TP. Nanoparticle polymer composites: where two small worlds meet. Science 2006;314:1107–10.

[93] Pelton M, Aizpurua J, Bryant G. Metal-nanoparticle plasmonics. Laser Phot Rev 2008;2:136–59.

[94] Oates TWH. Real time spectroscopic ellipsometry of nanoparticle growth. Appl Phys Lett 2006;88:213115.

[95] Oren R, Liang Z, Barnard JS, Warren SC, Wiesner U, Huck WTS. Organization of nanoparticles in polymer brushes. J Am Chem Soc 2009;131:1670–1.

[96] Rauch S, Eichhorn K-J, Stamm M, Uhlmann P. Spectroscopic ellipsometry of superpara-magnetic nanoparticles in thin films of poly(nisopropylacrylamide). J Vac Sci Technol A 2012;30:041514.

[97] Yashchenok AM, Gorin DA, Badylevich M, et al. Impact of magnetite nanoparticle incorporation on optical and electrical properties of nanocomposite LbL assemblies. Phys Chem Chem Phys 2010;12:10469–75.

[98] Oates TWH, Christalle E. Real-time spectroscopic ellipsometry of silver nanoparticle formation in poly(vinyl alcohol) thin films. J Phys Chem C 2007;111:182–7.

[99] Qian XM, Nie SM. Single-molecule and single-nanoparticle SERS: from fundamental mechanisms to biomedical applications. Chem Soc Rev 2008;37:912–20.

[100] Kasputis T, Koenig M, Schmidt D, et al. Slanted columnar thin films prepared by glancing angle deposition functionalized with polyacrylic acid polymer brushes. J Phys Chem C 2013;117:13971–80.

[101] Schmidt D, Müller C, Hofmann T, et al. Optical properties of hybrid titanium chevron sculptured thin films coated with a semiconducting polymer. Thin Solid Films 2011;519:2645–9.

[102] Hawkeye MM, Brett MJ. Glancing angle deposition: fabrication, properties, and applications of micro- and nanostructured thin films. J Vac Sci Technol A 2007;25:1317–36.

[103] Seto MW, Robbie K, Brett MJ. Mechanical properties of microspring thin films fabricated by glancing angle deposition (GLAD). Electr Comp Eng 1999;3:1616–20.
[104] Wang S, Fu X, Xia G, Wang J, Shao J, Fan Z. Structure and optical properties of ZnS thin films grown by glancing angle deposition. Appl Surf Sci 2006;252:8734–7.
[105] Schmidt D, Schubert E, Schubert M. Generalized ellipsometry characterization of sculptured thin films made by glancing angle deposition. In: Losurdo M, Hingerl K, eds. Ellipsometry at the nanoscale. Berlin (Germany): Springer; 2013.
[106] Nejadnik MR, Garcia CD. Staining proteins: a simple method to increase the sensitivity of ellipsometric measurements in adsorption studies. Colloid Surf B 2011;82:253–7.
[107] Beyerlein D, Kratzmüller T, Eichhorn K-J. Study of novel polymer architectures on solid surfaces by variable angle spectroscopic and imaging ellipsometry. Vib Spectrosc 2002;29:223–7.
[108] Fleischmann S, Hinrichs K, Oertel U, Reichelt S, Eichhorn K-J, Voit B. Modification of polymer surfaces by click chemistry. Macromol Rapid Commun 2008;29:1177–85.
[109] Rauch S, Eichhorn K-J, Oertel U, Stamm M, Kuckling D, Uhlmann P. Temperature responsive polymer brushes with clicked rhodamine B: synthesis, characterization and swelling dynamics studied by spectroscopic ellipsometry. Soft Matter 2012;8:10260–70.
[110] Logothetidis S. Polymer blends and composites. In: Hinrichs K, Eichhorn K-J, eds. Ellipsometry of functional organic surfaces and films. Berlin (Germany): Springer; 2014.
[111] Zhokhavets U, Erb T, Hoppe H, Gobsch G, Sariciftci NS. Effect of annealing of poly(3-hexylthiophene)/fullerene bulk heterojunction composites on structural and optical properties. Thin Sol Films 2006;496:679–82.
[112] Campoy-Quiles M, Ferenczi T, Agostinelli T, et al. Morphology evolution via self-organization and lateral and vertical diffusion in polymer:fullerene solar cell blends. Nat Mater 2008;7:158–64.
[113] Müller C, Bergqvist J, Vandewal K, et al. Phase behaviour of liquid-crystalline polymer/fullerene organic photovoltaic blends: thermal stability and miscibility. J Mater Chem 2011;21:10676–84.
[114] Engmann S, Turkovic V, Gobsch G, Hoppe H. Ellipsometric investigation of the shape of nanodomains in polymer/fullerene films. Adv Energy Mater 2011;1:684–9.
[115] Engmann S, Turkovic V, Hoppe H, Gobsch G. Aging of polymer/fullerene films: temporal development of composition profiles. Synth Mater 2012;161:2540–3.
[116] Engmann S, Turkovic V, Denner P, Hoppe H, Gobsch G. Optical order of the polymer phase within polymer/fullerene blend films. J Polym Sci B 2012;50:1363–73.
[117] Engmann S, Singh CR, Turkovic V, Hoppe H, Gobsch G. Direct correlation of the organic solar cell device performance to the in-depth distribution of highly ordered polymer domains in polymer/fullerene films. Adv Energy Mater 2013;3:1463–72.
[118] Karagiannidis PG, Kalfagiannis N, Georgiou D, et al. Effects of buffer layer properties and annealing process on bulk heterojunction morphology and organic solar cell performance. J Mater Chem 2012;22:14624–32.
[119] Chuang S-Y, Chen H-L, Lee W-H, et al. Regioregularity effects in the chain orientation and optical anisotropy of composite polymer/fullerene films for high-efficiency, large-area organic solar cells. J Mater Chem 2009;19:5554–60.
[120] Karagiannidis PG, Georgiou D, Pitsalidis C, Laskarakis A, Logothetidis S. Evolution of vertical phase separation in P3HT:PCBM thin films induced by thermal annealing. Mater Chem Phys 2011;129:1207–13.
[121] Madsen MV, Sylvester-Hvid KO, Dastmalchi B, et al. Ellipsometry as a nondestructive depth profiling tool for roll-to-roll manufactured flexible solar cells. J Phys Chem C 2011;115:10817–22.

[122] Herrmann F, Engmann S, Presselt M, Hoppe H, Shokhovets S, Gobsch G. Correlation between near infrared-visible absorption, intrinsic local and global sheet resistance of poly(3,4-ethylenedioxy-thiophene) poly(styrene sulfonate) thin films. Appl Phys Lett 2012;100:153301.
[123] Logothetidis S, Laskarakis A. Organic against inorganic electrodes grown onto polymer substrates for flexible organic electronics applications. Thin Sol Films 2009;518:1245–9.
[124] Gravalidis C, Laskarakis A, Logothetidis S. Fine tuning of PEDOT electronic properties using solvents. Eur Phys J Appl Phys 2009;46:12505.
[125] Schubert M, Bundesmann C, Jakopic G, et al. Infrared ellipsometry characterization of conducting thin organic films. Thin Sol Films 2004;455:295–300.
[126] Ino T, Hiate T, Fukuda T, Ueno K, Shirai H. Real-time ellipsometric characterization of the initial growth stage of poly(3,4-ethylenedioxythiophene):poly(styrene sulfonate) films by electrospray deposition using N,N-dimethylformamide solvent solution. J Non-Cryst Sol 2012;358:2520–4.
[127] Mauger SA, Chang L, Rochester CW, Moulé AJ. Directional dependence of electron blocking in PEDOT:PSS. Org Electr 2012;13:2747–56.
[128] Campoy-Quiles M, Heliotis G, Xia R, et al. Ellipsometric characterization of the optical constants of polyfluorene gain media. Adv Funct Mater 2005;15:925–33.
[129] Campoy-Quiles M, Etchegoin PG, Bradley DC. Exploring the potential of ellipsometry for the characterisation of electronic, optical, morphologic and thermodynamic proeprties of polyfluorene thin films. Synth Mater 2005;155:279–82.
[130] Yap BK, Xia R, Campoy-Quiles M, Stavrinou PN, Bradley DDC. Simultaneous optimization of charge-carrier mobility and optical gain in semiconducting polymer films. Nat Mater 2008;7:376–80.
[131] Pearson AJ, Wang T, Lidzey DG. The role of dynamic measurements in correlating structure with optoelectronic properties in polymer fullerene bulk-heterojunction solar cells. Rep Prog Phys 2013;76:022501.
[132] Yang X, Loos J, Veenstra SC, et al. Nanoscale morphology of high-performance polymer solar cells. Nano Lett 2005;5:579–83.
[133] Russel TP. Surface-responsive materials. Science 2002;297:964–7.
[134] Dong R, Lindau M, Ober CK. Dissociation behavior of weak polyelectrolyte brushes on a planar surface. Langmuir 2009;25:4774–9.
[135] Toomey R, Tirrell M. Functional polymer brushes in aqueous media from self-assembled and surface-initiated polymers. Annu Rev Phys Chem 2008;59:493–517.
[136] Cohen Stuart MA, Huck WTS, Genzer J, et al. Emerging applications of stimuli-responsive polymer materials. Nature Mater 2010;9:101–13.
[137] Tokarev I, Minko S. Stimuli-responsive hydrogel thin films. Soft Matter 2009;5:511–24.
[138] Boudou T, Crouzier T, Ren K, Blin G, Picart C. Multiple functionalities of polyelectrolyte multilayer films: new biomedical applications. Adv Mater 2010;22:441–67.
[139] Brittain WJ, Minko S. A structural definition of polymer brushes. J Polym Sci A 2007;45:3505–12.
[140] Wu T, Gong P, Szleifer I, Vlček P, Šubr V, Genzer J. Behavior of surface-anchored poly(acrylic acid) brushes with grafting density gradients on solid substrates: 1. Experiment. Macromolecules 2007;40:8756–64.
[141] Kooij ES, Sui X, Hempenius MA, Zandvliet HJW, Vancso GJ. Probing the thermal collapse of poly(N-isopropylacrylamide) graftsby quantitative in situ ellipsometry. J Phys Chem B 2012;116:9261–8.
[142] Biesalski M, Johannsmann D, Rühe J. Synthesis and swelling behavior of a weak polyacid brush. J Chem Phys 2002;117:4988–15.

[143] Junk MJN, Anac I, Menges B, Jonas U. Analysis of optical gradient profiles during temperature- and salt-dependent swelling of thin responsive hydrogel films. Langmuir 2010;26:12253–9.
[144] Toomey R, Freidank D, Rühe J. Swelling behavior of thin, surface-attached polymer networks. Macromolecules 2004;37:882–7.
[145] Gramm S, Teichmann J, Nitschke M, Gohs U, Eichhorn K-J, Werner C. Electron beam immobilization of functionalized poly(vinyl methyl ether) thin films on polymer surfaces Towards stimuli responsive coatings for biomedical purposes. eXPRESS Polym Lett 2011;5:970–6.
[146] Hinrichs K, Aulich D, Ionov L, et al. Chemical and structural changes in a pH-responsive mixed polyelectrolyte brush studied by infrared ellipsometry. Langmuir 2009;25:10987–91.
[147] Aulich D, Hoy O, Luzinov I, et al. In situ studies on the switching behavior of ultrathin poly(acrylic acid) polyelectrolyte brushes in different aqueous environments. Langmuir 2010;26:12926–32.
[148] Furchner A, Bittrich E, Uhlmann P, Eichhorn K-J, Hinrichs K. In-situ characterization of the temperature-sensitive swelling behavior of poly(N-isopropylacrylamide) brushes by infrared and visible ellipsometry. Thin Solid Films 2013;541:41–5.
[149] Malmsten M. Ellipsometry studies of protein layers adsorbed at hydrophobic surfaces. J Colloid Interface Sci 1994;166:333.
[150] Arwin H. Application of ellipsometry techniques to biological materials. Thin Solid Films 2011;519:2589–92.
[151] Berlind T, Tengvall P, Hultman L, Arwin H. Protein adsorption on thin films of carbon and carbon nitride monitored with in situ ellipsometry. Acta Biomater 2011;7:1369–78.
[152] Berlind T, Poksinski M, Tengvall P, Arwin H. Formation and cross-linking of fibrinogen layers monitored with in situ spectroscopic ellipsometry. Colloids Surf B 2010;75:410–7.
[153] Arwin H, Askendahl A, Tengvall P, Thompson DW, Woollam JA. Infrared ellipsometry studies of thermal stability of protein monolayers and multilayers. Phys Stat Sol C 2008;5:1438–41.
[154] Reichelt S, Eichhorn K-J, Aulich D, et al. Functionalization of solid surfaces with hyperbranched polyesters to control protein adsorption. Colloid Surf B 2009;69:169–77.
[155] Xue C, Yonet-Tanyeri N, Brouette N, Sferrazza M, Braun PV, Leckband DE. Protein adsorption on poly(N-isopropylacrylamide) brushes: dependence on grafting density and chain collapse. Langmuir 2011;27:8810–8.
[156] Bittrich E, Uhlmann P, Eichhorn K-J, Hinrichs K, Aulich D, Furchner A. Polymer brushes, Hydrogels, Polyelectrolyte multilayers: stimuli-responsivity and control of protein adsorption. In: Hinrichs K, Eichhorn K-J, eds. Ellipsometry of functional organic surfaces and films. Berlin (Germany): Springer; 2014.
[157] De Feijter JA, Benjamins J, Veer FA. Ellipsometry as a tool to study the adsorption behavior of synthetic and biopolymers at the air-water interface. Biopolymers 1978;17:1759–72.
[158] Benesch J, Askendal A, Tengvall P. Quantification of adsorbed human serum albumin at solid interfaces: a comparison between radioimmunoassay (RIA) and simple null ellipsometry. Colloid Surf B 2000;18:71–81.
[159] Arwin H. Optical properties of thin layers of bovine serum albumin, g-globulin, and hemoglobin. Appl Spectrosc 1986;40:313–8.
[160] Burkert S, Bittrich E, Kuntzsch M, et al. Protein resistance of PNIPAAm brushes: application to switchable protein adsorption. Langmuir 2010;26:1786–95.

Index